Italy: The New Domestic Landscape

Italy: The New Domestic Landscape
Achievements and Problems of Italian Design

Edited by Emilio Ambasz

The Museum of Modern Art, New York
in collaboration with
Centro Di, Florence

Distributed by New York Graphic Society Ltd., Greenwich,
Connecticut

The Museum of Modern Art
11 West 53 Street, New York, N. Y., 10019
All rights reserved

Library of Congress Catalog Card Number 73-164878
Hardbound ISBN 0-87070-394-3
Paperbound ISBN 0-87070-393-5

Printed in Italy, April 1972
Distributed in non-English-speaking countries by
Centro Di, 1r piazza de' Mozzi, 50125 Florence

Produced by Centro Di, Florence
Coordinating editor: Helen M. Franc
Layout: Matilde Contri (Centro Di)
Cover design: Emilio Ambasz

Photolitho: Mani, Florence
Printing: Stiav, Florence
Binding: Olivotto, Vicenza
Jacket: Asea, Florence

The exhibition 'Italy: The New Domestic Landscape,' shown at
The Museum of Modern Art, New York, from May 26 to September 11,
1972, is presented under the sponsorship of the Italian Ministry for
Foreign Trade and ICE — Institute for Foreign Trade, and the
Gruppo ENI.

CONTRIBUTORS TO THE EXHIBITION

ABET-Print
Alitalia
ANIC and Lanerossi (of the Gruppo ENI)
Anonima Castelli
Fiat
Olivetti

COLLABORATORS IN THE EXHIBITION

Artemide
Boffi
Carrozzeria Boneschi
Carrozzeria Renzo Orlandi
C & B Italia
Centro Cassina
Citroën
Elco-Fiarm
Gondrand
Gufram
Ideal-Standard
Kartell
Pirelli
Saporiti
Sleeping International System Italia

The Museum of Modern Art acknowledges with gratitude a generous subvention from the Consorzio per le Opere Pubbliche and the Istituto di Credito per le Imprese di Pubblica Utilità toward the preparation of the historical and critical articles included in this publication.

CONTENTS

PREFACE /11
ACKNOWLEDGMENTS /13
INTRODUCTION, by Emilio Ambasz /19
OBJECTS
Objects selected for their formal and technical means /25
Objects selected for their sociocultural implications /93
Objects selected for their implications of more flexible patterns of use
and arrangement /111

ENVIRONMENTS
Introduction /137
Design Program /139
Design as postulation:
   Gae Aulenti /150
   Ettore Sottsass, Jr. /160
   Joe Colombo /170
   Alberto Rosselli /180
   Marco Zanuso and Richard Sapper /190
   Mario Bellini /200
Design as commentary:
   Gaetano Pesce /212
Counterdesign as postulation:
   Ugo La Pietra /224
   Archizoom /232
   Superstudio /240
   Gruppo Strum /252
   Enzo Mari /262
Winners of the competition for young designers:
   Gianantonio Mari /268
   Group 9999 /276

HISTORICAL ARTICLES
Introduction /285
Art Nouveau in Italy, by Paolo Portoghesi /287
The Futurist Construction of the Universe,
by Maurizio Fagiolo dell'Arco /293
The Beginning of Modern Research, 1930-1940,
by Leonardo Benevolo /302
Italian Design 1945-1971, by Vittorio Gregotti /315

CRITICAL ARTICLES
Introduction /343
Italian Design in Relation to Social and Economic Planning,
by Ruggero Cominotti /345
Housing Policy and the Goals of Design in Italy, by Italo Insolera /352
Ideological Development in the Thought and Imagery of Italian Design,
by Giulio Carlo Argan /358
The Land of Good Design, by Alessandro Mendini /370
Radical Architecture, by Germano Celant /380
Design and Technological Utopia, by Manfredo Tafuri /388
A Design for New Behaviors, by Filiberto Menna /405

SUMMARY, by Emilio Ambasz /419
CREDITS /424

'You become responsible, forever, for what you
have domesticated.'

'What does that mean — "domesticated"?'

'It is an act too often neglected. It means to
establish bonds.'

. . . . .

'Please domesticate me,' said the fox.

'I want to, very much,' the little prince replied.
'But I have not much time. I have friends to discover,
and a great many things to understand.'

'One only understands the things that one domesticates,'
said the fox. 'Men have no more time to understand
anything. They buy things all ready made at the shops.
But there is no shop anywhere where one can buy friendship,
and so we have no friends any more. If you want a friend,
domesticate me. . . .'

'What must I do, to domesticate you?' asked the little prince.

. . . . .

'... One must observe the proper rites ...'

'What is a rite?' asked the little prince.

'Those are actions too often neglected,' said the fox.
'They are what make one day different from other days,
one hour from other hours.'

— Antoine de Saint-Exupéry, *The Little Prince*

It has been a long-standing assumption of the modern movement that if all man's products were well designed, harmony and joy would emerge eternally triumphant. Many signs from different sources are making it evident that, although good design is a necessary condition, it is not by itself sufficient to ensure the automatic solution of all the problems that precede its creation and of those that may arise from it. Consequently, many designers are expanding their traditional concern for the aesthetic of the object to embrace also a concern for the aesthetic of the uses to which the object will be put. Thus, the object is no longer conceived as an isolated entity, sufficient unto itself, but rather as an integral part of the larger natural and sociocultural environment.

This phenomenon is affecting designers the world over, but nowhere is the situation so complex, so well crystallized, and so rich in examples as in Italy. During the last decade, Italy has become one of the dominant forces in the creation and criticism of design; and by focusing on it, The Museum of Modern Art has wished, first, to honor the specific achievement of Italian designers, and second, to examine in general terms some of the problems affecting design today, and the diverse approaches being developed for their solution.

The history of the Museum's relationship with Italian architects and designers has been long. It goes back to 1932, when — an innovation for an 'art' museum — a separate Department of Architecture, which also concerned itself with industrial design, was established within our institution. It was the result of the conviction of the Museum's first director, Alfred H. Barr, Jr., that the meanings implicit in works of architecture and design should be as consciously investigated and evaluated as are those in works of painting and sculpture. Subsequently, a separate Department of Industrial Design was founded; in 1949, the two were combined to form the present Department of Architecture and Design.

Throughout the years, many objects of Italian design have been added to the Museum's collection and included in its temporary exhibitions. For example, Italy's primacy in the field of automobiles has thrice been recognized. The Cisitalia car, designed by Pinin Farina in 1946 and produced in 1949, was included in the exhibition 'Eight Automobiles' that Philip Johnson and Arthur Drexler, the present Director of the Department of Architecture and Design, organized in 1951; the Lancia GT, designed by Pinin Farina in 1951, and the SIATA Daina 1400, for which he designed the bodywork, appeared in the show 'Ten Automobiles' that Mr. Drexler presented in the Museum's Garden in 1953; and in 1966, he selected Pinin Farina's PF Sigma Italy 63, De Tomaso's Vallelunga with bodywork by Ghia, and Lamborghini's P-400 Miura, with bodywork by Bertone, for an exhibition entitled 'The Racing Car: Toward a Rational Automobile.' In 1952, Leo Lionni served as guest director for the exhibition 'Olivetti: Design in Industry.' in 1954, 'The Modern Movement in Italy: Architecture and Design,' organized by Ada Louise Huxtable for the Department of Circulating Exhibitions, was shown at the Museum before traveling throughout the United States and Canada. In 1955, Bruno Munari (together with the American Alvin Lustig) was featured in the exhibition 'Two Graphic Designers,' directed by Mildred Constantine. The Museum in its turn was honored in 1956 by receiving the 'Gran Premio Internazionale Compasso d'Oro,' in recognition of the contribution made by the Department of Architecture and Design to the evaluation and propagation of contemporary design.

Although the present exhibition was first conceived in May, 1970, intensive work on it began only in January, 1971, after a number of administrative and financial problems had been resolved. The preparation time for an exhibition of this size and scope is relatively long, owing in part to its intrinsic complexities, and in part to the Museum's lack of adequate space for so large a show. This long period of gestation has a certain advantage in that, if the concept is to be kept fresh and vital, the material must be constantly reexamined, the premises set forth at the outset questioned, and even one's own motives for such questioning be questioned. If, at the beginning of research for the exhibition, Italian design seemed so dazzling that it was momentarily possible to assume that transplanting its most outstanding examples would be sufficient to recall the luminosity of their original breeding ground, deeper examination made it increasingly evident that the problem was far from simple. Design in Italy today does not present a consistent body of ideas, with respect to either form or ideology. Its complex marches and countermarches made it necessary to develop what might be called 'provocation' techniques that would result in an exhibition revealing the contradictions and conflicts underlying a feverish production of objects, which are constantly generated by designers, and which in turn generate a state of doubt among them as to the ultimate significance of their activity.

The subject of Italian design is too alive to permit dissection, and its elements are too contradictory to be fitted into any single scheme of classification that one might be so careless as to propose. In order to aid comprehension of what the living process of Italian design actually is, a dozen of Italy's most outstanding designers were selected, on the basis of the formal and ideological positions they represent, and each was invited to give a solo performance as a statement of his position. A competition for young designers was held to give them a similar opportunity to expound their stance.

Now the props have been set, and the stage is lighted. The seemingly solo presentations that we shall shall see (and which in the concluding section of this publication we shall attempt to summarize) are not neatly defined segments cut out of a preexisting reality; they were specially devised for this event. Each represents a reckoning, on the part of its designer, of his own past, and an attempt to define his present position. In most cases, the designers have chosen to pose new questions rather than to present a synthesis of their achievements to date. Their questions, if perhaps somewhat rhetorical, are nevertheless incisive; and the sum total of their despairs and hopes, although contingent on their individual choice of idiom, may nonetheless express doubts and expectations that we all share.

— E. A.

ACKNOWLEDGMENTS

Preparing this exhibition has engaged the imagination and resources of virtually hundreds of people in Italy and the United States: the designers who responded to our invitation to propose environments, and the technicians who built the matrices that produced them; those who generously underwrote their construction, and those who shipped them; the manufacturers and designers who kindly donated or lent objects; those who participated in producing the audio-visual materials; and a host of others involved in administrative responsibilities, whose assistance was both indispensable and greatly appreciated, even though they may remain anonymous here.

On behalf of the Trustees of The Museum of Modern Art, I wish to express particular gratitude to the Italian Ministry of Foreign Trade and the Institute for Foreign Trade (ICE) and to the Gruppo ENI, for the generous and enlightened support that they gave to this project from its very inception. Special recognition is due to Vieri Traxler, Consul General of Italy in New York, and to Ernesto Toti-Lombardozzi, formerly Commercial Attaché in New York, and now in London, for their fine comprehension of the meaning that this show might have for Italy, and for the skillful guidance that they provided through the maze of bureaucratic procedures.

Eugenio Cefis, former president of ENI, likewise had a clear vision of the potential that this project might have for Italian industry; and his successor, the present president, Raffaele Girotti, warmly endorsed this initiative. To both, we express our deep appreciation. We are likewise indebted to Franco Briatico of ENI for his energetic support, as well as to Messrs. Magini and Ascione of his office. Enzo Viscusi, representative of AGIP in the United States, has been a most incisive guide to the complexities of Italian industry and has provided invaluable aid in all matters related to this exhibition. A special note of thanks is due to Carlo Robustelli, responsible for liaison between ENI and the Museum; his deep sense of cultural and social commitment honor the institution he serves, while his administrative resourcefulness has on more than one occasion been decisive for the outcome of this projects.

Guido Jannon, advisor to ABET-Print, merits particular recognition for his understanding of the cultural role of industry and his generous support of artists and designers. In thanking Renzo Zorzi, we wish not only to indicate our respect for his sensitivity and intellectual integrity but also to acknowledge once more the bonds of friendship that this Museum has had for many years with Olivetti. I also extend my thanks to Oddone Camerana, who very ably presented the scope of this show to the Board of FIAT and enlisted their support; to Paolo Conti and Giuliano di Somerville of Alitalia, who have been most understanding and always ready to assist with any means at their disposal; to Leonida Castelli, president of Anonima Castelli, who patiently followed the course of successive designs for the exhibition containers and most generously underwrote them; and to Messrs. Conversano, San Marco, and Salvadori, together with the technical staff of ANIC-Lanerossi, for so competently coordinating and supervising the production of five of the environments.

The response of the designers of the environments to our invitation to participate in this exhibition has been most gratifying. They and their collaborators spared no pains and effort, committing great portions of their time to the detriment of certainly more lucrative undertakings, and graciously subjecting themselves to many anxieties and frustrations during the course of executing their prototypes. We are grateful, also, for the descriptions and statements they have provided to elucidate their general positions and these specific environments. The sudden and untimely death of Joe Colombo, one of Italy's best-known designers, took place during the summer of 1971, while his project was in preparation. On the basis of detailed drawings that he made for it, however, his prototype was built under the supervision of Ignazia Favata and other members of his studio, to whom we extend our special thanks.

On a previous page, we have listed more than a dozen industries who collaborated in support of this exhibition; they are, however, only part of a still more extensive list. In singling out for special mention Giulio Castelli of Kartell; Cesare Cassina and Rodrigo Rodriguez of Cassina; Messrs. Bellato, Sr. and Jr., of Elco; and Alberto Vignatelli of Sleeping International System Italia (SISI), we wish symbolically to express our gratitude to all the numerous companies and individuals charged with supervising and executing the environments and thus giving form to the designers' ideas. To all the workers and subcontractors, listed institutions, and supervisors unnamed here, go our sincerest thanks for their cooperation in this endeavor.

Elsewhere, we have listed the manufacturers who generously donated or lent their products for the section of the exhibition devoted to objects. We are especially indebted to Ernesto Gismondi and the capable technical staff of Artemide, who produced the lamps for the containers in which the objects are displayed. In designing these containers, I benefited greatly from the enthusiasm and structural imagination of Giancarlo Piretti; they were constructed by Eckol Containers Systems, Inc., of Glenside, Pennsylvania. I am most grateful to Juhani Pallaasma for the time and effort he expended in helping to explore the possibility of having them constructed in Finland, and also to Renzo Piano, who attempted to devise a lightweight structure to cover the upper and lower terraces of the Museum's Garden, where the objects are displayed; regrettably, the scarcity of time for experimentation, as well as of available funds, made it impossible to carry out his elegant design.

With great patience and dedication, Vittorio Conti and Mr. Giuseppe Gusmaroli of Gondrand checked the more than eighteen hundred items coming from over a hundred manufacturers in many localities, to ensure that they somehow found their way to the collection warehouse and, even more miraculously, were containerized and shipped rigorously on schedule.

The audio-visual and graphic systems that introduce the various sections of the exhibition were underwritten by Olivetti. It is a pleasure to express my gratitude not only to Mr. Zorzi, already mentioned, but also to Riccardo Felicioli and Pierre Denivelle, who supervised and coordinated the different projects. Umberto Bignardi gave poetic visualization to the text of the critical section, and Giacomo Battiato showed sensitivity and faithfulness to the text and script presented to him in making the film that introduces the exhibition. Oliva di Collobiano executed the sets, and the film producer Sergio Lentati deserves our particular gratitude for the capability and affability, surpassed only by his superlative skill, with which he coordinated all material aspects of the film. We are indebted to Paolo Pillitteri, member of the City Council of Milan in charge of cultural affairs, for his enthusiastic interest and assistance in obtaining the cooperation of Milan's police force and the management of the Galleria Vittorio Emanuele in clearing the areas required, during the shooting of the introductory film. We likewise express special thanks to the municipal authorities of Piacenza and the management of its Municipal Theater for making their facilities available for shooting the last part of this film. In the United States, Robert Whitman of Experiments in Art and Technology (EAT) was most helpful in advising on the setup for the film's projection system.

The orientation leaflet distributed gratis to visitors was contributed and produced by Olivetti; Franco Bassi of the graphics section of the firm's public information office in Milan, and his staff, executed it with true professional competence and dedication.

Our heartfelt admiration and thanks go to Aldo Ballo, who photographed most of the objects for this book and made the enlargements shown in the exhibition containers. Valerio Castelli very sensitively photographed many of the environments. Those by Alberto Rosselli and by Studio Zanuso were skillfully photographed by Leombruno, Bodi, and Lami; while Klaus Zaugg's photographs imaginatively interpreted the environment by Gaetano Pesce.

I am obliged to Peter Eisenman and my fellow colleagues at the Institute for Architecture and Urban Studies for having granted me a year's leave of absence to work on the preparation of this exhibition. During the many months spent in Italy for this purpose, I met with cordial cooperation from far more people than can be enumerated here. I am particularly grateful to Lisa Licitra Ponti and Marianne Lorenz of the editorial board of *Domus*, who repeatedly called my attention to matters that they thought merited investigation. Mrs. Inge Feltrinelli was a most gracious cicerone amid the social intricacies of Milan. Tomás Maldonado once again, as at Ulm several years ago, assumed the role of older brother, giving generously of his advice and directing me toward intellectual resources.

The kindness of Bartlett Hayes, Jr., in consenting to have the American Academy in Rome serve as clearing house for all the entries submitted to the competition that the Museum sponsored for young designers is much appreciated.

The historical and critical essays contained in this book were in part underwritten by a generous subvention from the Consorzio di Credito per le Opere Pubbliche and the Istituto di Credito per le Imprese di Pubblica Utilità. I deeply appreciate the aid given by Tomaso Carini, Executive Director of the Credito Italiano in Rome, in obtaining this grant. We are most grateful to the authors of the articles for their incisive and thoughtful contributions. We also express our thanks to those who assisted in translating these texts and those accompanying the environments: James Pallas, Barbara Angelillo, Susan Contini, Angela Gibbon, Felicity Lutz, and Angela Redini.

This publication is issued by The Museum of Modern Art in collaboration with Centro Di of Florence, where it was designed and produced. Ferruccio Marchi was a most generous and gracious host to all my graphic requirements and devoted himself with constant patience and understanding to the difficult problems that producing it in a very short time presented. I am greatly indebted to him, the designer Matilde Contri, and her assistant Simonetta Doni, for all that may be graphically good in the book. Alessandra Marchi provided invaluable assistance in bibliographical research and placed the extensive resources of the Centro Di library at our disposal. Helen M. Franc, formerly Editor-in-Chief of the Museum's Department of Publications, served as coordinating editor; I am grateful alike for her high professional standards and skill, and for her unflagging enthusiasm for the task, despite its complexity and exceptionally arduous production schedule. With great good humor, the staffs of Tipolitografia Stiav, Fotolito Mani, and Asea of Florence exerted special efforts to meet this schedule without loss of quality.

I am indebted in many ways to numerous colleagues at the Museum, not least to John Szarkowski, Director, Department of Photography; William Rubin, Chief Curator of Painting and Sculpture Collections; and Alicia Legg, Associate Curator, Painting and Sculpture, who kindly consented to changes in the scheduling of exhibitions of their own to accommodate the dates for this show; and to Betsy Jones, Associate Curator, Painting and Sculpture Collections, for removing a number of sculptures from the Garden in order to provide space. Richard Palmer, Assistant to the Director in charge of exhibitions, most competently supervised the very complicated administrative aspects, and Betty Burnham, Associate Registrar, assumed charge of the intricacies of registration. Matthew Donepp, Building Operations Manager, and Charles Froom, Production Manager, concerned themselves with problems relating to the show's installation — by far the most complex that the Museum has undertaken in many years.

Within my own Department of Architecture and Design, I am especially grateful to its Director, Arthur Drexler, who courageously allowed me to undertake this project, though fully aware of the difficulties and risks involved. He has spent much effort in helping to resolve many of the problems that inevitably arose. I also wish to thank John Garrigan, Assistant Curator of Graphic Design, for his friendly cooperation, particularly with regard to all graphic matters related to the Design Program and the exhibition labels; and my resourceful secretary, Camilla Martellini.

This exhibition would have been virtually impossible without the help of Thomas Czarnowski and Anna Tucci. For more than a year, they worked seven days a week, and fourteen hours a day, Mr. Czarnowski coordinating, and Mrs. Tucci assembling the objects selected and gathering the necessary documentation. For their energy, perseverance, and wholehearted support, I express the deepest gratitude on my own behalf and that of the Museum.

— Emilio Ambasz, Director of the Exhibition

INTRODUCTION

INTRODUCTION
Emilio Ambasz

The emergence of Italy during the last decade as the dominant force in consumer-product design has influenced the work of every other European country and is now having its effect in the United States. The outcome of this burst of vitality among Italian designers is not simply a series of stylistic variations of product design. Of even greater significance is a growing awareness of design as an activity whereby man creates artifacts to mediate between his hopes and aspirations, and the pressures and restrictions imposed upon him by nature and the manmade environment that his culture has created.

Italy has become a micromodel in which a wide range of the possibilities, limitations, and critical issues of contemporary design are brought into sharp focus. Many of the concerns of contemporary designers throughout the world are fairly represented by the diverse and frequently opposite approaches being developed in Italy. The purpose of this exhibition, therefore, is not only to report on current developments in Italian design, but to use these as a concrete frame of reference for a number of issues of concern to designers all over the world.

It is possible to differentiate in Italy today three prevalent attitudes toward design: the first is conformist, the second is reformist, and the third is, rather, one of contestation, attempting both inquiry and action.

By the first, or conformist, approach, we refer to the attitude of certain designers who conceive of their work as an autonomous activity responsible only to itself; they do not question the sociocultural context in which they work, but instead continue to refine already established forms and functions. As a group, they may be distinguished by certain characteristics (also found, however, among other Italian designers): their bold use of color, and their imaginative utilization of the possibilities offered by the new hard and soft synthetic materials and advanced molding techniques. Their work, which constitutes the most visible part of Italian design production, is mainly concerned with exploring the aesthetic quality of single objects — a chair, a table, a bookcase — that answer the traditional needs of domestic life.

The second, or reformist, attitude is motivated by a profound concern for the designer's role in a society that fosters consumption as one means of inducing individual happiness, thereby insuring social stability. Torn by the dilemma of having been trained as creators of objects, and yet being incapable of controlling either the significance or the ultimate uses of these objects, they find themselves unable to reconcile the conflicts between their social concerns and their professional practices. They have thus developed a rhetorical mode to cope with these contradictions. Convinced that there can be no renovation of design until structural changes have occurred in society, but not attempting to bring these about themselves, they do not invent substantially new forms; instead, they engage in a rhetorical operation of redesigning conventional objects with new, ironic, and sometimes self-deprecatory sociocultural and aesthetic references.

In their ambiguous attitude toward the object, these designers justify their activity by giving to their designs shapes that deliberately attempt a commentary upon the roles that these objects are normally expected to play in our society. The diversity of these rhetorical operations permits us to recognize at least half a dozen different procedures for recharging known forms with altered meanings.

Some of these designers are involved in a process of revival. For the most part, they restate forms created by the earliest modern movements in design — mainly, Art Nouveau and the Bauhaus — since these forms have by now become understood, and the complex set of ideas that they once connoted have by now become explicit. Sometimes, however, in their nostalgia for the past, they reach back even to medieval times.

Other groups are more concerned with ironic manipulation of the sociocultural meanings attached to existing forms, rather than with changing those forms. Specifically, they design deliberately kitsch objects, as a way of thumbing their noses at objects created to satisfy the desire for social status and identification. Some, taking their cue from Pop art, adopt forms from the manmade elements that compose our milieu, presenting them transformed in little else but scale.

Other designers seek neither to add anything to, nor to alter, the profile of our environment; they use the device of giving their designs the guises of nature.

Conversely, others satisfy the same intention by assembling their designs solely from already existing industrial elements, recovered from the surrounding industrial landscape; by this recycling, they avoid the proliferation of new formal matrices.

For a few designers, the cultural premises predominant today have no validity and can therefore provide only a false basis for any formal inquiry. Devoid of any firm referents, they return, in a somewhat self-deprecatory attempt at purification, to the human figure as the source of all formal truth.

Recognizing that the object in our society often serves as a fetish, some designers underscore that quality by assigning to their designs an explicitly ritualistic quality. The object is given sculptural form and conceived as an altarpiece for the domestic liturgy.

For some designers, the object can be stripped of its mystique only if it is tamed, if it is made to assume the role of house pet. Reduced to graspable size, the object no longer intimidates us; endowed with the stablility of inert matter, and created for no specific function, such objects can be allowed into our homes in the certainty that they will never make evident the passage of biological or social time.

Confronted with the erosion of the simplistic doctrine of functionalism, some designers produce objects whose function is not evident from their form, and whose structural properties, in fact, contradict the behavior one would expect from that form. In such cases, no longer does 'form follow function' but, on the contrary, it aggressively conceals it.

The distinction between the two main approaches so far discussed, the conformist and the reformist, is in reality not so clear-cut. The oscillations of designers between these two attitudes reflect the contradictions and paradoxes that result from simultaneously doubting the benefits of our consumer society, and at the same time enacting the role of voyeurs of the technological dream.

The third approach to design, which we have designated as one of contestation, attempts to deal with such a situation. This attitude reveals itself in two main trends in Italy today, each trying to get to the root in very different ways. The first is by a commitment to a 'moratorium' position and an absolute refusal to take part in the present socioindustrial system. Here, 'antiobject' literally means 'not making objects,' and the designers' pursuits are either confined to political action and philosophical postulation, or else consist of total withdrawal.

Those following the second tendency share with the preceding group the disbelief that an object can be designed as a single, isolated entity, without regard for its physical and sociocultural context. Their reaction to the problem, however, is not one of passive abstention but rather one of active critical participation. They have thus come to conceive of objects and of their users as an ensemble of interrelated processes, whose interaction results in constantly changing patterns of relationships. To the traditional preoccupation with aesthetic objects, these contemporary designers have therefore added a concern for an aesthetic of the uses made of these objects. This holistic approach is manifested in the design of objects that are flexible in function, thus permitting multiple modes of use and arrangement. To one accustomed to dealing with finite shapes that can act as points of reference, such objects can be offensive, because they refuse to adopt a fixed shape or to serve as reference to anything. In contrast to the traditional object, these objects, in some instances, assume shapes that become whatever the users want them to be, thereby providing an open-ended manner of use. Objects of this sort are conceived as environmental ensembles and permit different modes of social interaction, while at the same time they allow the user to make his own statement about both privacy and communality.

The products of this mode of Italian design do, in fact, seem to correspond to the preoccupations of a changing society. It has therefore been the environmental approach that the present exhibition has been particularly concerned with exploring. Accordingly, The Museum of Modern Art invited a number of well-known Italian designers to propose environmental concepts and translate them into physical designs; at the same time, it conducted a competition in the same terms for young Italian designers under the age of thirty-five. Both the commissioned designers and those entering the competition were requested to explore the domestic landscape with a sense for its 'places,' and to propose the spaces and artifacts that give them form, the ceremonies and behaviors that assign them meaning. Special attention was paid to the individual's need for spaces both of an adaptable and fixed nature, in which previously unrealized and unthought-of relationships might be openly expressed.

To complement these concrete environments, a number of designers who believe that no substantial solution can emerge from physical design, but only from social and political involvement, were also invited to present their points of view.

The objects shown in this exhibition, and illustrated in its catalogue, therefore, serve to provide a cultural context for the environments. These examples of design produced in Italy in the last decade have not been selected with an historical intent, but rather with the purpose of indicating the different design positions now evolving in Italy. The two parts of the exhibition — objects and environments — are thus complementary.

This publication documents the exhibition and follows its scheme of organization. In the first section, the objects selected are presented in three distinct groups, according to the three main tendencies in Italian design discussed above. The second section presents the environments and the Design Program submitted both to the designers invited by the Museum and to the young designers entering the competition it sponsored. The articles in the two sections that follow provide, respectively, historical and critical frames of reference for the ideas presented by the designers in the preceding sections.

All designers are Italian unless another nationality is specified. A list of designers and manufacturers appears on pages 424-27.

The date of design of each object is followed by the date of its production, in parentheses. All objects are mass produced unless otherwise stated. In dimensions, height precedes width, followed by depth.

Every object in the exhibition is reproduced; therefore, no separate checklist has been provided, the captions giving the necessary data. In those few instances in which the caption may not seem to describe accurately what is visible in the illustration, it should be understood that the data apply to what is exhibited, rather than to what can be seen in the reproduction.

All color and black-and-white photographs of objects in this section of the catalogue were specially taken by Aldo Ballo, Milan, with the exception of those indicated in the list of photograph credits on page 429.

Objects selected for their formal and technical means

Cesare Leonardi and Franca Stagi
Dondolo rocking chair. 1969 (1969)
Ribbed fiber glass, 30 3/4x68 7/8x15 3/4
inches (78x175x40 cm)
Elco. Gift of the manufacturer

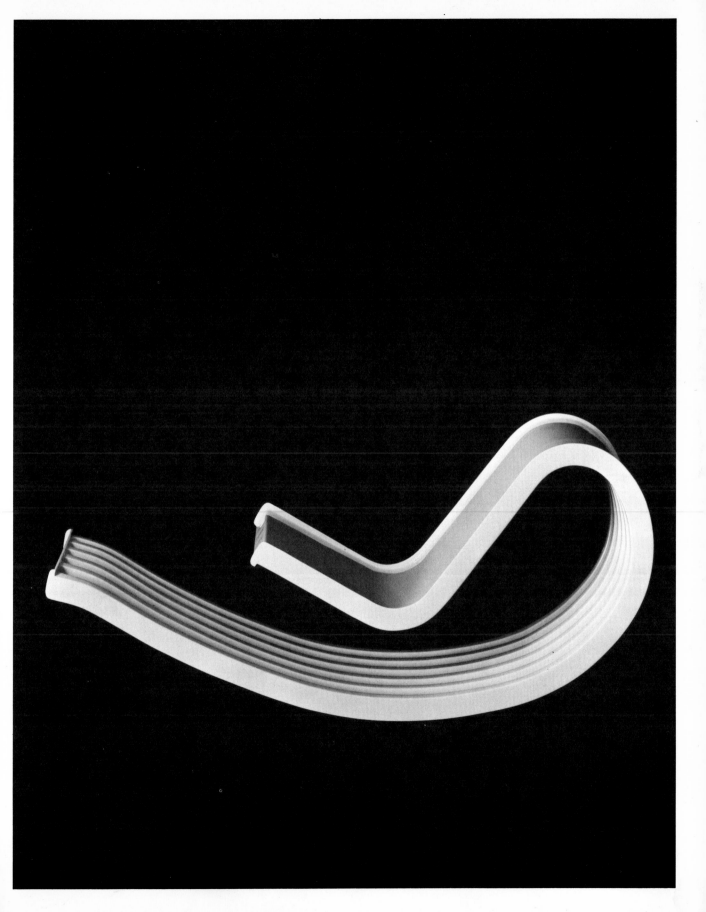

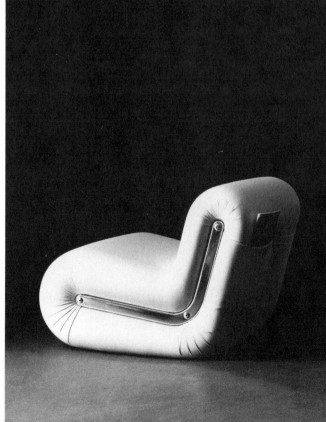

Tobia and Afra Scarpa
Soriana lounge chair. 1970 (1970)
Polyurethane and dacron covered in fabric or
leather, with external chromed metal frame,
28x35 1/2x41 3/8 inches (71x90x105 cm)
Cassina. Gift of the manufacturer

Rodolfo Bonetto
Boomerang lounge chair. 1969 (1970)
Polyurethane covered in fabric, with external
chromed plastic braces, 26x29 1/8x37 inches
(66x74x94 cm)
Flexform. Lent by the manufacturer

Tobia and Afra Scarpa
Ciprea armchair. 1968 (1968)
Self-supporting polyurethane, slip-covered in fabric,
31 1/2x36 1/4x34 5/8 inches (80x92x88 cm)
Cassina. Gift of the manufacturer

Cini Boeri
Bobolungo lounge chair. 1969 (1969)
Self-supporting polyurethane covered in fabric,
27 1/2x24 3/4x52 3/4 inches (70x63x134 cm)
Arflex. Gift of the manufacturer

29

Joe Colombo
Armchair, model 4801/5. 1965 (1966)
Three molded plywood elements, slip-jointed,
painted, 23 1/4x24xx25 1/4 inches (59x61x64 cm)
Kartell. Gift of the manufacturer

Tobia and Afra Scarpa
Lounge chair, model 925. 1966 (1966)
Back and seat molded plywood, covered in
leather, legs wood, 27 1/4x26x26 inches
(69x66x66 cm)
Cassina. Gift of the manufacturer

Mario Bellini
Armchair, model 932/2 — single-seat version
1967 (1967)
Four upholstered cushions, covered in leather,
belted, 24 3/8x36 5/8x33 1/2 inches (62x
93x85 cm)
Cassina. Gift of the manufacturer

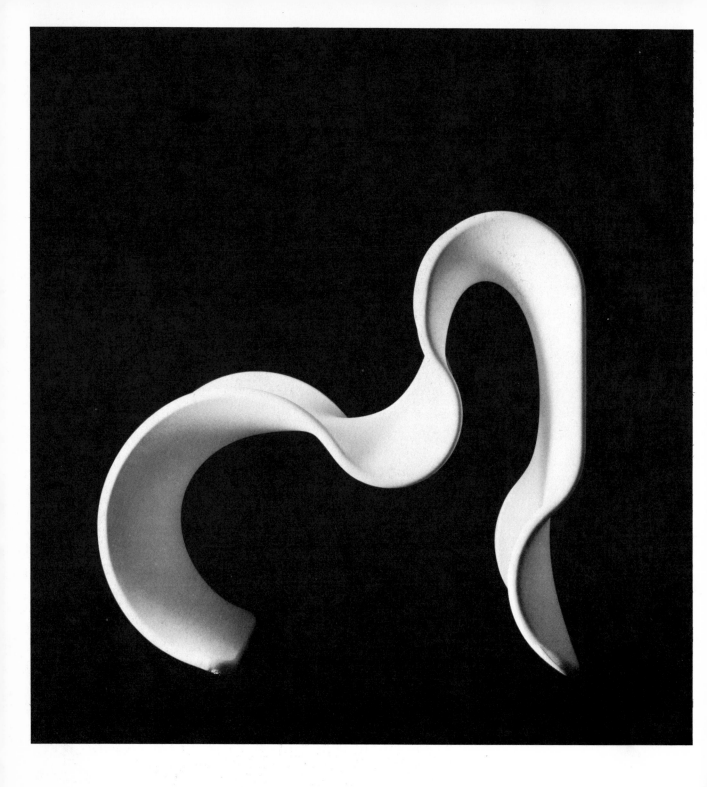

Group G 14 (Gianfranco Facchetti, Umberto Orsoni,
Gianni Pareschi, Pino Pensotti, Roberto Ubaldi)
Fiocco ('Yib') armchair. 1970 (1970)
Iron tube structure, covered with tensed stretch
fabric, 40 1/2x27 1/2x46 7/8 inches
(103x70x119 cm)
Busnelli. Gift of the manufacturer
32

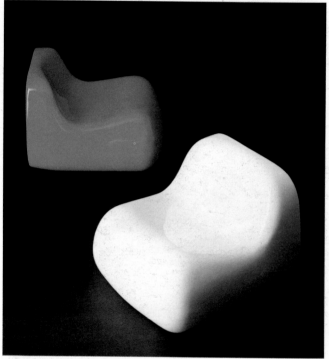

Rodolfo Bonetto
Melaina armchair. 1968 (1969)
Fiber glass, 24x30 3/4x26 inches (61x78x66 cm)
Driade. Gift of the manufacturer

Alberto Rosselli
Jumbo lounge chair. 1968 (1968)
Fiber glass, 17 1/4x20 7/8x26 inches (44x
53x66 cm)
Saporiti. Gift of the manufacturer

Vico Magistretti
Vicario ('Vicar') armchair. 1970 (1971)
Reinforced polyester, 26 3/8x25 5/8x28 3/8
inches (67x65x72 cm)
Artemide. Gift of the manufacturer

33

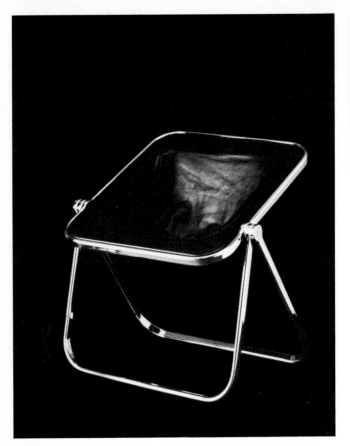
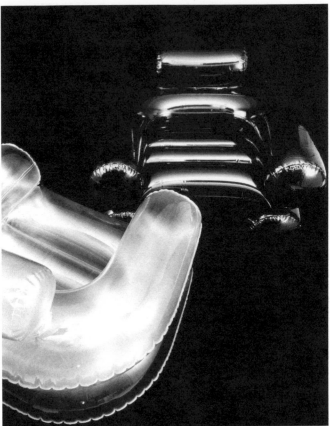

Giancarlo Piretti
Plona folding armchair. 1970 (1971)
Polished aluminum structure, leather seat,
28 3/8x27x21 1/4 inches (72x68,5x54,5 cm)
Anonima Castelli. Gift of the manufacturer

Paolo Lomazzi, Donato D'Urbino, and Jonathan
De Pas
Blow inflatable armchair. 1967 (1967)
Pneumatic structure, PVC plastic, 23 5/8x39 3/8
x39 3/8 inches (60x100x100 cm)
Zanotta. Gift of the manufacturer

34

Gaetano Pesce
UP 2 armchair. 1969 (1969)
Polyurethane, covered in stretch fabric (expands
on release from compression within package),
26 3/8x39 3/8x39 3/8 inches (67x100x100 cm)
C & B. Gift of the manufacturer

Marco Zanuso
Springtime armchair. 1965 (1966)
Demountable wood structure, upholstered, covered
in corduroy, 26x25 1/4x29 1/2 inches
(66x64x75 cm)
Arflex. Gift of the manufacturer

Sergio Mazza and Giuliana Gramigna
Poker demountable armchair. 1970 (prototype;
not yet in production)
Four identical elements of ABS plastic, filled
with soft polyurethane, metal joints, 26x30 3/4
x30 inches (66x78x76 cm)
Cinova. Gift of the manufacturer

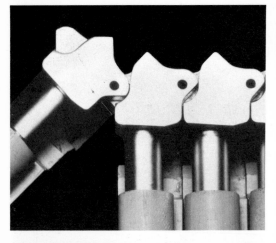

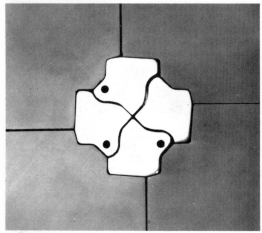

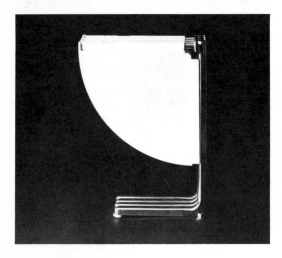

Giancarlo Piretti
Plano folding table. 1970 (1971)
Aluminum base, jointed, top reinforced polyester
in four sections, 28 3/4x37 3/4 inches diameter
(73x96 cm)
Anonima Castelli. Gift of the manufacturer

a. detail showing joints in process of opening
b. detail view showing position of joints when
table is open
c. side view of table when folded

Giancarlo Piretti
Plia folding and stacking chair. 1969 (1970)
Aluminum frame, back and seat transparent plastic,
30 1/4x18 1/2x18 1/2 inches (77x47x47 cm)
Anonima Castelli. Gift of the manufacturer

Alberto Rosselli
Jarama chair and table. 1969 (1969)
Fiber glass, chair 28x17x19 3/4 inches
(71x43x50 cm), table 29 1/8x63x76 3/4
inches (74x160x95 cm)
Saporiti. Gift of the manufacturer

Vico Magistretti
Stadio 80 table. 1967 (1968)
Reinforced polyester, 28 3/8x32 1/4x32 1/4
inches (72x82x82 cm)
Artemide. Gift of the manufacturer

Vico Magistretti
Gaudi armchair. 1970 (1971)
Reinforced polyester, 29 1/8x21 5/8x23 5/8
inches (72x55x60 cm)
Artemide. Gift of the manufacturer

Joe Colombo
Stacking chair, model 4860 - 61/5. 1967 (1968)
ABS plastic, 14 1/4 - 17x17x20 1/2 inches
(35 - 43x43x12 cm)
Kartell. Gift of the manufacturer

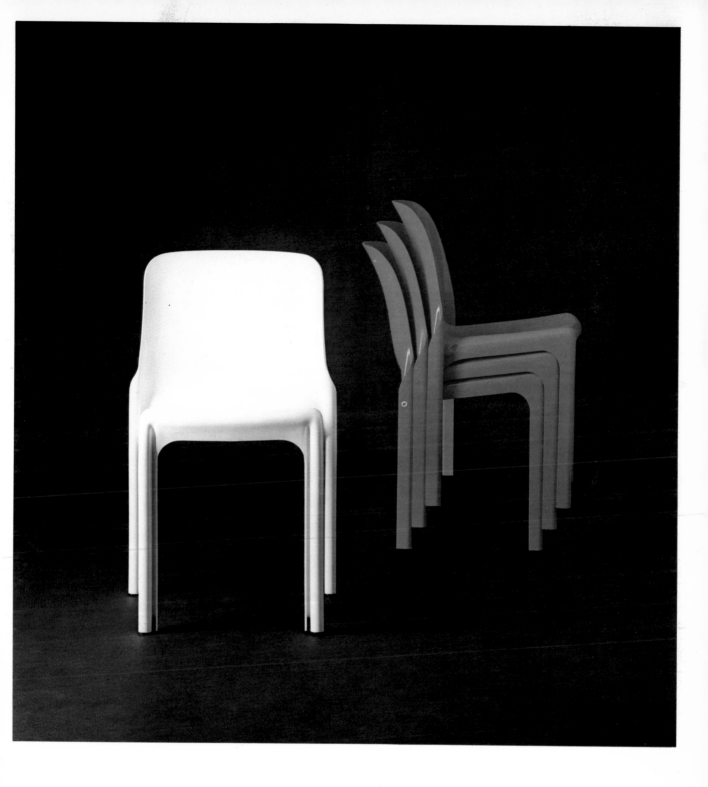

Vico Magistretti
Selene stacking chair. 1961 (1966)
Reinforced polyester, 30x18x19 1/2 inches
(76x47x50 cm)
Artemide. Gift of the manufacturer

41

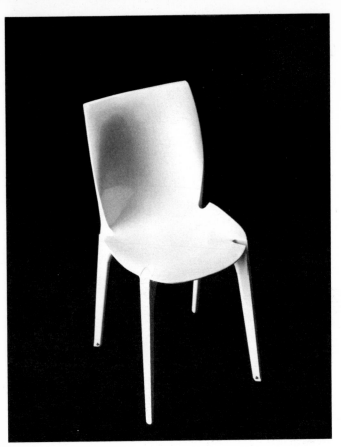 

Marco Zanuso
Lambda chair. 1963 (1963)
Lacquered sheet metal, 30 3/4x15 3/4x18 1/2
inches (78x40x47 cm)
Gavina. Gift of the manufacturer

Silvio Coppola
Gru ('Crane') chair. 1970 (1970)
Lacquered metal tube, seat and arm rest covered
in fabric, 27 1/2x19 3/4x20 7/8 inches
(70x50x53 cm)
Bernini. Gift of the manufacturer

Gae Aulenti
Aprile ('April') folding chair. 1964 (1966,
1969; this version 1971)
Stainless steel structure, seat covered in
artificial leather, 33 7/8x17 7/8x17 3/8
inches (86x45,5x44 cm)
Zanotta. Gift of the designer

Angelo Jacober and Pierangela d'Aniello
Fiera di Trieste folding chair. 1966 (1966)
Wood structure with caned seat, 28x21 1/4x
18 inches, open; 21 1/4x30 3/4x2 inches,
closed (71x54x46 cm; 54x78x5 cm)
Bazzani. Gift of the manufacturer

43

Marco Zanuso and Richard Sapper (Italian, born
Germany)
Small child's chair, model 4999/5. 1961 (1967)
Polyethylene, 19 3/4x11x11 inches (50x28x28 cm)
Kartell. Gift of the manufacturer

Giorgio Decursu, Jonathan De Pas, Paolo Lomazzi,
Donato d'Urbino
Chica demountable child's chair. 1971 (prototype)
ABS plastic, 19 1/4x13x13 inches (49x33x33 cm)
BBB Bonacina. Gift of the designers

Joe Colombo
Poker card table with demountable legs
1968 (1968)
Laminated wood and stainless steel, 27 1/2x39 3/8
x39 3/8 inches (70x100x100 cm)
Zanotta. Gift of the manufacturer
45

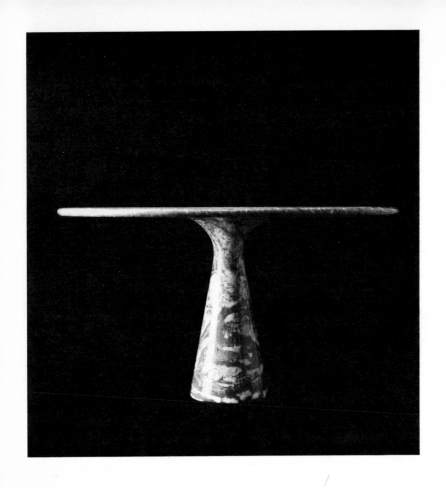

Angelo Mangiarotti
M 1 table. 1961 (1969)
Marble, 28 3/8x51 1/4 inches diameter
(72x130 cm)
Tisettanta. Gift of the manufacturer

Anna Castelli and Ignazio Gardella
Table, model 4997/5. 1969 (1970)
Reinforced polyester resin, 6 feet inchesx51 1/4
x28 3/8 inches (200x130x72 cm)
inches (200x130x72 cm)
Kartell. Gift of the manufacturer
47

Giotto Stoppino
Nesting tables, model 4905-6-7/5. 1967 (1968)
ABS plastic, 12 5/8, 14 1/2, and 16 1/2x17 inches
diameter (32, 37, and 42x43 cm)
Kartell. Gift of the manufacturer

Mario Bellini
4 Gatti ('4 Cats') stacking tables. 1966 (1967)
Fiber glass, 9 7/8x19 3/4x19 3/4 inches
(25x50x50 cm) each
C & B. Gift of the manufacturer

Sergio Asti
Trentatré ('Thirty-three') low tables. 1970 (1970)
Fiberglass, (square table: 31 1/2x31 1/2 inches,
rectangular table: 13 3/4x31 1/2 inches
13 3/4x31 1/2x31 1/2 inches (35x80x80 cm)
Sintesis. Gift of the manufacturer

48

Rodolfo Bonetto
Quattro quarti ('Four quarters') combinable low
tables. 1969 (1969)
ABS plastic, 11 3/4x39 3/8 inches diameter
assembled (30x100 cm)
Bernini. Gift of the manufacturer
49

Ettore Sottsas, Jr. (Italian, born Austria)
Nefertiti desk. 1969 (1969)
Plywood covered with plastic laminate striped
white and green, 43 1/4x51 1/4x13 3/4 inches
(110x130x35 cm)
Poltronova. Gift of the manufacturer

Giancarlo Piretti
Platone ('Plato') folding desk. 1971 (1972;
prototype)
Chromed steel tubing and ABS plastic, 28x32 7/8
x25 1/2 inches (71x83,5x65 cm)
Anonima Castelli. Gift of the manufacturer
51

Anna Castelli Ferrieri
Stacking storage units, models 4953-54-55-56-
5960/5. 1969 (1970)
ABS plastic, 9, 10 5/8, 15 3/8, and 17x16 1/2
inches diameter (23, 27, 39, and 43x42 cm)
Kartell. Gift of the manufacturer

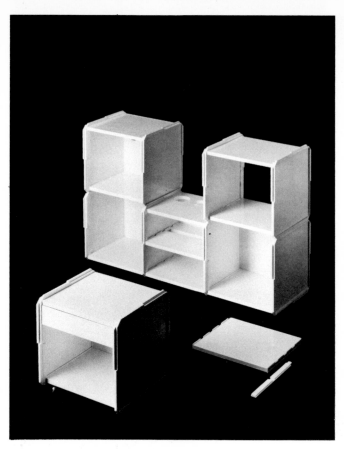

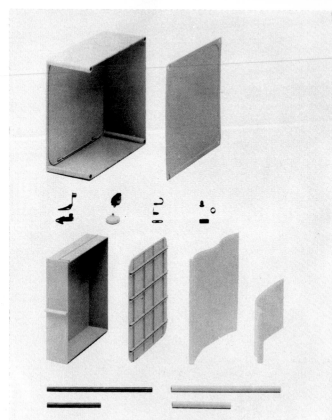

Franco Cattelan
Cubo Idea demountable storage units. 1969 (1970)
ABS plastic, 15 3/4x15 3/4x15 3/4 inches each
(40x40x40 cm)
Xilema. Gift of the manufacturer

Joe Colombo
Square Plastic System demountable storage cubes
1970 (1971)
ABS plastic, 18 1/2x47 1/4x16 1/4 inches
(47x120x42 cm), 33 1/2x49 5/8x16 1/2
inches (85x126x42 cm), and 17 3/4x
16 1/2x16 1/2 inches (45x42x42 cm)
Elco. Gift of the manufacturer

53

Enzo Mari
Glifo ('Glyph') ·demountable storage cubes
1969 (1969)
Plastic, 14 1/2x14 1/2x14 1/4 inches each
(37x37x36 cm)
Gavina. Gift of the manufacturer
54

Ugo La Pietra
Uno sull'altro ('One on Top of the Other')
stacking shelves. 1970 (1971)
Wood, 19 1/4x39 3/8, 31 1/2, 19 3/4, and 15 3/4
x21 1/2 inches (49x100, 80, 50, and 40x54,5 cm)
Poggi. Gift of the manufacturer

55

Sergio Mazza
Sergesto stacking bookshelves. 1969 (1970)
ABS plastic, 70 7/8x57 1/2x10 1/4 inches
(180x146x26 cm)
Artemide. Gift of the manufacturer
56

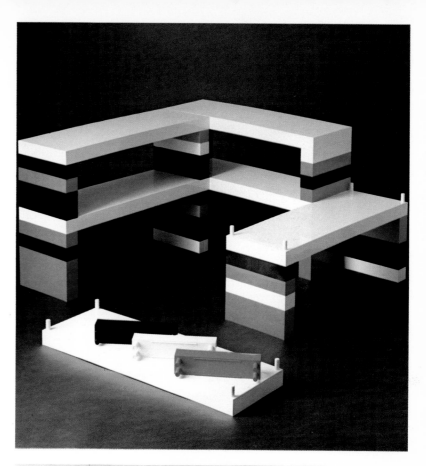

Jonathan De Pas, Donato D'Urbino, Paolo Lomazzi
Brick modular shelving system. 1970 (1971)
Plastic, shelves 2 3/8x12 5/8x37 3/4 inches each
(6x32x96 cm), 'bricks' 2 3/8x2 3/4x12 5/8 inches
each (6x7x32 cm)
Longato. Gift of the manufacturer

Marcello Siard
Wall brackets, model 4949-50-51/5. 1966 (1968)
ABS plastic, 13 3/4x17 3/4, 22, and 31 1/2x
11 3/4 inches (35x45, 56, and 80x30 cm)
Kartell. Gift of the manufacturer

Nanda Vigo
Utopia table lamp. 1970 (1971)
Stainless steel, 20x20x3 1/2 inches
(51x51x9 cm)
Arredoluce. Gift of the manufacturer
58

Vico Magistretti
Eclisse (Eclipse) table lamp with adjustable
shade. 1965 (1965)
Lacquered aluminum, 8 5/8x4 3/4 inches diameter
(22x12 cm)
Artemide. Gift of the manufacturer
59

Antonio Macchi Cassia
Table lamp 541. 1968 (1969)
Two independent lacquered metal spheres, one
magnetized, the other containing the bulb,
permitting random orientation, 9 7/8x5 1/8 inches
diameter (25x13 cm)
Arteluce. Gift of the manufacturer

Giotto Stoppino, Lodovico Meneghetti, Vittorio
Gregotti
Table lamp with rotating head, model 537. 1967
(1968)
Lacquered aluminum, 7 7/8x3 3/4 inches diameter
(20x9,5 cm)
Arteluce. Gift of the manufacturer

60

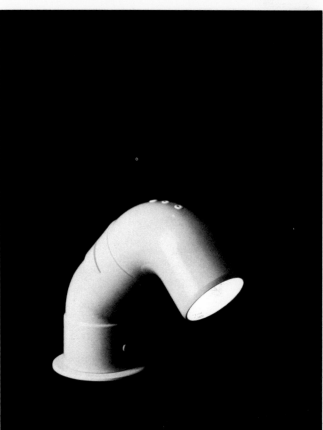

Gino Sarfatti
Lamps, 600 G (grande) and 600 P (piccolo). 1965
(1966)
Artificial leather base filled with heavy pellets
permitting random orientation, metal shade;
G: ,7 1/8x2 3/8 inches (18x6 cm), P: 5 7/8x2 3/8
inches (15x6 cm)
Arteluce. Gift of the manufacturer

Cini Boeri
Lamps with rotating heads, models 602 and 1098
1968 (1969)
PVC plastic, 602:47 1/4x4 3/8 inches diameter
(120x11 cm), 1098: 13 3/4x5 7/8 inches diameter
(35x15 cm)
Arteluce. Gift of the manufacturer

61

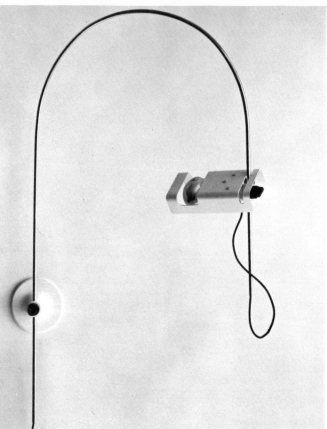

Danilo and Corrado Aroldi
Periscopio ('Periscope') adjustable table lamp
1966 (1966)
Lacquered metal, 18 1/2x4 inches diameter
(47x10 cm)
Stilnovo. Gift of the manufacturer

Joe Colombo
Spider wall lamp with adjustable height, swinging
stem, rotating reflector. 1965 (1965)
Metal, base and reflector lacquered, 39 3/8x23 5/8
inches (100x60 cm)
O-Luce. Gift of the manufacturer

Marcello Cuneo
Longobardo table lamp. 1967 (1967)
Ceramic, 6 1/4x3 7/8 inches diameter (16x10 cm)
Gabbianelli. Gift of the manufacturer

Gruppo Architetti Urbanisti Città Nuova
Nesso ('Nèssus') table lamp. 1962 (1963)
ABS plastic, 13 3/4x21 5/8 inches diameter
(35x55 cm)
Artemide. Gift of the manufacturer

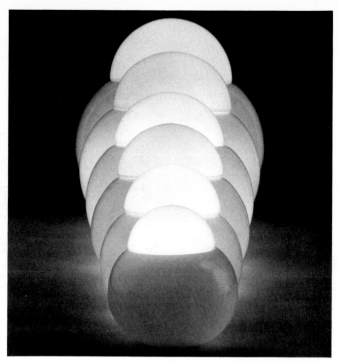

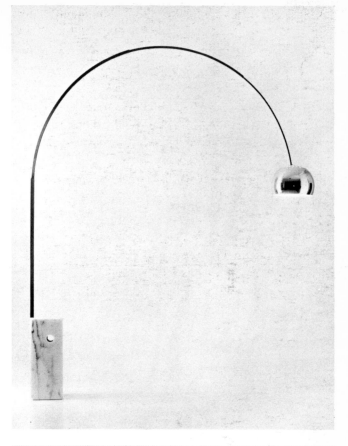

Opposite:

Giancarlo Mattioli
MT lamp. 1969 (1970)
Lacquered metal, 18 1/2x17 3/4 inches diameter
(47x45 cm)
Sirrah. Gift of the manufacturer

Sergio Asti
Daruma lamp (set of six). 1968 (1969)
Opaline glass, 5 7/8, 7 1/2, 9, 10 7/8, 13 1/4,
and 15 3/4 inches diameter (16, 19, 23, 27,5,
33,5, and 40 cm)
Candle. Gift of the manufacturer

Giampiero and Giovanni Bassi
Crack lamp with glass top functioning as switch
1968 (1968)
Lacquered metal and opaline glass, 3 1/2x8 5/8
inches diameter (9x22 cm)
Studio Luce. Gift of the manufacturer

Eleonore Peduzzi-Riva (Italian, born Switzerland)
Vacuna floor lamp. 1965 (1965)
Blown glass, 16 1/2 inches diameter (42 cm)
Artemide. Gift of the manufacturer

Above and right:

Achille and Pier Giacomo Castiglioni
Arco ('Arc') adjustable floor lamp. 1962 (1962)
Marble base, stainless steel shade, aluminum
reflector, 8 feet 2 1/2 inchesx6 feet 7 inches
(250x200 cm) maximum
Flos. Gift of the manufacturer

Vico Magistretti
Giunone ('Juno') floor lamp with four rotating
shades. 1969 (1970)
Lacquered metal, 7 feet 6 1/2 inchesx29 7/8
inches diameter (230x76 cm)
Artemide. Gift of the manufacturer

65

Achille and Pier Giacomo Castiglioni
Taraxacum small hanging lamp. 1960 (1960)
Metal structure, cocoon-sprayed with synthetic
fiber, 20 1/2x25 1/4 inches diameter (52x64 cm)
Flos. Gift of the manufacturer

Achille and Pier Giacomo Castiglioni
Splügen Bräu/C hanging lamp with adjustable
head. 1961 (1961)
Aluminum with varnished counterweight, 8 1/8x
14 1/8 inches diameter (205x36 cm)
Flos. Gift of the manufacturer

Gianfranco Frattini and Livio Castiglioni
Boalum segmental flexible lamp. 1969 (1970)
Plastic and metal, each segment 35 1/2x2 3/8
inches diameter (90x6 cm)
Artemide. Gift of the manufacturer

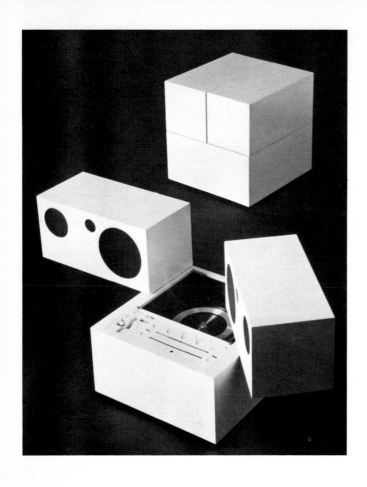

Mario and Dario Bellini
Totem Hi-Fi set with detachable speakers, model
RR 130 Fo-St. 1970 (1972)
Metal case, 20 1/2x20 1/8x20 1/8 inches
(52x51x51 cm)
Brionvega. Gift of the manufacturer

Marco Zanuso and Richard Sapper (Italian, born Germany)
Doney television set. 1961 (1962)
Plastic case, 11 3/4x13 3/4x11 3/4 inches (30x36x30 cm)
Brionvega. Gift of the manufacturer

Marco Zanuso and Richard Sapper (Italian, born Germany)
Black television set. 1969 (1970)
Semitransparent black acrylic plastic case, 10 5/8x12 5/8x11 3/4 inches (27x32x30 cm)
Brionvega. Gift of the manufacturer

69

Mario Bellini
GA 45 Pop automatic record player. 1968 (1969)
ABS plastic case, 9x8 1/4x3 1/4 inches
(23x21x8 cm)
Minerva. Gift of the manufacturer
70

Marco Zanuso and Richard Sapper (Italian, born Germany)
TS 502 portable radio. 1964 (1965)
ABS plastic case in two equal parts, hinged, 5 1/4x8 5/8x5 1/4 inches (13x22x13 cm)
Brionvega. Gift of the manufacturer

Rodolfo Bonetto
Magic Drum portable radio, tuned by turning upper half, hinged annular FM antenna. 1968 (1969)
ABS plastic case, 4 3/4x4 inches (12x10 cm)
Autovox. Gift of the designer

71

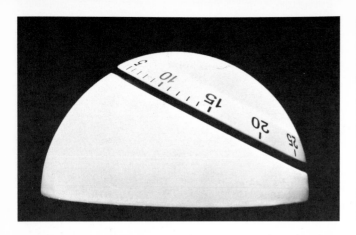

Rodolfo Bonetto
Contaminuti ('Minute counter') timer. 1962 (1963)
ABS plastic. 2 1/8x4 1/2 inches diameter
(5,2x11,5 cm)
Veglia Borletti. Gift of the designer

Rodolfo Bonetto
Sfericlock alarm clock. 1963 (1963)
Plastic case, 3 3/4 inches diameter (8,5 cm)
Veglia Borletti. Gift of the designer

Gino Valle
Cifra 3 ('Figure 3') synchronized digital clock
1965 (1965)
Plastic case, 3 3/4x7 1/8x3 3/4 inches
(9,5x18x9,5 cm)
Solari. Gift of the manufacturer

Pio Manzù
Battery clock. 1968 (1968)
ABS plastic case, 3 1/8x2 3/4 inches diameter
(8x7 cm)
Italora. Gift of the manufacturer

Marco Zanuso and Richard Sapper (Italian, born
Germany)
Knife-sharpener. 1963 (1964)
ABS plastic case, 4 7/8x3 5/8 inches diameter
Necchi. Gift of the manufacturer

Marco Zanuso.
Scale. 1969 (1969)
Lacquered metal, foam pad covered in artificial
leather, 3 1/4x11 3/4x13 inches (8x30x33 cm)
Terraillon. Gift of the manufacturer

73

Marco Zanuso and Richard Sapper (Italian, born
Germany)
Grillo ('Cricket') telephone, with built-in dial. 1965 (1967)
ABS plastic, 2 3/4x6 1/4x3 1/4 inches (7x16x8 cm)
closed
Sit-Siemens. Gift of the manufacturer

Ettore Sottsass, Jr. (Italian, born Austria)
Valentine portable typewriter. 1969 (1969)
Plastic housing and carrying case, 4 1/2x13 1/2
x13 3/4 inches (11,4x34,4x35,1 cm)
Olivetti. Gift of the manufacturer

75

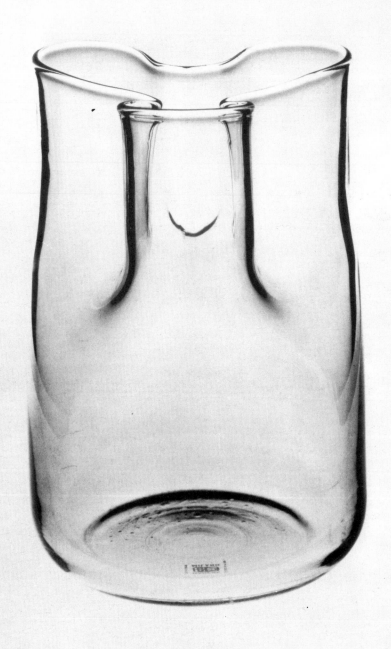

Enzo Mari
Trilobed carafe, model 3094. 1969 (1971)
Blown glass, 6 3/4x4 1/2 inches diameter
(17x11,5 cm)
Danese. Gift of the manufacturer

Massimo Vignelli
Max 1 stacking dishes. 1964 (1964-71)
Melamine, 9 7/8x9 7/8x15 1/2 inches
(25x25x38,75 cm)
Heller. Gift of the manufacturer

Giuliana Gramigna
Pomona place settings and serving dishes
1969 (1969)
Ceramic, plates 9 3/4, 11, and 16 1/8 diameter
(25, 28, and 41 cm), soup plate 8 5/8 inches
diameter (22 cm), serving dish 20 inches
diameter (51 cm)
Gabbianelli. Lent by the manufacturer

Enzo Mari
Table bowls, models 3089 A, B, C, D. 1969 (1969)
Melamine, 2 3/8, 3 1/2, 4 3/4, and 3 1/2x4 3/4,
6 1/4, 9 1/2, and 11 3/4 inches diameter
(6, 9, 12, 24 and 30x12, 16, 24, and 30 cm)
and 30 cm)
Danese. Gift of the manufacturer

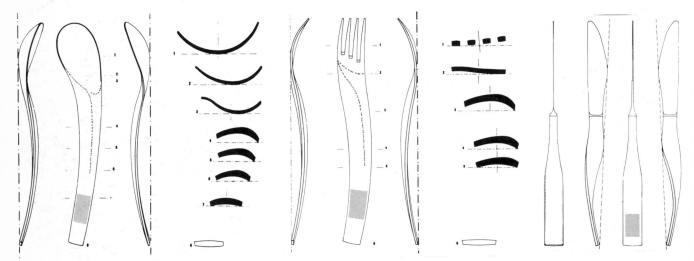

Roberto Mango
Flatware. 1959 (1960-68; prototypes)
Silver.
Reed and Barton. Lent by the manufacturer

Designs for knife, fork, and spoon submitted in
Reed & Barton design competition for Italy,
December 1959

Roberto Sambonet
Center Line set of cooking utensils. 1965 (1971)
Stainless steel, largest 6 3/4x12 1/4 inches
diameter (17x31 cm)
Sambonet. Gift of the manufacturer

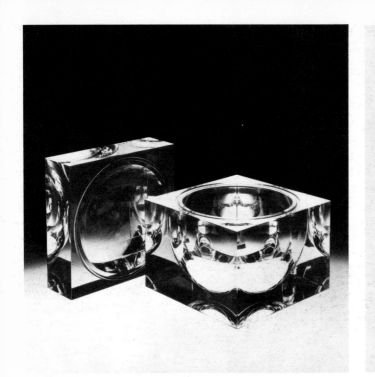

Studio TG
Bombo ('Bumble-bee') ice bucket. 1970 (1970)
Transparent acrylic plastic, 8 5/8x8 5/8x8 5/8
inches (22x22x22 cm)
Guzzini. Gift of the manufacturer

Studio OPI
Ice-bucket cube, model 123. 1971 (1971)
Plastic, 6 1/2x6 1/2x6 1/2 inches diameter
(16,5x16,5x16,5 cm)
Cini & Nils. Gift of the manufacturer

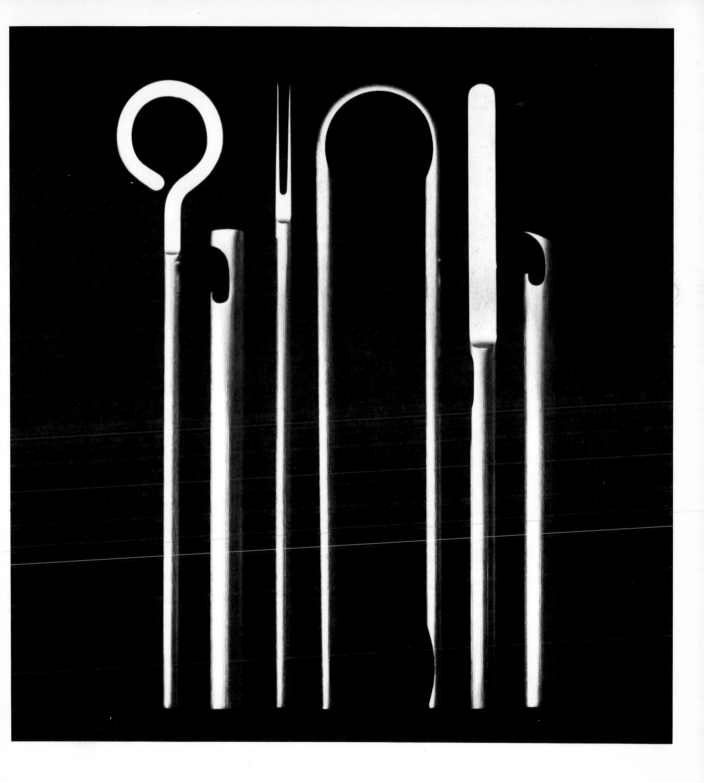

Studio OPI
Cylindrical bar set 1969 (1969)
Stainless steel, 8 1/4 and 6 1/4 inches high (210
and 160 cm)
Cini & Nils. Gift of the manufacturer
81

Angelo Mangiarotti
Ashtray, model 4000 B. 1963 (1964)
Ceramic, 2x7 7/8 inches diameter (5x20 cm)
Danese. Gift of the manufacturer

Roberto Arioli
DA 8 ashtray. 1965 (1965)
Ceramic, 1 1/2x9 1/2 inches diameter (4,5x24 cm)
Gabbianelli. Gift of the manufacturer

Ennio Lucini
Mangiafumo ('Smoke-eater') ashtray. 1968 (1968)
1968 (1968)
Ceramic, 2 3/8x4 3/4 inches diameter (6x12 cm)
Gabbianelli. Gift of the manufacturer

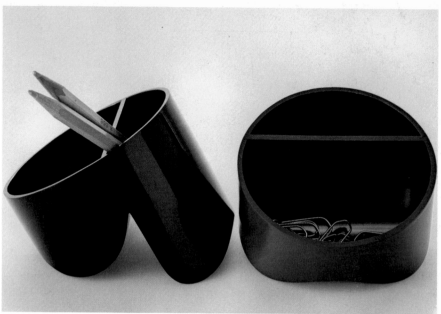

Giorgio Soavi
Paperweight with ball. 1964 (1964)
Stainless steel, 2 3/4x2 7/8 inches diameter
(7x7,5 cm)
Olivetti. Gift of the manufacturer

Enzo Mari
Colleoni pen and pencil holder. 1969 (1970)
ABS plastic, 3 1/4x3 3/4 inches diameter
(8x10 cm)
Danese. Gift of the manufacturer

Gian Nicola Gigante, Marilena Boccato,
and Antonio Zambusi
Mattia umbrella stand. 1970 (1970)
Ceramic, 6 4/4x15 3/4 inches diameter (17x40 cm)
SICART. Gift of the manufacturer

Studio OPI
Revolving magazine rack. 1968 (1969)
ABS plastic, 12 1/4x11 3/4 inches diameter
(31x30 cm)
Cini & Nils. Gift of the manufacturer

Roberto Arioli
Uno Team 11 ashtray. 1970 (1970)
Marble, 3 7/8x10 5/8 inches diameter (10x27 cm)
Gabbianelli. Gift of the manufacturer

Roberto Arioli
Uno team 12 marble dish. 1970 (1970)
Marble, 2 3/4x18 7/8 inches diameter (7x48 cm)
Gabbianelli. Gift of the manufacturer

Eleonore Peduzzi-Riva (Italian, born Switzerland)
Fruit bowls, models 619 and 618. 1971 (1971)
Blown glass with stainless steel insert slotted
for drainage, 4 3/8x15 and 5 1/8x11 3/8 inches
diameter (11x38 and 13x29 cm)
Vistosi. Gift of the manufacturer

Detail: draining insert seen from above

 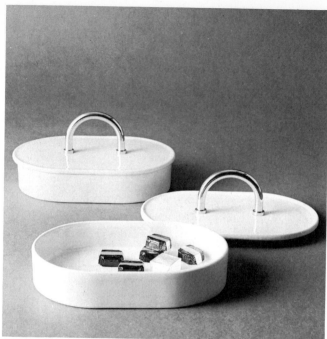

Eleonore Peduzzi-Riva (Italian, born Switzerland)
Large dish, model S 621. 1971 (1971)
Blown glass, 3 7/8x19 5/8 inches diameter
(10x50 cm)
Vistosi. Gift of the manufacturer

Roberto Arioli
Containers. 1968 (1968)
Ceramic with metal handles, 3 1/2x8 1/2x5 5/8
inches (9x21,5x14,5 cm)
Gabbianelli. Gift of the manufacturer

Sergio Asti
Collina ('Hill') sculptural flower holder in four
sections. 1968 (1968)
9 1/2x15 inches diameter overall (24x38 cm)
Cedit for Gavina. Gift of the manufacturer

Sergio Asti
Marco vase. 1962 (1962)
Glass, 8 5/8x4 1/4 inches diameter (22x10,5 cm)
Salviati. Gift of the manufacturer

88

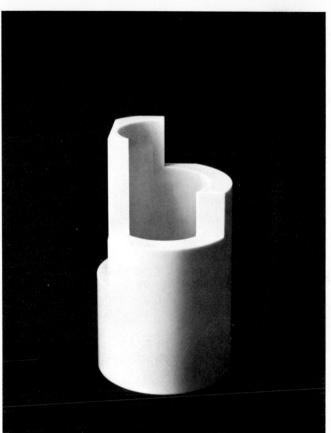

Enzo Mari
Set of six single-flower vases. 1961 (1962)
Marble, each 5 7/8x2 inches each side (15x5 cm)
Danese. Gift of the manufacturer

Enzo Mari
Marmo H ('Marble H') vase. 1963 (1964)
11 3/4x7 1/8 inches diameter (30x18 cm)
Danese. Gift of the manufacturer

Enzo Mari
Reversible vase, model 3087. 1969 (1969)
ABS plastic, 11 3/4x7 7/8 inches diameter
(30x20 cm)
Danese. Gift of the manufacturer

Enzo Mari
Tortiglione ('Spiral') set of two vases, model
3083 B. 1968 (1969)
PVC plastic, 7 1/8 and 14 1/4x9 7/8 inches
diameter (18 and 36x25 cm)
Danese. Gift of the manufacturer

Enzo Mari
Bambù set of five vases, models 3084 A, B, C, D
1968 (1969)
PVC plastic, 15 3/4, 14 1/2, 13 3/8, 8 5/8, and
11 3/4x4 7/8 inches diameter (40, 37, 34, 22, and
30x12,5)
Danese. Gift of the manufacturer

91

Objects selected for their sociocultural implications

The objects within this group are those whose formal characteristics
are derived from, or motivated by, the semantic manipulation of
established sociocultural meanings.

'Some, taking their cue from Pop art, adopt forms from the manmade elements that compose our milieu, presenting them transformed in little else but scale' (pp. 95-97).

'For a few designers, the cultural premises predominant today have no validity and can therefore provide only a false basis for any formal inquiry. Devoid of any firm referents, they return, in a somewhat self-deprecatory attempt at purification, to the human figure as the source of all formal truth' (p. 98).

'Other designers seek neither to add anything to, nor to alter, the profile of our environment; they use the device of giving their designs the guises of nature' (pp. 99-101).

'Conversely, others satisfy the same intention by assembling their designs solely from already existing industrial elements, recovered from the surrounding industrial landscape; by this recycling, they avoid the proliferation of new formal matrices' (p. 102).

'Confronted with the erosion of the simplistic doctrine of functionalism, some designers produce objects whose function is not evident from their form, and whose structural properties, in fact, contradict the behavior one would expect from that form. In such cases, no longer does "form follow function" but, on the contrary, aggressively conceals it' (p. 103).

'Recognizing that the object in our society often serves as a fetish, some designers underscore that quality by assigning to their designs an explicitly ritualistic quality. The object is given sculptural form and conceived as an altarpiece for the domestic liturgy' (pp. 104-106).

'For some designers, the object can be stripped of its mystique only if it is tamed, if it is made to assume the role of house pet. Reduced to graspable size, the object no longer intimidates us; endowed with the stability of inert matter, and created for no specific function, such objects can be allowed into our home in the certainty that they will never evidence the passage of biological or social time' (p. 107).

'Other groups are more concerned with ironic manipulation of the sociocultural meanings attached to existing forms, rather than with changing those forms. Specifically, they design deliberately kitsch objects, as a way of thumbing their noses at objects created to satisfy the desire for social status and identification' (p. 108).

'Some of these designers are involved in a process of revival. For the most part, they restate forms created by the earliest modern movements in design — mainly, Art Nouveau and the Bauhaus — since these forms have by now become understood, and the complex set of ideas that they once connoted have by now become explicit. Sometimes, however, in their nostalgia for the past, they reach back even to medieval times' (p. 109).

Paolo Lomazzi, Donato D'Urbino, and Jonathan
De Pas
Joe sofa. 1970 (1971)
Polyurethane covered in leather, 33 7/8x65 3/4
x41 3/8 inches (86x167x105 cm)
Poltronova. Gift of the manufacturer
95

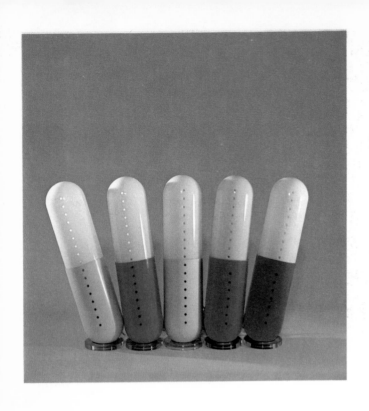

Cesare Casati and Emanuele Ponzio
Pillola ('Pill') lamp. 1968 (1969)
21 5/8x5 1/8 inches diameter (55x13 cm)
ABS plastic and acrylic, 21 5/8x5 1/8 inches
diameter (55x13 cm)
Ponteur. Gift of the manufacturer

96

Gaetano Pesce
Moloch floor lamp, articulated to flex and swing
1970/71 (1972); design derived from four-time
enlargement of original by Naska Loris, Sweden
Metal alloy shaft, steel base and shade,
7 feet 6 1/2 inchesx9 feet 5 inches (230x312 cm),
base 33 7/8 inches diameter (86 cm), shade
25 1/2 inches diameter (65 cm)
97

Alfredo Pizzo Greco
Venere 70 ('Venus 70') whatnot. 1970 (1970)
Transparent acrylic plastic, 59x17x24 3/8 inches
(150x43x62 cm)
Acerbis. Gift of the manufacturer
(The object at right in form of a foot is a stand
in expanded polyurethane by Gaetano Pesce; not
in exhibition)

Gaetano Pesce
UP 5 Donna ('Lady') armchair with UP 6 hassock
connected by chain. 1969 (1969)
Polyurethane covered in stretch fabric, 40 1/2x
47 1/4x51 1/8 inches, open (103x120x130 cm)
C & B. Gift of the manufacturer

Piero Gilardi
I Sassi ('The Rocks') set of seats. 1967 (1968)
Polyurethane, varying sizes, largest 17 3/4x23 5/8
inches diameter (45x60 cm), smaller 9 7/8x13 3/4
(25x35 cm) and 7 7/8x5 7/8 (20x15 cm)
Gufram. Gift of the manufacturer

Right, above, and opposite, above:

Giuseppe Raimondi
Cirro ('Cirrus') triple mirror, model 3127
1970 (1970)
Two-colored glass, 21 1/4x21 1/4 inches
(54x54 cm)
Cristal Art. Gift of the manufacturer

Marcello Pietrantoni and Roberto Lucci
Nuvola ('Cloud') ceiling lamp. 1966 (1966)
Plexiglas, two elements, a, 29 1/8x7 7/8x57 1/8
inches (74x20x145 cm), b, 29 1/8x29 3/4x57 1/8
inches (74x30x145 cm)
Stilnovo. Gift of the manufacturer

Gino Marotta
Dalia ('Dahlia') wall or ceiling lamp. 1968 (1968)
Plastic, 11x21 5/8 inches diameter (26x55 cm)
Poltronova. Gift of the manufacturer

Right, below, and opposite, below:

Superstudio (Adolfo Natalini, Cristiano Toraldo
di Francia, Roberto and Alessandro Magris,
Piero Frassinelli)
Passiflora ('Passion-flower') floor lamp. 1968
(1968)
Plastic, 15x11 3/4x9 inches (38x30x23 cm)
Poltronova. Gift of the manufacturer

Bracciodiferro
Archizoom (Andrea Branzi, Gilberto Corretti,
Dario and Lucia Bartolini, Massimo Morozzi,
Paolo Deganello)
Sanremo floor lamp. 1968 (1968)
Lacquered metal and perspex, 7 feet 10 3/4
inches x37 3/8x37 3/8 inches (240x95x95 cm)
Poltranova. Gift of the manufacturer

Gruppo Strum (Giorgio Ceretti, Piero Derossi,
Riccardo Rosso)
Pratone ('Big Meadow') mat. 1970 (1971)
Polyurethane, 37 3/8x57 1/2 inches
(95x145x136 cm)
Gufram. Gift of the manufacturer

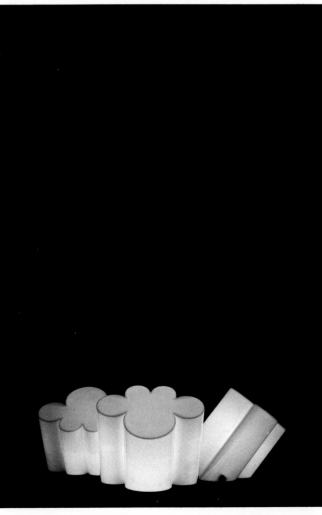

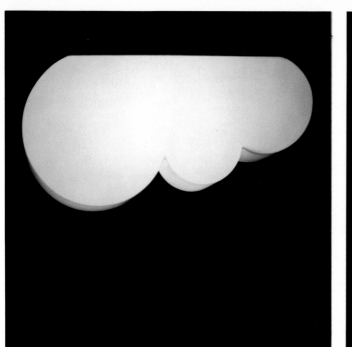
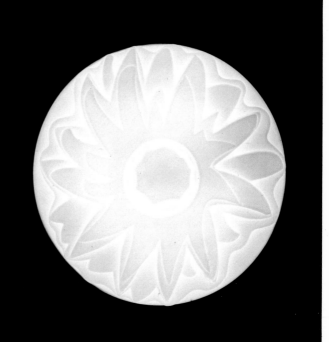
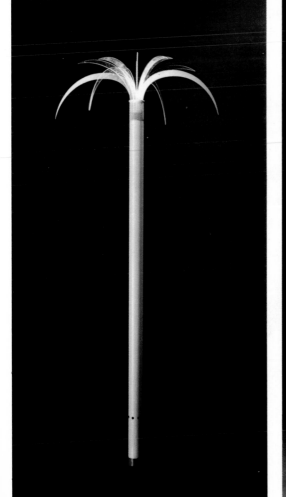
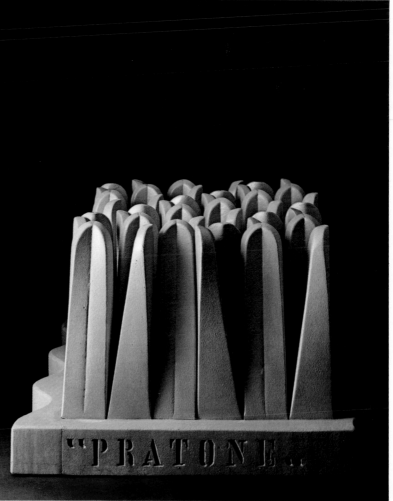

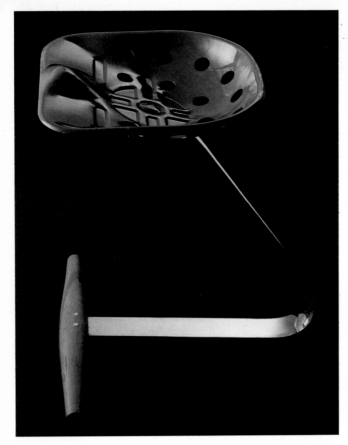

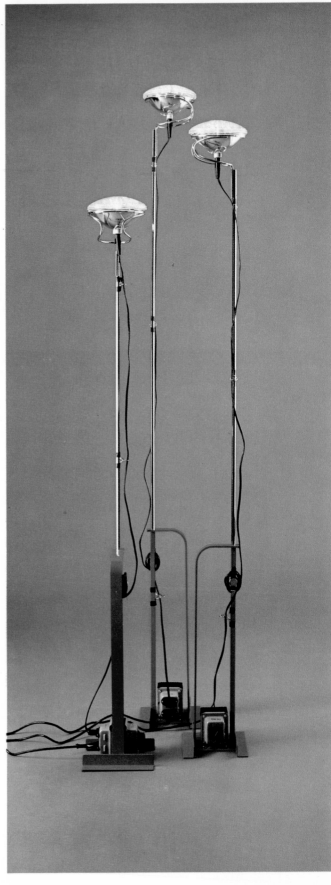

Achille and Pier Giacomo Castiglioni
Mezzadro seat. 1955 (1970)
Metal and wood, 20x19 3/4 inches diameter
(51x50 cm)
Zanotta. Gift of the manufacturer

Achille and Pier Giacomo Castiglioni
Toio adjustable-height floor lamp. 1962 (1962)
Lacquered steel base, nickeled brass shaft,
6 feet 7 inches maximum x8 1/4 inches (200x21 cm)
Flos. Gift of the manufacturer

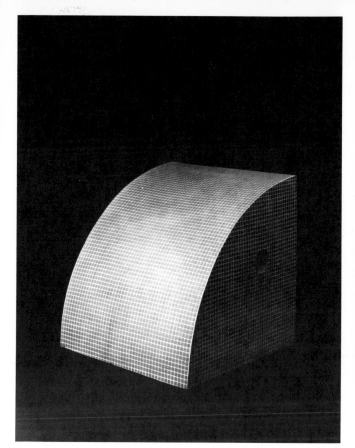 

Giorgio Ceretti, Piero Derossi, Riccardo Rosso
Torneraj ('You'll Come Back') armchair
1969 (1969)
Polyurethane, 35 1/2x35 1/2x35 1/2 inches
(90x90x90 cm)
Gufram. Gift of the manufacturer

Archizoom (Andrea Branzi, Gilberto Corretti,
Dario and Lucia Bartolini, Massimo Morozzi, and
Paolo Deganello)
Mies armchair. 1969 (1969)
Chromed metal and rubber, 31 1/2x29 1/8x51 1/2
inches (80x74x131)
Poltronova. Gift of the manufacturer

Ettore Sottsass, Jr. (Italian, born Austria)
Models of wardrobes. 1966 (1966; prototypes)
(not in exhibition)

104

Ettore Sottsass, Jr. (Italian, born Austria)
Cupboards. 1966 (1971; prototypes)
Plywood and plastic laminate, 33x11 3/4
x12 1/4 inches (84x30x31 cm)
ABET-Print. Gift of the manufacturer

Opposite, above:

Ettore Sottsass, Jr. (Italian, born Austria)
Yantra vases, Y 23, Y 15, Y 37. 1970 (1970)
Ceramic, Y 23: 9 5/8x9 5/8 inches
(24,1x24,5x24,5 cm), Y 15: 9 1/2x4 3/2x20 1/2
inches (24x12x52 cm), Y 37: 8 1/4x8 1/4x20 7/8
inches (21x21x53 cm)
Poltronova. Gift of the manufacturer

Below:

Ettore Sottsass, Jr. (Italian, born Austria)
Asteroide ('Asteroid') lamp. 1968 (1968)
Perspex and wood, 27 1/2x11x6 1/4 inches
(70x28x16 cm)
Poltronova. Gift of the manufacturer

Above:

Hans von Klier. Italian, born Czechoslovakia
Gli Animali ('The Animals') miniature drawers,
min. 7, 2, and 6. 1969 (1969)
Lacquered wood, min. 7: 41 3/8x15 3/4x12 1/4
inches (105x40x31 cm), min 2: 15 3/4x15 3/4x
x18 7/8 inches (40x40x31 cm), min. 6: 15 3/4x
x15 3/4x7 1/8 inches (41x40x18 cm)
Planula. Gift of the manufacturer

 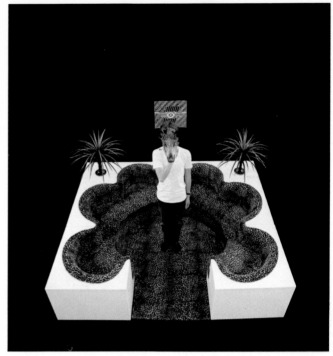

Archizoom (Andrea Branzi, Gilberto Corretti,
Dario and Lucia Bartolini, Massimo Morozzi,
Paolo Deganello)
Superonda ('Superwave') convertible sofa bed
1966 (1967)
Expanded resin with plastic fabric covering,
39 3/8x7 feet 10 3/4 inches x15 inches
(100x240x38 cm)
Poltronova. Gift of the manufacturer

Archizoom (Andrea Branzi, Gilberto Corretti,
Dario and Lucia Bartolini, Massimo Morozzi,
Paolo Deganello)
Safari two-part composite sofa. 1968 (1968)
Lacquered wood covered in artificial leopard
skin, each element 25 5/8x33 1/2x33 1/2 inches
(65x85x85 cm)
Poltronova (not in exhibition)

Archizoom (Andrea Branzi, Gilberto Corretti,
Dario and Lucia Bartolini, Massimo Morozzi,
Paolo Deganello)
Beds. 1967 (miniature models; never produced)
(not in exhibition)

108

Vico Magistretti
Golem chair. 1969 (1969)
Lacquered wood, 49 1/4x19 3/4x22 inches
(125x50x56 cm)
Poggi. Gift of the manufacturer

Sergio Asti
Charlotte chair, 1968 (1968)
Chromed steel, upholstered seat covered in velvet,
25 5/8x23 5/8x20 1/2 inches (65x60x52 cm)
Zanotta. Gift of the manufacturer

Sergio Asti
Démodé vase. 1969 (1969)
Opaque veined glass, 5 7/8x10 3/4 inches
diameter (12,5x30 cm)
Venini. Gift of the manufacturer

Gae Aulenti
Pipistrello ('Bat') adjustable lamp. 1965 (1966)
Lacquered aluminum base, telescopic shaft
stainless steel, shade perspex, 27 1/2x20 7/8
inches diameter (70x53 cm)
Martinelli-Luce. Gift of the manufacturer

109

Objects selected for their implications of more flexible patterns of use and arrangement

The objects in this section are flexible in function, permit multiple modes of arrangement and use, and propose more informal patterns of behavior in the home than those currently prevailing.

Piero Gatti, Cesare Paolini, Franco Teodoro
Sacco ('Sack') beanbag lounge chair. 1969 (1969)
Polyurethane pellets contained in leather or
plastic sack, 45 1/4x19 3/4 inches diameter,
variable (115x50 cm)
Zanotta. Gift of the manufacturer

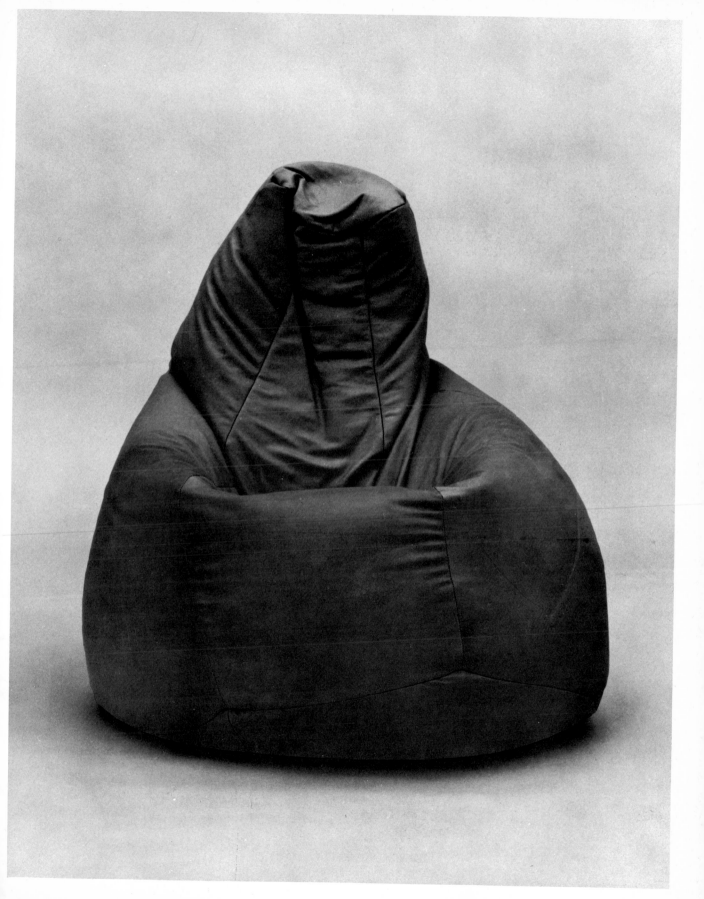

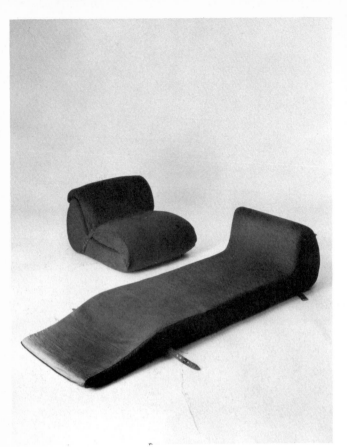 

Umberto Catalano and Gianfranco Masi
Ghiro ('Dormouse') convertible mattress-lounge
chair, belted. 1967 (1967)
Polyurethane covered in fabric, 22x31 1/2x8 feet
6 1/2 inches open (56x80x260 cm), 25 5/8x
31 1/2x45 1/4 closed (65x80x115 cm)
NY Form. Gift of the manufacturer

Jonathan De Pas, Donato D'Urbino, and Paolo
Lomazzi
Galeotta ('Galley') three-position lounge chair
1967 (1968)
Polyurethane covered in fabric, 25 1/4x25 5/8x
37 3/8 inches (64x65x95 cm)
BBB Bonacina. Gift of the manufacturer

114

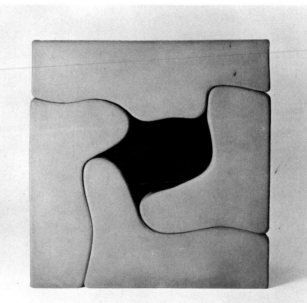

Sebastiano Matta (Chilean, active in Italy)
Malitte cushion system of four combinable seats
and ottoman. 1966 (1967)
Polyurethane upholstered in fabric, 63x63x24 3/4
inches assembled (160x160x63 cm)
Gavina. Gift of the manufacturer

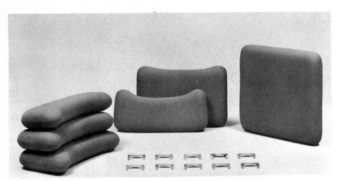

Joe Colombo
Tube Chair of nesting and combinable elements
1969 (1970)
PVC plastic tubes padded with polyurethane,
covered in fabric, 25 1/4x19 3/8 inches diameter
nested (64x50 cm)
Flexform. Gift of the manufacturer

Joe Colombo
Additional System combinable lounge chair and
ottoman. 1968 (1968)
Interchangeable polyurethane blocks covered in
stretch fabric, metal clamps, chair 27 1/2x30 3/4
x30 3/4 inches (70x78x78 cm), ottoman
15 3/4x16 1/2x30 3/4 inches (40x42x78 cm)
Sormani. Gift of the manufacturer

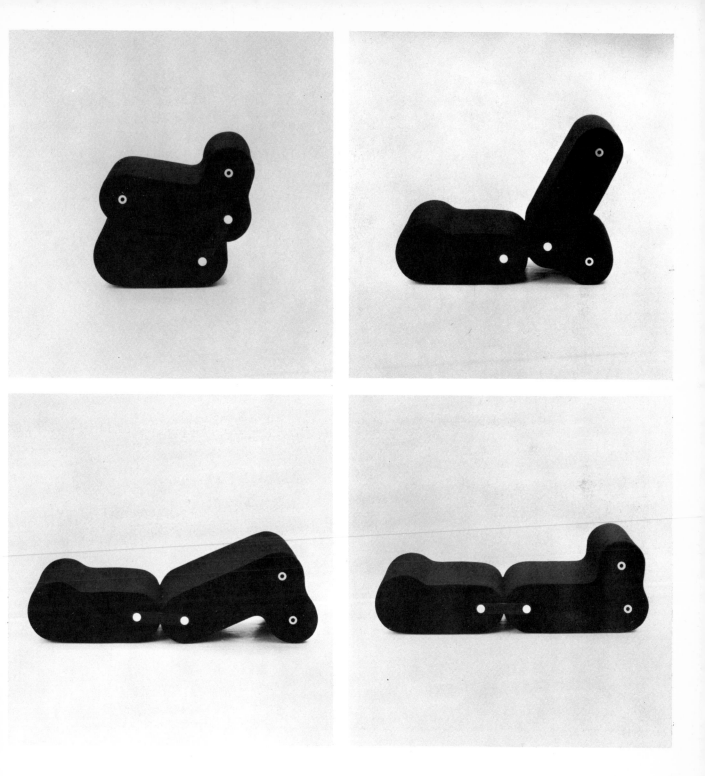

Joe Colombo
Multichair two-element adjustable chair. 1969
(1970)
Polyurethane covered in cloth, 27 1/2x22 3/4x
42 1/2 inches (70x58x108 cm)
Sormani. Gift of the manufacturer
117

Achille Castiglioni
Primate kneeling bench. 1970 (1970)
Artificial leather, 18 1/2x18 7/8x31 inches
(47x48x80 cm)
Zanotta. Gift of the manufacturer

Cesare Casati and C. Emanuele Ponzio
Rocchetto ('Bobbin') three-position armchair
1965 (1966; prototype)
Fiber glass, 27 1/2x31 1/2x33 1/2 inches (70
x80x85 cm)
Arflex. Gift of the manufacturer

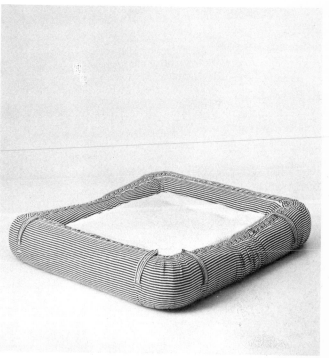

Alessandro Becchi
Anfibio ('Amphibious') three-position convertible
couch. 1971 (1971)
Polyurethane, covered in fabric or leather, 25 1/2
x73x38 1/2 inches, closed (65x184x98 cm)
Giovannetti. Gift of the manufacturer

Mario Bellini
Camaleonda ('Chameleon') unlimited cushion
system. 1970 (1971)
Polyurethane covered in fabric, base cushion
16 1/8x35 1/2x35 1/2 inches (41x90x90 cm),
longer cushion 11 3/4x13 3/4x35 1/2 inches
(30x34x90 cm), shorter cushion 11 3/4x7 7/8x
35 1/2 inches (30x20x90 cm)
C & B Gift of the manufacturer

Cini Boeri
Serpentone ('Jumbo Snake') seat of unlimited
length. 1970-71 (1971)
Polyurethane, each section 14 1/8x14 1/2x35 1/2
inches, variable (36x37x90 cm)
Arflex. Gift of the manufacturer

 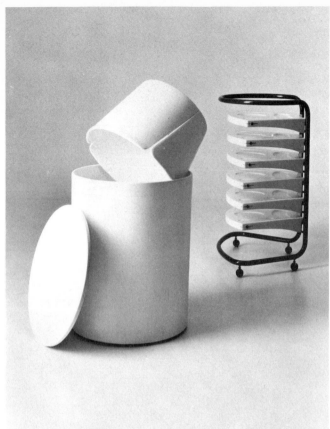

Fabio Lenci
Table and chair set, model 230/1/2. 1970 (1970)
(The central cylinder stores six folding chairs;
the cart on castors holds six trays)
Polyurethane and metal, table 29 7/8x59 inches
diameter, open (76x150 cm), cart 44 7/8x30 1/4
x22 7/8 (114x77x58 cm)
Bernini. Gift of the manufacturer

Achille and Pier Giacomo Castiglioni
Rampa ('Ramp') double-faced stair-shaped unit on
castors, containing desk and shelves. 1963 (1963)
Wood, 51 1/4x30 3/4x59 inches (130x78x150 cm)
Bernini. Gift of the manufacturer

Joe Colombo
Minikitchen on castors. 1963 (limited production)
Wood and stainless steel, 37 3/8x23 5/8x43 1/4
inches (95x60x100 cm)
Boffi. Gift of the manufacturer

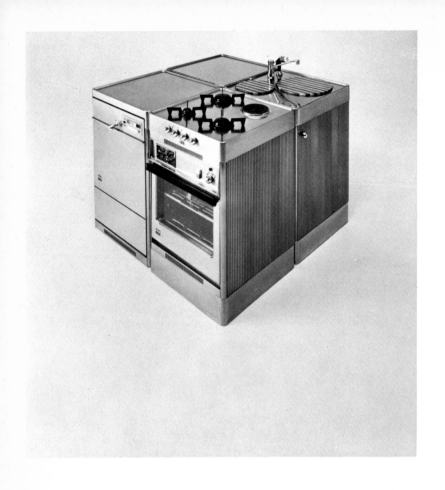

Giancarlo Iliprandi
Arcipelago ('Archipelago') combinable four-piece
kitchen unit. 1970 (1972)
Stainless steel and anodized aluminum, 35 1/2x
47x47 inches overall, closed (90x120x120 cm)
RB. Gift of the manufacturer

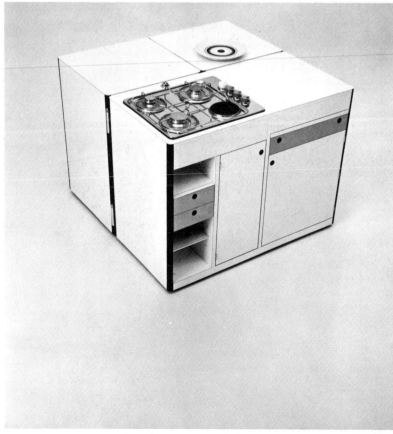

Ufficio Tecnico Snaidero
Centralblock hinged and folding kitchen on
castors. 1969 (1969)
Steel and plywood covered with plastic laminate,
36 1/4x48 7/8x48 7/8 inches, closed
(92x124x124 cm)
Snaidero. Gift of the manufacturer

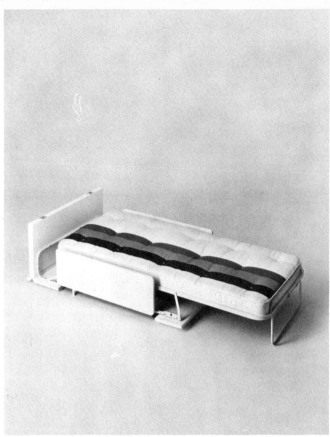

Alberto Salvati and Ambrogio Tresoldi
Tavoletto ('Little Table-Bed') low table on
castors containing folding bed. 1967 (1969)
Lacquered wood, 15x35 3/4x35 3/4 inches, closed
(38x91x91 cm)
Campeggi. Gift of the manufacturer

 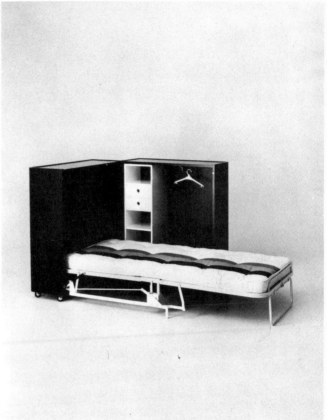

Alberto Salvati and Ambrogio Tresoldi
Armadio-letto ('Wardrobe bed') wardrobe on
castors containing folding bed. 1967 (1969)
Lacquered wood, 37 3/8x25 1/4x35 1/2 inches,
closed (95x64x90xcm)
Campeggi. Grift of the manufacturer

127

Angelo Mangiarotti
Cub 8 component wall system with closets, doors,
shelves, drawers, bed, folding desk, and bar
1966 (1967)
(The name derives from the maximum number of
cubes that can be composed by using elements of
the system meeting simultaneously at one point)
Painted wood, interiors plastic, joints PVC plastic,
each unit 7 feet 10 1/2x17 1/2, 23 5/8, or 35 1/2
x23 5/8 inches (240x45, 60, or 90x45 cm)
Poltronova. Gift of the manufacturer
Right: Diagrams of system of interlocking elements

128

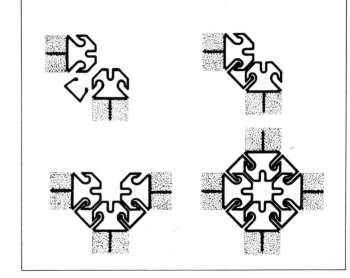

Luigi Massoni (Studio BMP)
A 1 component wall system, wardrobe. 1970 (1970)
Laminated wood, in modular sizes, each element
13 1/4x5 3/4, 17 1/2, 23 5/8, 35 1/2, or 23 5/8
inches (33,5x15,5, 45, 60, 90, or 120x60 cm)
Boffi. Gift of the manufacturer

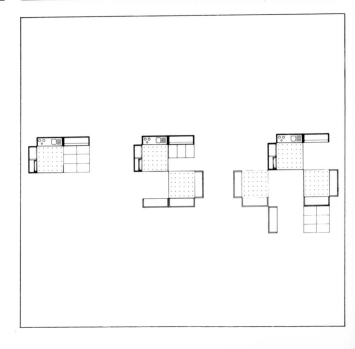

Giancarlo and Luigi Bicocchi and Roberto Monsani
Component wall and ceiling system with kitchen,
convertible bed, cupboard, bookshelves, ceiling
unit with lighting and loud speaker
1971 (1972; prototype)
Plywood and plastic laminate, each unit
6 feet 9 inchesx70 7/8x25 3/8 inches
(210x180x60 cm)
I.C.F. De Padova. Gift of the manufacturer and of
ABET-Print

Bruno Munari
Abitacolo ('Cockpit') habitable structure. 1971
(1971)
Welded steel, varnished, 7 feet 6 inches x35 1/2
x7 feet 1 inch (200x90x190 cm)
Robots. Gift of the manufacturer

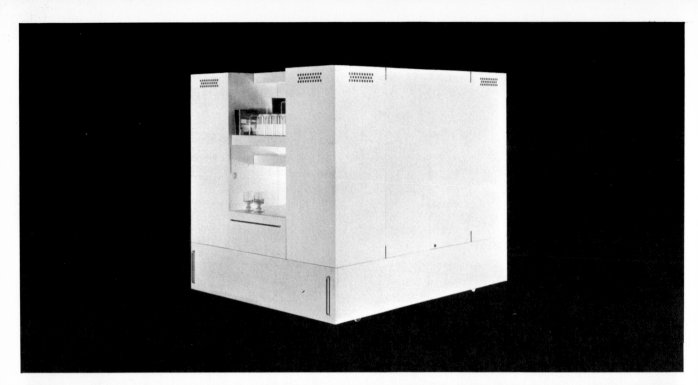

Alberto Seassaro
Central Block, containing bed, table, wardrobe,
toilet, shelves. 1968 (prototype 1970)
Painted wood and chromed steel, 7 feet 6 /34
inchesx6 feet 7 inchesx6 feet 7 inches, closed
(230x200x200 cm), 22 feet 8 3/4 inchesx11 feet
2 inchesx6 feet 7 inches, open (660x345x200 cm)
Gift of the designer

Internotredici (Carlo Bimbi, Gianni Ferrara,
Nilo Gioacchini)
Tuttuno ('All-in-One') single-block unit with bed,
table, sofa, drawers, and shelves. 1969, redesigned
1971 (limited production)
Plywood covered with plastic laminate, 51 1/4x
7 feet 2 1/4x7 feet 2 1/4 inches, closed (130
x222x222 cm)
Giosué Turri. Lent by the manufacturer

ENVIRONMENTS

The environments illustrated in this section have been specially researched, designed, and produced for the exhibition 'Italy: The New Domestic Landscape' by Italian designers and manufacturers. They represent two opposite attitudes to environmental design currently prevalent in Italy.

The first attitude involves a commitment to design as a problem-solving activity, capable of formulating, in physical terms, solutions to problems encountered in the natural and sociocultural milieu. The opposite attitude, which we may call one of counterdesign, chooses instead to emphasize the need for a renewal of philosophical discourse and for social and political involvement as a way of bringing about structural changes in our society.

In order to bring these two design attitudes into focus, the director of the exhibition prepared a special Design Program, with specific and general considerations to be borne in mind by designers of microenvironments and microevents to be presented in the show. It was intended to provide the greatest possible freedom of inquiry. This program was submitted to a number of invited designers of established reputation and also formed the basis of a competition, open only to Italian designers under the age of thirty-five, so that they might have an opportunity of expressing their points of view.

The Design Program (pp. 139-46) asked the participants to consider the recent history of design, in whose first, heroic period modern architects and designers were mainly concerned with arriving at the 'prototypical solution,' that impeccable conceptual vision that animated one generation's long journey from an imperfect *today* to a harmonious *tomorrow*.

But, in their quest for the prototypical solution that might justify the long journey, the pioneer modern architects and designers overlooked the fact that the succession of new perceptual experiences that occur between today and tomorrow would inevitably modify those constants with which they were preoccupied.

It becomes evident that neither the experiences of today nor the vision of tomorrow can be emphasized at the expense of one another, and that the search for quality in daily existence cannot afford to ignore the concomitant problems of pollution, the deterioration of our cities and institutions, and poverty. The task, therefore, is to reconcile the overview with the exigencies of the moment, to be aware of both the goal of the long journey and of the day-to-day activities along the way.

The Design Program for this exhibition was prepared with the specific intention of exploring possible approaches to these problems. One approach is to search for the long-range meanings of the rituals and ceremonies of the twenty-four hours of the day, in order to design the spaces and artifacts that give it structure. The complementary approach is to divest ourselves of the spaces and artifacts inherited by our present culture, in order to arrive at a redefinition of the ideal way to live.

The group representing designers who believe in design as a positive activity were asked to explore the domestic landscape with a sense for its places, and to postulate the spaces and artifacts that give them form, the ceremonies and behaviors that assign them meaning. Special attention was to be paid to new forms and patterns of use emerging as a result of changing life styles, more informal social and family relationships, and evolving notions of privacy and territoriality; as well as to the exploration of new materials and techniques of production. The Design Program also asked the designers to conceive of their environments not as self-contained units, but to explore their ideas further in terms of dwellings, with explicit concern for the context of housing.

By contrast, the exponents of the counterdesign position were invited to elaborate on their designs for destruction of the object and to expound their strategies for cultural change. They were then provided with a platform from which to present their philosophical and political manifestos.

I. Specific Considerations

Swiftly the years, beyond recall.
Solemn the stillness of this spring morning.
— Anonymous Chinese poem, translated by Arthur Waley

The human mind has two main scales on which to measure time. The large one takes the length of a human life as its unit, so that there is nothing to be done about life, it is of an animal dignity and simplicity, and must be regarded from a peaceable and fatalistic point of view. The small one takes as its unit the conscious moment, and it is from this that you consider the neighbouring space, an activity of the will, delicacies of social tone, and your personality. The scales are so far apart as almost to give the effect of defining two dimensions; they do not come into contact because what is too large to be conceived by the one is still too small to be conceived by the other. Thus, taking the units as a century and the quarter of a second, their ratio is ten to the tenth and their mean is the standard working day; or taking the smaller one as five minutes, their mean is the whole summer. The repose and self-command given by the use of the first are contrasted with the speed at which it shows the years to be passing from you, and therefore with the fear of death; the fever and multiplicity of life, as known by the use of the second, are contrasted with the calm of the external space of which it gives consciousness, with the absolute or extra-temporal value attached to the brief moments of self-knowledge with which it is concerned, and with a sense of security in that it makes death so far off.

Both these time-scales and their contrasts are included by these two lines in a single act of apprehension, because of the words *swift* and *still*. Being contradictory as they stand, they demand to be conceived in different ways; we are enabled, therefore, to meet the open skies with an answering stability of self-knowledge; to meet the brevity of human life with an ironical sense that it is morning and springtime, that there is a whole summer before winter, a whole day before night.

I call *swift* and *still* here ambiguous, though each is meant to be referred to one particular time-scale, because between them they put two time-scales into the reader's mind in a single act of apprehension. But these scales, being both present, are in some degree used for each adjective, so that the words are ambiguous in a more direct sense; the *years* of a man's life seem *swift* even on the small scale, like the mist from the mountains which 'gathers a moment, then scatters'; the *morning* seems *still* even on the large scale, so that this moment is apocalyptic and a type of heaven.

Lacking rhyme, metre, and any overt device such as comparison, these lines are what we should normally call poetry only by virtue of their compactness; two statements are made as if they were connected, and the reader is forced to consider their relations for himself. The reason why these facts should have been selected for a poem is left for him to invent; he will invent a variety of reasons and order them in his own mind. This, I think, is the essential fact about the poetical use of language.
— William Empson, *Seven Types of Ambiguity*

In their first, heroic period, modern architects and designers were mainly concerned with arriving at 'the prototypical solution,' that impeccable conceptual model that would lead us slowly but surely from today to tomorrow.

In their quest for that conceptual ideogram that would insure the success of 'the long journey,' they neglected to consider the succession of constant and new experiences, conceptual and perceptual, that occur and recur between today and tomorrow.

Our task, therefore, is to reconcile one time scale with another. One possible approach may be to search for the meanings of the *rituals and ceremonies* of the twenty-four hours of the day, and to design the *artifacts and spaces* that give it structure.

From the point of view of the competition, the particular objective of this Design Program is the designing of a domestic environment, adaptable enough to permit the enactment of different private and communally imagined new events, but at the same time sufficiently fixed to permit the reenactment of those constant aspects of our individual and social memory. The competitor is thus asked to propose microenvironments and microevents: he is *to design the spaces and artifacts* that, singly and collectively, support domestic life; and he is also *to demonstrate the ceremonial and ritual patterns* in which they may be used.

A. Programmatic Considerations

1. The possible activities and ceremonies that customarily constitute domestic life are too numerous to list completely; any one of them represents the continual reinterpretation of man's acts and, even taken singly, cannot be complete, for it remains in the memory to restructure future interpretation and attribution of meanings. Similarly, the physical apparatus that supports the ceremonials and rituals of domestic life occupies not only physical space but also psychological space.

In order to describe the experiential quality of the enactment of domestic ceremonies and rituals in relation to the spaces and artifacts of their settings, the designer should bear in mind at least two considerations. First, he must take into account the mode of participation: whether the ceremony is *private* or *communal*. Second, he must define the character of the space or artifact: whether the space or artifact is *fixed* — that is, specific to only a certain ceremony or ritual — or whether it is *adaptable* and capable of sustaining several ceremonial meanings and roles.

In order to illustrate these categories by example, they may be set out in tabular form:

	fixed	fixed-adaptable	adaptable
private	1	4	7
private-communal	2	5	8
communal	3	6	9

In order to discover the manner of use and the behavioral character of an artifact, it is important to remember that its status will vary according to its context.
(In order not to prejudice the designer, the following examples have been drawn from the exterior domestic landscape.)

Thus, for example, a treehouse is a fixed private domain, until its owner invites his friend, at which point it becomes a communal place; we may thus place it in 1-2-3.
Or the shade of a tree, an eminently adaptable space, can serve as the setting for personal reverie and meditation, for quiet conversation with a friend, or for a picnic; we may thus place it in 7-8-9.
In between fixed and adaptable artifacts we find the sandbox, circumscribed in space, but eminently adaptable to provide physical support to a child's fantasies. A child playing alone finds it a very personal place, although a group of children may enact a communal ceremony of very private meaning; we must thus place the sandbox in 4-5-6.
And so we may list the tool shed as 4; the front steps as 3; a grotto as 1, while the lawn would be 7-8-9, and an ambulatory 1-2-3.

2. Programmatic Option

The competitor is invited to select only one of the two following programmatic options, which describe in turn the *use*, the *role*, and the *ceremonial function* of the object.

Users in option 1 shall be a young couple, represented by M/W.
In option 2, a couple with a child shall be considered, represented by M/W (or M/W w or m).

	OPTION 1	OPTION 2
USERS	M/W	M/W w or m
ROLE	communal-personal (a fusion of the Italian *stanza* and *camera*)	communal (*stanza*)
CEREMONIAL FUNCTION	1. *living*: conversation relaxation work play reception entertaining ___ ___ ___ 2. *cooking-eating* 3. *sleeping*	1. *living*: conversation relaxation work play reception entertaining ___ ___ ___ 2. *cooking-eating* 3. *sleeping* (optional)

Note: Bathroom functions will be assumed to be satisfied outside of the given 'spatial boundaries' of the environment. They are, therefore, *not* to be designed nor included in the general scheme. This rule can be broken *only* if the designer is convinced that its inclusion is essential to his proposal.

Owing to the spatial limitations of the environment, its envisioned *role* will vary according to its *users*. Thus, in option 1, the environment should satisfy both personal and communal needs — the roles traditionally performed by *stanza* (communal) and *camera* (personal); and, in option 2, the environment should restrict itself to the satisfaction of the communal role (traditionally *stanza*).

The ceremonial functions listed here are to serve only as a guide; the list is left open-ended to allow the competitor's own contributions and proposals.

3. Manufacture

The sponsors encourage exploration into the potentialities of the synthetic materials and fibers.

It is the hope of The Museum of Modern Art that the winning environment may be chosen by Italian industry as the basis for the industrial development of an environmental system. It should thus be borne in mind by the competitors that the environments they propose are to be capable of being manufactured in large quantity, and therefore should be envisioned as industrial prototypes. Materials and methods of joining the elements should be selected accordingly.

4. Cost

With the goal of being economically available to low- to middle-income families (Italian average income), the environments should be designed with attention to cost of materials and ease of manufacturing.

B. Exhibition Considerations

1. Spatial limitations

The environments will be presented in a large exhibition gallery at The Museum of Modern Art. This gallery will be painted matte black and will not be lit.

Each exhibited environment will be granted spatial boundaries described by the dimension 3.60 meters high x 4.80 meters width x 4.80 meters depth (12x16x16 feet).

The competitor should note that the assigned spatial boundaries do not correspond to standard construction dimensions (e.g., commonly used floor-to-ceiling heights), and that they have been established for exhibition purposes only. While not giving sufficient room for a full double height, the size of the gallery nevertheless allows the designer to adopt more than one level. This is important to note, for although the competitor is given a large space in which to present his prototype, the proposed environment, once produced industrially, should be intended for the dimensions of housing, whether high- or low-rise. Thus, the elements are to be seen not as isolated pieces of furniture, but as constituents, at a microenvironmental scale, of a range of housing elements to be built by a building industry.

The competitor is thus free to present his proposed microenvironment and microevent in any manner he wishes, within the limits of this spatial boundary. He can occupy *part* or *all* of it; he may use stairs, should he introduce more than one level; but he may *not*, under any circumstances, exceed the assigned boundaries.

On the other hand, should the designer's concept so demand, he may disregard the spatial requirements and present his proposal solely by audio-visual means (films, stage sets, sounds, written texts, etc.). If the proposed environment has vertical planes, the exterior faces of the outermost planes (i.e., those facing the main exhibition space) will be painted black to correspond to the color of the exhibition gallery.

The competitor shall include within his design a supporting podium 40 cm. high (16 inches), plus any other surface required for the physical support of elements: vertical surface(s) or horizontal covering surface(s).

Should the competitor wish to allow the visiting public access within the spatial boundary allotted to him, he may do so by cutting a suitable passage through the podium, which is provided to protect the exhibits from the 3,000 persons who, on the average, visit The Museum of Modern Art daily.

The competitor shall include within the spatial boundaries of his environment, and possibly as an integral part of it, space for at least one 23-inch television screen, located in such a fashion as to be visible from outside the spatial boundary by visitors to the exhibition. This television set will be used to present a film depicting the varying patterns of the ceremonies, rituals, and uses of space of the proposed environment.

2. Light sources

All light sources are to be provided for by the competitor in his submission. As noted, the exhibition hall will be painted matte black and will not be lit.

3. Construction and shipment of the exhibition prototype

It is envisioned that the winning environment prototype to be presented at the exhibition will be constructed in Italy. Once manufactured in Italy, the environment will be assembled for testing and photographing before being shipped to the United States.

The proposed environments are to be shipped to the USA in containers measuring 5.9x2.3x2.2 meters internally and having an access door measuring 2.2x2.1 meters. The environments must therefore be demountable and/or composed of elements of a size small enough to permit their entrance into, and packing within, these containers.

II. General Considerations

The following General Considerations are listed to provide a context to the Specific Considerations. They regard problems of a universal nature, not exclusive to Italy. They are listed here, nevertheless, not only because they do affect and are related to contemporary problems of Italian design, but also because we believe that although some of these problems can certainly not even begin to be solved within the scope of this exhibition and the competition, they ought to form part of the designer's largest frame of reference.

As the designer will soon observe, these General Considerations are not part of a precise ideological sequence, but rather a collage of related thoughts. Not all of them can, therefore, be consistently reconciled, because satisfying the spirit of one consideration may entail excluding others. In short, the intention is not to list a number of precise design requirements that must be satisfied, but rather to render an *impression* of the attitudes that animate this exhibition and competition.

A. The Domestic Landscape as Urban Society

1. By the year 2000, a very high percentage of the world will be living in urban conditions. The processes ensuing from such urban phenomena will not become comprehensible nor manageable solely by resorting to the methods applied to industrial phenomena; any attempt at language renovation and invention can result only from a radical critique and postulation of urban practices and theory.

2. The questions of urban society is to be taken seriously because we should not take it for granted that we shall live in a true urban environment in the sense we take it now. The domestic environment appears to be more readily the basic unit or *individual cell*, for it acts as the direct *connection* between the individual and urban environment. It is *into* this more or less closed individual environment that the individual is able to introduce 'News of the World,' i.e., messages from the immediate socio-cultural world (and from remote historical worlds) which cross (independently) time and space, influencing his ways of thinking, judging and decisions, only to the extent that he wants to pay attention to them.
— Letter from Abraham A. Moles to Emilio Ambasz, January 10, 1971

The domestic environment becomes, then, the theater for the reenactment of forms and scripts introduced from the outside. But in this process, we inevitably assign to these importations other levels of meaning and thereby modify their structure. This procedure allows us to develop different modes of beholding the reality of these forms and scripts. These new modes are, in turn, exported to the outside world, i.e., to other domestic environments. Once taken from the private domain in which their original meaning was reelaborated, and introduced into the social domain (i.e., into the society of domestic environments), these new or modified modes of beholding reality begin to affect the way in which society may act upon that particular reality. Thus, a domestic environment equipped for introducing and transforming 'news from the world' may contribute to the process of making and breaking urban patterns and meanings.

3. An urban society cannot be reduced to its economic base or social *structures. It includes superstructures as well.* It requires not only institutions [for the long journey] but also 'values' and ideas [artifacts and ceremonies], ethics and aesthetics for everyday existence,
— Henry Lefebvre, *The Explosion, Marxism, and the French Upheaval*

Everyday existence is the terrain that supports the social structure.

Contestation... rejects the ideology which views the passive acts of consumption as conducive to happiness, and the purely visual preoccupation with pure spectacle as conducive to pleasure.
— Lefebvre, ibid.

Substitution of this ideology, and the giving of new forms to the terrain of daily existence upon which it presently rests, demands supplementing political imagination with the designing of artifacts and processes which are legitimate extensions of our desire for continuous participation in the self-shaping and self-management of everyday existence.

> Self-design and self-management indicate the road toward the transformation of everyday existence. Language plays in this a necessary role — but it does not suffice; life is not changed miraculously by magic words.
> — Lefebvre, ibid.

The partial contributions that the designer — as one among many others — can make to the transformation of everyday life require developing possibilities of practice and modes of thought capable of creating institutions, artifacts, and patterns of behavior that can be integrated in an individual manner rather than being collectively dissociative.

B. The Domestic Landscape as Family Environment

> ... the modern nuclear family is based on maintaining the incompleteness of its individual members so that 'mother,' 'father' and 'child' have become relatively simple and unyielding roles.
> A member assigned a role cannot [without revolt] be anything but what that role dictates and what its concommitantly assigned space allows.
> — David Cooper, *The Death of the Family*

Complementing the psychological mechanisms that the family as an entity employs to maintain its domination over its members, the present spatial layout and hierarchies of the family environment are the primary physical devices to reinforce present family role assignments.

> In conventional social analysis, the family is characterized as that social institution most resistant to change.
> — Cooper, ibid.

Introducing changes into the present psychological structure of the family, although of capital importance, may not be enough. If we do not, at the same time, change the corresponding structure of the physical environment to make it more adaptive, so that each member may educate and learn from the other, assuming and expressing all the roles that his own inventiveness may suggest, little will really change, since family relations, and, by extension, the physical organization of the house provide models for the structure of many non-familial institutions.

Each family member should have the widest range of roles to express vis-à-vis himself, the family and, by extension, everyone else outside. Suggested solutions to the deformations which occur inside the family have, until now, come mainly from the psychosocial domain, and are not too radical at that: enforcement of parent-child emotional contacts; abolition of the rigid structure and hierarchy of the family, and the consequent abandonment of authority as the principal coordinator of human and family action.

To complement these objectives, any proposed design would have to contemplate individual needs for adaptable and fixed spaces in which to enact private roles, as well as communal spaces in which to express openly previously unrealized and unthought-of relationships. For example, the particular case of the kitchen, the heart of the home in certain cultures, can be reinterpreted by modification of traditional spatial definitions and by the introduction of new home hardware (for example, totally mobile equipment).

C. The Domestic Landscape as Private Domain

1. The psychological properties of the nonhuman environment: influence of built form on behavior.
(See recommended background readings on this subject, p. 146).

2. The metaphysics of the nonhuman environment

The Greeks filled the landscape with psychic and religious meanings: they

gave a spiritual import to rocks, mountains, and sea.
— Siegfried Giedion, *The Architecture of Transition*

As the natural landscape replaced the divine milieu in becoming
Renaissance man's primary area of concern, so a newly emerging
manmade landscape — part sociocultural environment and part
nonhuman environment — has become the all-pervading framework
of contemporary man's thought and imagery.

... the human being is engaged, throughout his life span, in an unceasing
struggle to differentiate himself increasingly fully, not only from his human,
but also from his non-human environment, while developing, in proportion
as he succeeds in these differentiations, an increasingly meaningful relatedness
with the latter environment as well as with his fellow human beings.
— Harold F. Searles, *The Non-Human Environment*

Man assigns values and functions to the objects that constitute the
nonhuman environment that surrounds him, endowing them with the
properties of *intermediaries* that may help him to reconcile the
confrontation between his fears and desires, and the constraints
imposed upon him by the natural, the human, and the nonhuman
environments.

Oh, my beautiful vacuum cleaner who served me so well and whom I so much
wanted and cared for; it does not work anymore. Shall I deposit it into the
cellar, or shall I send it, with the other things to the attic?
— Freely quoted statement from seminar given by Abraham A. Moles,
Princeton University School of Architecture, April 1967

Once a need or a desire is satisfied, the respective artifact or space
that served it begins to undergo a process of stylization.

The artifacts, spaces, and ceremonies of the private domain can thus
be seen, on the one hand, as directly determined by physiological
needs (for example, food), but, on the other hand, they can be seen
as icons whose symbolic contents are, in part, intrinsic to them and, in
part, culturally assigned as layers of semantic change. These
assigned levels of meaning proceed from different sources. One
of the sources is social and comes from values and meanings
extracted from the outside. The other is private and stems from
experiences and ideas that we recover from our individual memory.

In waking life, as well as in dreams, human beings frequently undergo
temporary 'regressions' or adopt new roles, as a means of gaining
release from the demands of interpersonal living and a means of gaining a
restoration of emotional energy so that, refreshed now, they can participate
in more strictly human interpersonal relatedness with new freshness and vigor.
— Searles, ibid.

A radical design operation would, thus, imply not so much returning
to the protohistorical and functional reasons for the artifact's
existence and stripping it bare of any social and private meanings it
may have been assigned, but, rather, becoming conscious of this
meaning-accrual process and designing the domestic environment in
such an adaptive way that it may satisfy the requirements for the
enactment of any play, regardless of its origin — whether Proustian or
Strindbergian.

Furthermore, the environment should be so designed as to permit us
to improvise ceremonies and gestures, the meaning of which we
need not be conscious of at the time of their spontaneous performance,
but of which we may become aware afterward, when we reinterpret
them. Underlying the proposal of such a feasible, nonrepressive
environmental arrangement is the basic premise that man's actions
and visions are irrational; only after the word is pronounced and the
deed committed can we assign it a possible logical structure to
describe its purposes and explain its laws.

Man's action transforms reality as he is triggered and transformed by it. The
organism imposes a schema on the surrounding world. As it develops, as its
relations to the environment become more intricate and creative, that schema
is modified. Thus the fascinating hypothesis that our memory code, far from

being fixed and essentially automatic (as is that of a computer), is itself in a constant process of restructuring. We repack the past for our new need as we travel ahead.
— George Steiner, review of *Beyond Reductionism* (edited by Arthur Koestler and J. R. Smythies)

... the capacity of living things to alter while retaining their identity seems to depend on a subtle play between indeterminacy in the small and determinacy in the gross.
— Ibid.

Suggested Reading

To provide a frame of reference for the designers' considerations, the following texts, reprinted by permission of the respective publishers, accompanied the Design Program in a separate volume. The first two selections were included to define the present 'state of the art,' as an indication of how much (or how little) is known about the relation of the nonhuman environment to human behavior. The other selections were intended to provide a poetical insight into the meanings and images of the domestic environment.

Environmental Psychology: Man and His Physical Setting, edited by Harold M. Proshansky, William H. Ittelson, and Leanne G. Rivlin. New York, Chicago, San Francisco: Holt, Rinehart and Winston, 1970.
1. Passages from the General Introduction and from the Introductions to Parts One, Three, Four, and Five.
2. Chapter 2: Edward T. Hall, 'The Anthropology of Space: An Organizing Model'; Chapter 3: Harold M. Proshansky, William H. Ittelson, and Leanne G. Rivlin, 'The Influence of the Physical Environment on Behavior'; Chapter 4: David Stea, 'Space, Territory and Human Movements'; Chapter 6: Raymond G. Studer, 'The Dynamics of Behavior — Contingent Physical Systems'; Chapter 7: James Marston Fitch, 'Experimental Bases for Esthetic Decision'; Chapter 16: Harold M. Proshansky, William H. Ittelson, and Leanne G. Rivlin, 'Freedom of Choice and Behavior in a Physical Setting'; Chapter 19: René Dubos, 'The Social Environment'; Chapter 26: Roger Sommer, 'The Ecology of Privacy'; Chapter 27: Oscar Lewis, 'Privacy and Crowding in Poverty'; Chapter 28: Alexander Kira, 'Privacy and the Bathroom'; Chapter 34; Alvin L. Schorr, 'Housing and Its Effects'; Chapter 36: Oscar Lewis, 'A Poor Family Moves to a Housing Project'; Chapter 40: Elizabeth Richardson, 'The Physical Setting and Its Influence on Learning'; Chapter 42: P. Sivadon, 'Space as Experienced: Therapeutic Implications'; Chapter 55: Humphry Osmond, 'Function as the Basis of Psychiatric Ward Design.'

Harold F. Searles, M. D. *The Nonhuman Environment in Normal Development and in Schizophrenia.* (Monograph Series on Schizophrenia No. 5). New York: International Universities Press, 1960. Selected passages.

W. H. Auden. *About the House.* New York: Random House, 1959. 'Thanksgiving for a Habitat.'

George Matoré, 'Existential Space' (excerpt). *Landscape* (Blair M. Boyd, editor). Spring 1966.

CIAM 1959 in Otterlo (Documents of Modern Architecture, Vol. I), edited by Jurgen Joedicke. London: Alec Tiranti; Stuttgart: Karl Kramer, 1961.
Passage by Aldo van Eyck.

Martin Pawley, 'The Time House.' *Architectural Design* (London), September 1968.

MANHATTAN:
CAPITAL OF THE TWENTIETH CENTURY
Emilio Ambasz

... once I have grasped it, then an old, as it were rebellious, half apocalyptic province of my thoughts will have been subdued, colonized, set in order.
— Walter Benjamin, Letter to Gerhard Scholem, April 23, 1928

Manhattan, unencumbered by permanent memory, and more interested in becoming than in being, can be seen as the city of that second technological revolution brought about by the development of processes for producing and controlling information rather than just energy. It has, after all, incorporated the worship of communication with the idolatry of the industrial product and, by so doing, has provided the ground for supporting any infatuation with the now as the ultimate configuration of reality. However, seen in a different light, Manhattan may reveal an unforeseen potential for conceiving of a quite different notion of the city.

Manhattan is, in essence, a network. If beheld as an infrastructure for the processing and exchange of matter, energy, and information, Manhattan may be seen either as the overwrought roof of a subterranean grid of subway tunnels and train stations, automobile passages, postal tubes, sewage chambers, water and gas pipes, power wires, telephone, telegraph, television, and computer lines; or, conversely, as the datum plane of an aerial lattice of walking paths, automobile routes, flight patterns, wireless impulses, institutional liaisons, and ideological webs. In any of these roles, the points of Manhattan's network have repeatedly been charged, on and off, with different meanings. Entire systems and isolated elements have been connected to, and processed by, these networks, only to be later removed and replaced by new ones.

Were we willing, for the sake of argument, to suspend disbelief, forget coordinates, and imagine that all present constructions had been completely removed, Manhattan's infrastructure would emerge — in all the complexity of its physical organization, the capacity of its input-output mechanism, and the versatility of its control devices — as the most representative urban artifact of our culture.

Once having freed it in this manner from its current limitations, we might, to further this transfer operation, remove Manhattan's infrastructure from its present context and place it, for example, in the center of San Francisco Bay, on the plains of Africa, among the chateaux of the Loire Valley, along the Wall of China....

Manhattan's infrastructure, thus liberated, belongs to us all. But an infrastructure, though necessary, is not sufficient to make a city. The next step, then, is for us all to undertake the postulation of its possible superstructures. The method may belong either to remembrance or to invention, for — conceived as the idea rather than as the actual configuration — Manhattan's infrastructure provides the framework in which all crystallized fragments rescued from the city of the memory, and all figments envisioned for the city of the imagination may dwell together in an ensemble, if not by reason of their casual or historical relationships (since no reconstruction is hereby intended), then by grace of their affinities. The outcome of such an undertaking may be provocative, yielding, if not actual proposals of superstructures, at least an explicit Inventory of Qualities of everyday existence toward a yet to be defined 'City of Open Presents.'

In a first, retrospective phase, we may, as one of many possible approaches, assemble piecemeal any surviving fragments of memory on the infrastructure:

bologna's arcades ... mandelstam's st. petersburg ... john nash's regent's park ... gabriel's petit trianon ... katsura's promenades to observe the sunset ... mies' barcelona pavilion ... wallace stevens'

Editor's note: This article, originally written in 1969 and first printed in Casabella, XXXV, no. 359-60, 1971, pp. 93-94, was included with the Design Program as an encouragement to the designers to make incursions into imaginary realms.

wind on a wheatfield ... john soane's house ... frank zappa's los angeles ... baudelaire's fleeting instants ... debussy's submerged cathedral ... michael heizer's landmarks ... joan littlewood's fun palace ... ray bradbury's brown clouds ... le notre's gardens of chantilly....

This tearing of the fragment from its former context, this rescuing of the irreducible word from its sentence, involves not only the usual process of design by discriminate selection, but suggests a further process of bringing together whereby, instead of establishing fixed hierarchies, the fragments rescued from tradition are placed on the same level in ever-changing juxtapositions, in order to yield new meanings and thereby render other modes of access to their recondite qualities.

In a second, prospective phase, the form of any superstructure to be assembled on the infrastructure should come from the domain of invention.

But envisioned qualities do not come in wholes. They must be apprehended as they rush by — partially denoting an inverted tradition, or possible states that may become; once grasped, they must be dialectically confronted with the many meanings that can be temporarily assigned to our fragmentary experience of the Present.

If beheld as icons, the architectural and ceremonial forms that these constantly changing superstructures may adopt represent an instance of the perennial state of transaction between the fears and desires underlying the individual's aspirations and the assembled forces of his natural and his sociocultural milieu. If unfolded, these superstructures would provide an insight into the goals and the values of their designer: man, the private being and the member of society.

Expanding the Inventory of Qualities of urban existence by this process of interpreting the meaning of the individual values and goals underlying the invention of superstructures involves bringing the subjective content of these individual values up to a communal, objective level, so that they may be accepted or rejected by the community.

This process of expanding the community's ethical framework involves observing or projecting the possible effects that these values, if implemented, may have on the community, and assessing communally whether these effects should be enhanced or reduced. As the meaning of these superstructures can be interpreted only in the context of the patterns of relationships established with other superstructures, this process generates new meanings, which in turn will require further intepretation. By this reiterative process, the envisioned superstructures assume constructive powers. Insofar as they question the context of the Present, they assign it new meanings; insofar as they propose alternative states, they restructure it.

The environments within this category fall into two groups:
'House Environments' (Gae Aulenti, Ettore Sottsass, Jr., Joe Colombo)
and 'Mobile Environments' (Alberto Rosselli, Marco Zanuso and
Richard Sapper, Mario Bellini).

GAE AULENTI

Gae Aulenti, active as a designer in Milan, took her degree in architecture at the Politecnico in Milan, and has been assistant lecturer on architectural composition at the universities of Venice and Milan. From 1954 to 1962, she was a member of the editorial staff of *Casabella Continuità*; and from 1966 to 1969 served as vice-president of the Association for Industrial Design (ADI). Her projects and designs have been published in both Italian and foreign publications. For her contribution to the Italian Pavilion at the XIII Triennale, Milan, in 1964, she received the First International Prize.

Designer: Gae Aulenti
Patrons: ANIC-Lanerossi; Kartell
Producers: Kartell; with the assistance of Zanotta
Film: directed by Massimo Magri

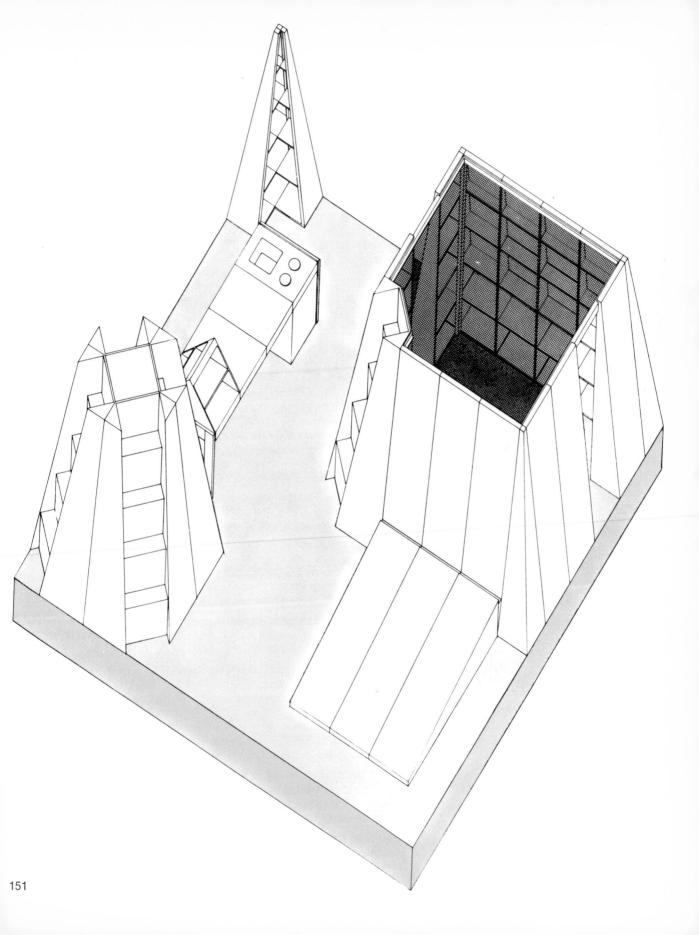

'Nothing is built on stone, all is built on sand, but we must build as if the sand were stone.'
— Jorge Luis Borges, *Fragments of an Apocryphal Gospel*

Architecture is designed beyond the strife of governments, wars, and hunger. Architecture is concrete space, a positive thing that has as its substance the city, in which both private and collective factors join to transform nature through the exercise of reason and memory.

None of man's objects, whether monument or den, can escape its relationship to the city, which is the place where the human condition is manifested. It is only possible, therefore, to analyze the object if we can define it as a dialectic form of the whole; if we can demonstrate how it finds its place, and the laws that bring it into being.

The existence of the object is defined by the actual circumstances of its own relationship to the city, that is, by the relations established between economic and social processes, forms of behavior, norms, techniques — characteristics that, even if they are not expressed by or inherent in the object itself, allow it to come into being and discover its relationship to other objects — to find its place.
This premise is the consolation and hope of reality, the only key to a possible interpretation of the object, eliminating ambiguities, contradictions, discords, and the distortions of our history.

The objects with which we generally deal are extremely numerous; new, for the most part, but also rather unstable and changeable, and at least some of them are doomed to disappear quickly.
Some of them seem to have possible uses that have not been exploited hitherto; they distort behavior, they unleash anxieties and psychoses, are prone to injure the nervous system and cause casualties in mental adjustment, even criminality. Their appearance is based on wrong principles and indifference regarding beliefs and traditions; it is the outcome of a desire 'not to see,' a desire not to interpret the world objectively.

If the condition or general principle for designing an object is as we have described it, this design can find its proper relationship only in a field of which it does not constitute the center.

A domestic environment should be designed in its general form, for its positive qualities can reside only in the sum of the conditions in accordance with which its spatial elements and attribution of meanings approach a synthesis, which is possible only by using and testing all the criteria applied to defining a city.

Our concern, then, is to make things appear in all their complexity and density, even if the result can represent only a limited part of the whole field, a utilization of fragments only.

In the present case, our choice has been restricted to recovering the positive significance of man, who finds fulfilment through creating for himself an artificial atmosphere with an aesthetic intention. This choice, which looks forward hopefully to a more authentic existence for man, the rediscovery of his stable and permanent values, has been a poetic one — an arbitrary selection; it therefore has an emblematic value, alluding to the exorcism of a new reality.

The conscious principle in this design has been to achieve forms that could create experiences, and that could at the same time welcome everyone's experiences with the serenity of an effortless development, determined by an independent critical faculty.

The predominant object of the design is made up of elements so composed as always to make their original purpose evident, while at the same time remaining open to a determination of their future purposes. Use has been made of arrangements of concave and convex spaces, the integration of different types of space; its validity can be checked by comparison with other kinds of values, by the rules for its use, and the areas that it can complement.

Three Possible Interpretatiions

1. Objective description

a. System of three different elements, of which one is linear and two angular. By arranging them differently, they can create areas with the following uses: bed, cupboard, bookcase (either longitudinal or free standing), shelves, seats.
Material: fiber glass, rubber joints, cages of structural metal.
(To be prepared for production)
b. Extendable table with modular service units: a plain surface, a storage surface. Material: steel structure with ABS plastic facing.
(Ready for production)
c. Chair, a corrugated form. Material: fiber glass or rigid polyurethane.
(Ready for production)
d. Lamp, eight rotating elements. Material: metal.
(To be prepared for production)

2. Subjective description

a. Pyramids: form as a precise sign to be placed within reality and serve as the measure for the process of transformation.
b. Ruler and triangle: geometry as the matrix of everyday mutability.
c. Shell: nature as the generating force whereby things come into being, and of their properties and qualities.
d. Fire: allegory as a synthetic and comprehensive representation of an idea through images.

3. Critical description

a) and b) are single elements that express in themselves values that are not self-sufficient, symbolic of a will to create experiences.
c) and d) are two objects that express autonomous values, a self-sufficiency symbolic of a rational will.

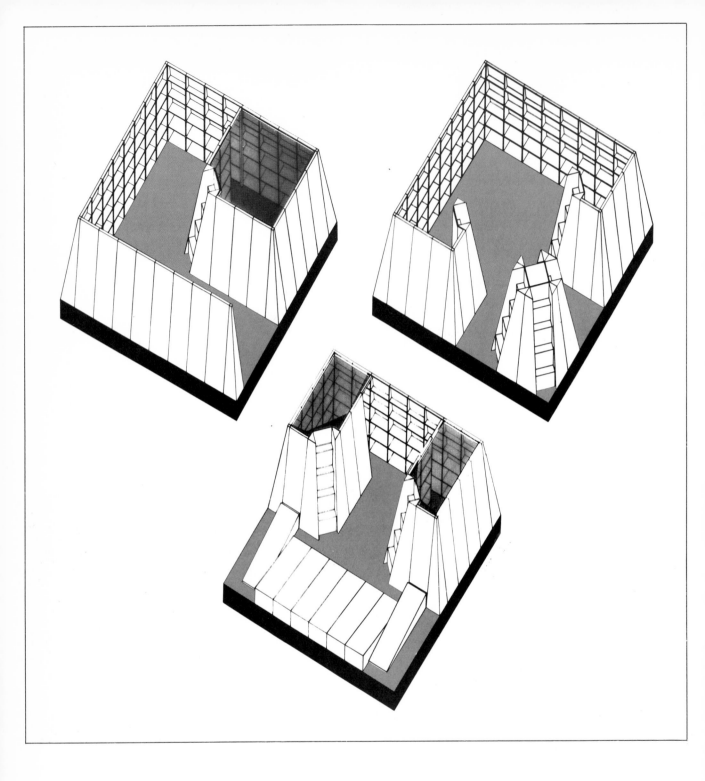

154

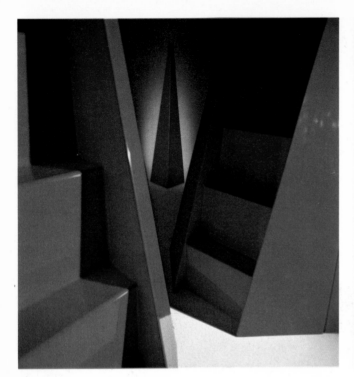

159

ETTORE SOTTSASS, JR

Born in Innsbruck, Austria, in 1907, Ettore Sottsass, Jr., came to Italy in the early 1930s, receiving his professional training at the Architectural School of the University of Turin. He has worked in a wide variety of mediums and materials, including painting, architecture, ceramics, jewelry, furniture, and tapestries, as well as in interior furnishings and industrial design, in which he has specialized since 1955. He has served as consultant for several industries, including Olivetti and Arredoluce, designing for the former a notable series of office machines and electronic equipment. Recently, he has been particularly active in ceramics.

Designer: Ettore Sottsass, Jr.
Patrons: ANIC-Lanerossi, Kartell, Boffi, Ideal-Standard
Producers: Kartell, Boffi, Ideal-Standard; with the assistance of Tecno
Film: directed by Massimo Magri

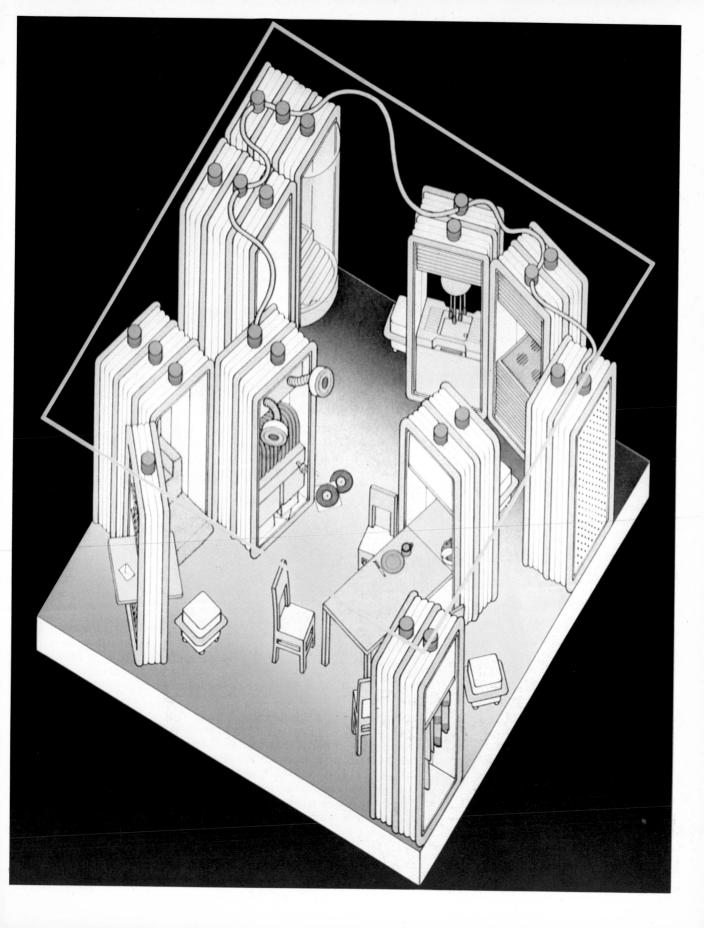

To Nanda, who explained everything to me

Given the time and conditions, and given the general views held by people as well, my pieces of furniture on view in this exhibition can be nothing more than prototypes, or perhaps even pre-prototypes, and thus, if you approach them, you realize that hardly anything really 'works.' You realize that no water flows through the pipes, that the stove doesn't heat, that the refrigerator isn't cold, and so forth; you realize that no 'product engineering' (as they say in industry) has been done. These pieces of furniture, in fact, represent a series of ideas, and not a series of products to be put on the market this evening or tomorrow morning. So I hope no one will wonder how much they cost and where they can be bought, because obviously they are not priced, and they are not on sale anywhere.

The point should be made very clear that the aim of the project was not to achieve a product, but to state and provoke ideas. I wasn't in the least concerned with making furniture, or an elegant, 'cute,' sweet, or amusing environment, and still less was I concerned with designing silent things that would allow the spectator to remain calm and happy within his psychic and cultural status quo (which may nonetheless turn out to be extremely complicated). But perhaps what I really did was the opposite. The form is not cute at all. It is a kind of orgy of the use of plastic, regarded as a material that allows an almost complete process of deconditioning from the interminable chain of psycho-erotic self indulgences about 'possession.' I mean the possession of objects, I mean the pleasure of possessing something that seems to us precious, that seems to us precious because it is made out of a precious material, it has a precious form, or perhaps because it was difficult to make, or may be fragile, etc.

The form isn't cute but rather brutal and even, maybe, rough, and the expected deconditioning process, even if it works in a negative direction, I mean in the direction of eventually eliminating the self-indulgence of possession, will certainly impose a responsibility upon whoever ventures to use these objects. Eliminating the protective layer of alibis we build around ourselves always necessitates great commitment.

To explain this more simply, let's say that the idea is to succeed in making furniture from which we feel so detached, so disinterested, and so uninvolved that it is of absolutely no importance to us. That is, the form is — at least in intention — designed so that after a time it fades away and disappears.

Inside these pieces of furniture, which therefore become mere equipped 'containers' — ordinary boxes — are placed all the other elements that have been invented to supply as efficiently as possible the traditional catalogue of needs our industrial-productive society has drawn up one by one. There is the stove to cook on, the refrigerator to keep the food in, the cupboard for storing clothes, the shower under which to bathe, a place to sit and read a book, a juke-box to emit sounds, a bookcase for books, and so on. The catalogue of needs grows or diminishes according to the culture of the ethnic group to which the user belongs, but the containers remain impassive. They have no formal link with the owner's ethnic group. He will use more or less containers, own more or less boxes, and will finally resolve the problem in terms of quantity rather than quality (as the current phrase goes).

Smoothly running wheels have then been attached to the pieces, so that, even when they are weighed down by whatever is inside them, a child can move them about as he likes with complete ease. The idea is that the elements can be moved closer or farther away from oneself, one's friends, or one's relatives, however and whenever the fancy strikes. So everybody, either as an individual or as the representative of a group, can indicate through his furniture the different situations through which he passes during his private or communal adventure; because the states of need, tragedy, joy, illness, birth, and death always take place within the given area. The pieces of furniture move like beasts of the sea, they diminish or increase, they go to right or left, up or down, they coalesce into colonies, dissolve into dust, solidify into rocks, or soften to plankton, and so on.

Thus, given our advanced technologies, or at least those technologies of which we speak so much, I thought of the possibility of eliminating a series of rigid elements in the building of houses, above all of eliminating the installation systems for services and utilities. It is possible to imagine each person having his personal stock — of liquids, heat, air, refuse, words, sounds, or anything else, which he could carry about with him wherever and whenever he liked.
For such an idea to work, however, we must be able to envisage a society, or groups of people, not inclined to barricade themselves within great walled fortresses; people who don't wish to hide, people who don't feel the need, or perhaps even the unavoidable necessity, to demonstrate continually their imagined status, nor to live in houses that are nothing other than cemeteries containing the tombs of their memories. An idea like this can only work on the assumption that the 'rite' of life, as Emilio Ambasz has called it, can begin a new morning with a new awareness of existence, and on the assumption that our memories (which we know no one can dispense with or eliminate from his own consideration) can remain as memories, without necessarily solidifying into emblems; rather, they can become a sort of living plasma with which, day after day, we can always start over again from the beginning.

The pieces of furniture can then either be joined together with demountable hinges, or detached, or connected with the various outlets — for electric power, water, and air, that are likewise demountable. So not only can these containers be grouped together or dispersed, but they can also continually assume new forms, sinuous as a snake or rigid as the Great Wall of China, they can create areas that are either transparent or closed, deep and narrow, or wide and open. In other words, they can at any moment provide the most suitable setting for the drama that is about to take place, or is already in progress.

I have often wondered what the relation is between an environment and the events that originate and take place in that environment.
There would certainly seem to be some relation between environment and events; and indeed, if there is such a relation, then the idea of this environment of furniture on wheels is that through its neutrality and mobility, through being so amorphous and chameleonlike, through its ability to clothe any emotion without becoming involved in it, it may provoke a greater awareness of what is happening, and, above all, a greater awareness of our own creativity and freedom.

Whether this idea or ambition, this proposal of mine, may be successful or not this time, is another matter. But there is no doubt that, sooner or later, something will be done so that one can put on one's own house every day as we don our clothes, as we choose a road along which to walk every day, as we choose a book to read, or a theater to go to; as we daily choose a day to live, within the limits that destiny or fate impose upon us.

I have only wished to suggest such thoughts, without the slightest intention either of aesthetics or, as it is called, design.

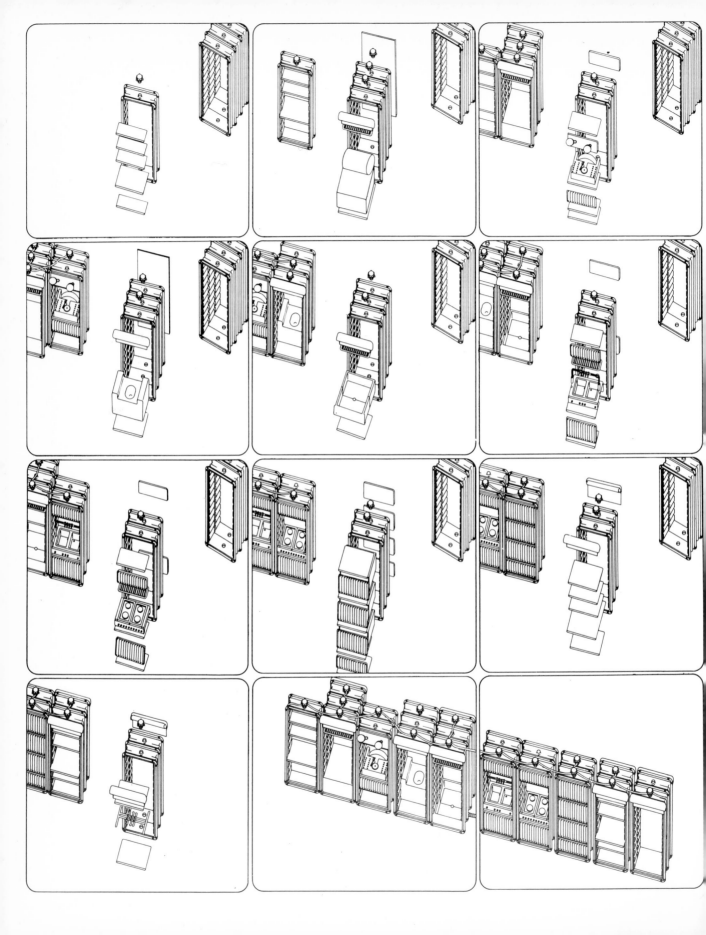

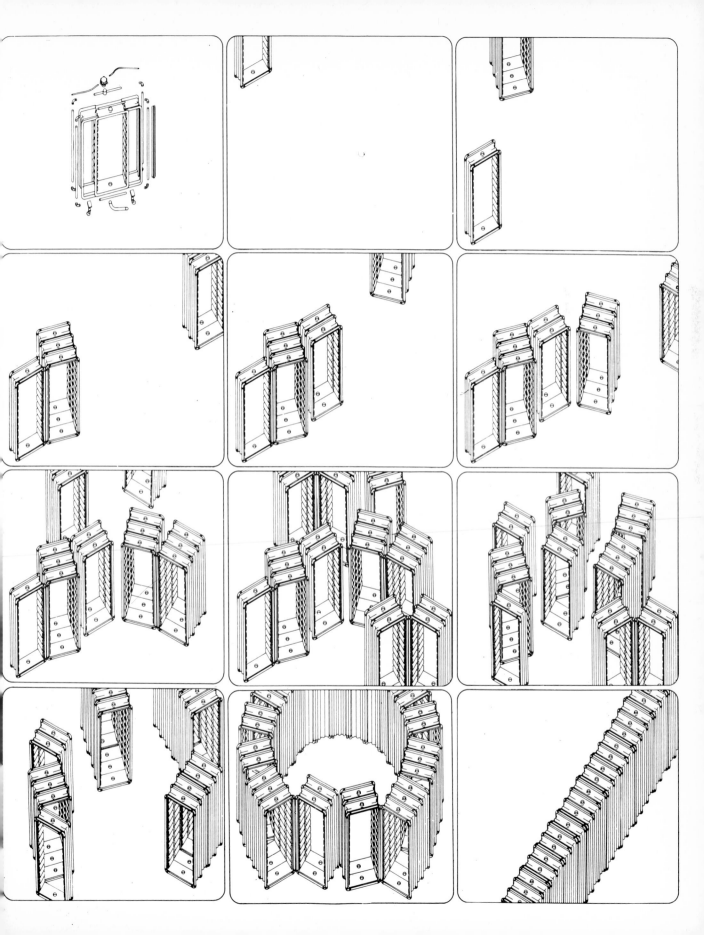

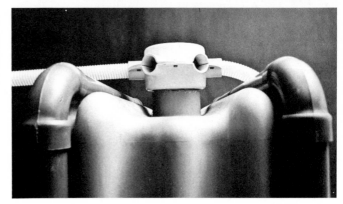

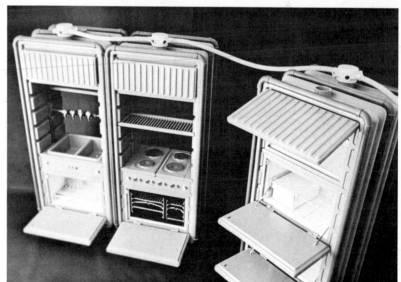

JOE COLOMBO

Joe Colombo was born in Milan in 1930 and died in 1971 at the early age of forty-one, having attained international renown both for his design and his research. Following his early activity in architecture and painting, he concentrated definitively on design from 1962 on, and soon thereafter, working in collaboration with industry and department stores, he began intensive efforts to make good design widely available to all social classes. Although the objects he designed were acclaimed both in Italy and abroad, during the last years of his life he concentrated particularly on problems relating to man's habitat. His researches in ecology and ergonomics led him increasingly to view the individual habitat as a microcosm, which should serve as the point of departure for a macrocosm attainable in the future by means of coordinated structures created through programmed systems of production.

Designer: Joe Colombo; collaborator Ignazia Favata
Patrons: ANIC-Lanerossi, Elco-FIARM, Boffi, Ideal-Standard
Producers: Elco-FIARM, Boffi, Ideal-Standard; with the assistance of Sormani
Film: directed by Gianni Colombo and Livio Castiglioni

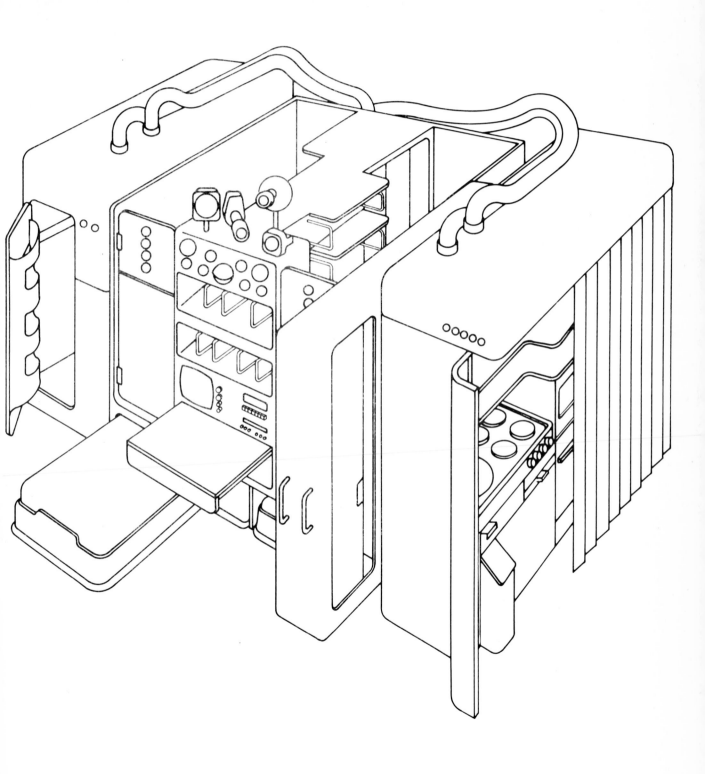

If we accept the idea that homogeneity is the basic premise underlying our designs, then the methodology that has determined them may be explained according to the following relationships:

a. The relation between the city and the dwelling unit
b. The relation between green spaces and the dwelling unit
c. The relation between man and the dwelling unit

The fact that urban centers are not integrated with green spaces, and that they are expanding in a way that is increasingly chaotic and that is conditioning our lives to an ever greater degree, does not mean that our choices must be of a compensatory nature. Indeed, if our homes were to become merely places of refuge, we should be obliged to face every day the nightmare of trying to accommodate ourselves to the outside world.

It therefore becomes necessary to create a dwelling unit that more closely approximates the actual life style of today and tomorrow, but that is also closer to man's true requirements, and thus less restricting and less representative of taste, prestige, and so forth. On the contrary, the dwelling should be adapted more and more to man, rather than the other way around.

It is valid, therefore, to conceive of a unit planned in accordance with mass production, as the facts of life demand; but it should be so defined that all its functions become as nearly perfect as possible.

The space within this unit should be dynamic; that is, it should be in a continual state of transformation, so that a cubic space smaller than the conventional norm can nevertheless be exploited to the maximum, with a maximum economy in its interior arrangement. At this point, one can easily envisage the form that such a proposal should take: a series of suitably equipped 'furnishing units,' freely placed within their allocated areas.

Such furnishing units, developed to serve the various functions of the home and private life, have been differentiated so that, in turn, they may be flexible enough to be adapted to various kinds of space or to differing requirements.

Four different kinds of units are proposed:
Kitchen
Cupboard
Bed and privacy
Bathroom

They function according to the various moments in which they are in use. It is worth noting that, while the kitchen and bathroom serve no other purposes than those for which they were originally intended, the cupboard also acts as a screen between the entrance and the area that will eventually be used at night, or at any rate, it separates the two areas; whereas the day-and-night (bed and privacy) unit comprises within itself all the functions of living — from sleeping, eating, reading, receiving friends, etc., to withdrawing privately to an inner space specifically designed for this purpose.

PATTERNS OF USE

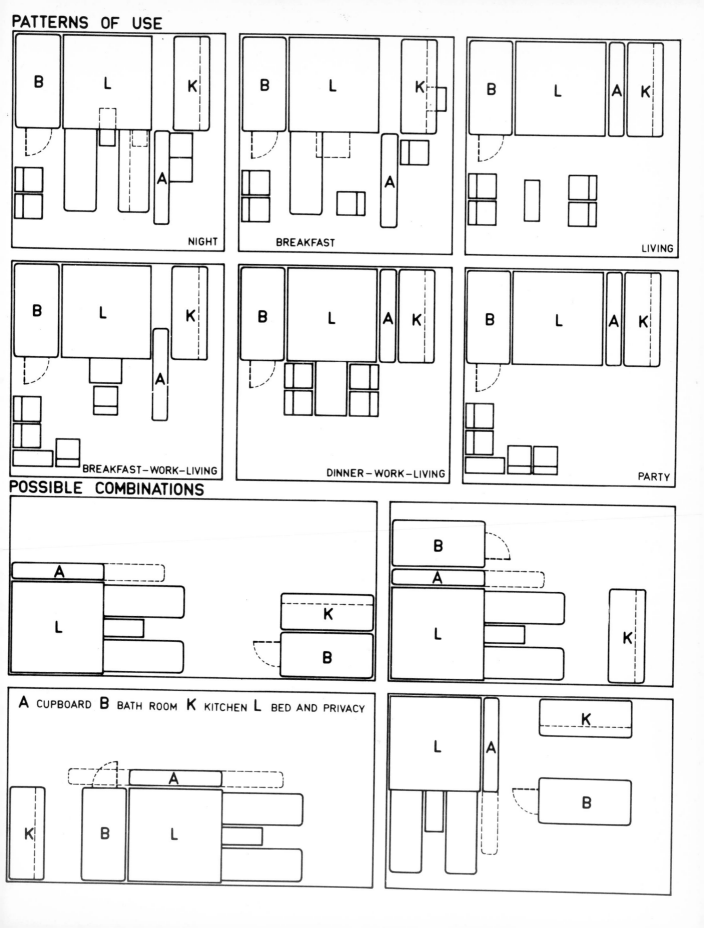

NIGHT

BREAKFAST

LIVING

BREAKFAST—WORK—LIVING

DINNER—WORK—LIVING

PARTY

POSSIBLE COMBINATIONS

A CUPBOARD B BATH ROOM K KITCHEN L BED AND PRIVACY

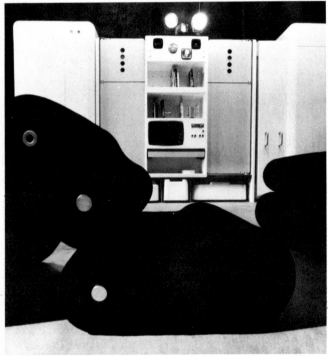

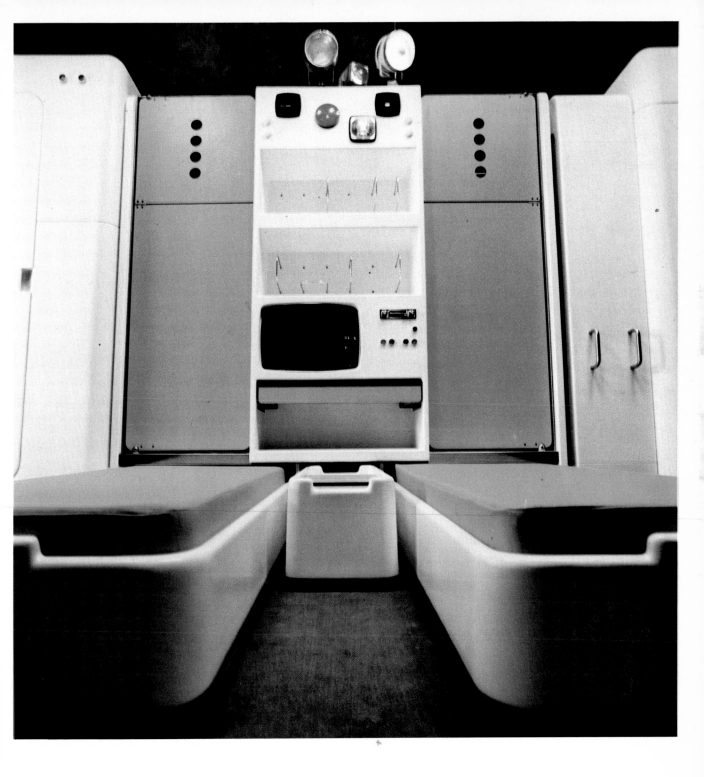

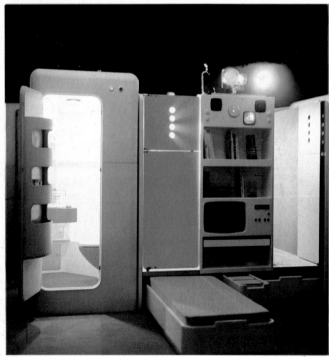

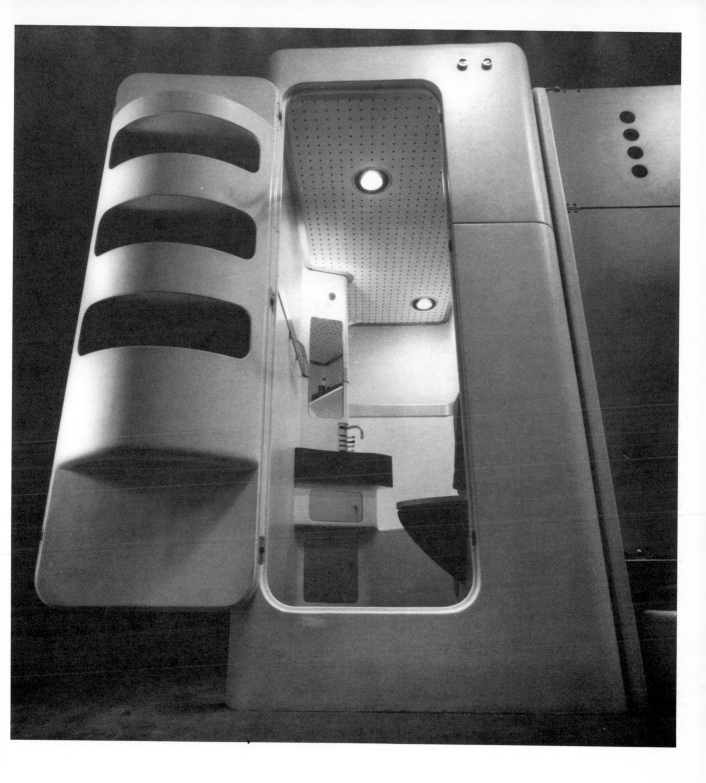

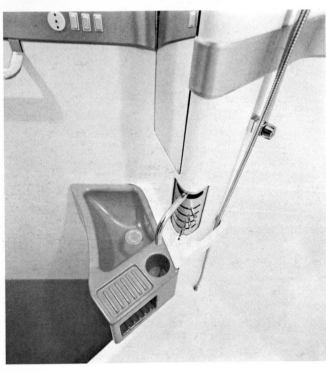
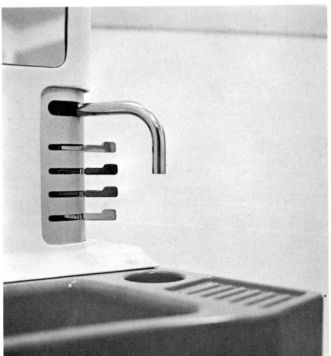

ALBERTO ROSSELLI

Alberto Rosselli, born in 1921, completed his study of architecture at the Politecnico of Milan, and since 1950 has been active in that city as designer as well as architect. In 1953, he founded the periodical *Stile Industria*, which he continued to direct until 1963. He has served as president of the ADI (Association for Industrial Design) and as a member of its Council. Since 1963, he has been professor of industrial design in the Architectural School of the Politecnico. His activity as designer has included serving as consultant to various branches of industry, including product design, furnishings, and transport. In recent years, he has devoted himself particularly to the designing of transport vehicles and works for mass production, especially furniture and sanitary equipment.

Designer: Alberto Rosselli; collaborator Isae Hosoe
Patron: FIAT
Producers: Carrozzeria Renzo Orlandi, Carrozzeria Boneschi, Industria Arredamenti Saporiti, Boffi; with the assistance of Valenti, Nonwoven, Rexedil
Film: CINEFIAT (Ernesto Prever and Osvaldo Marini)

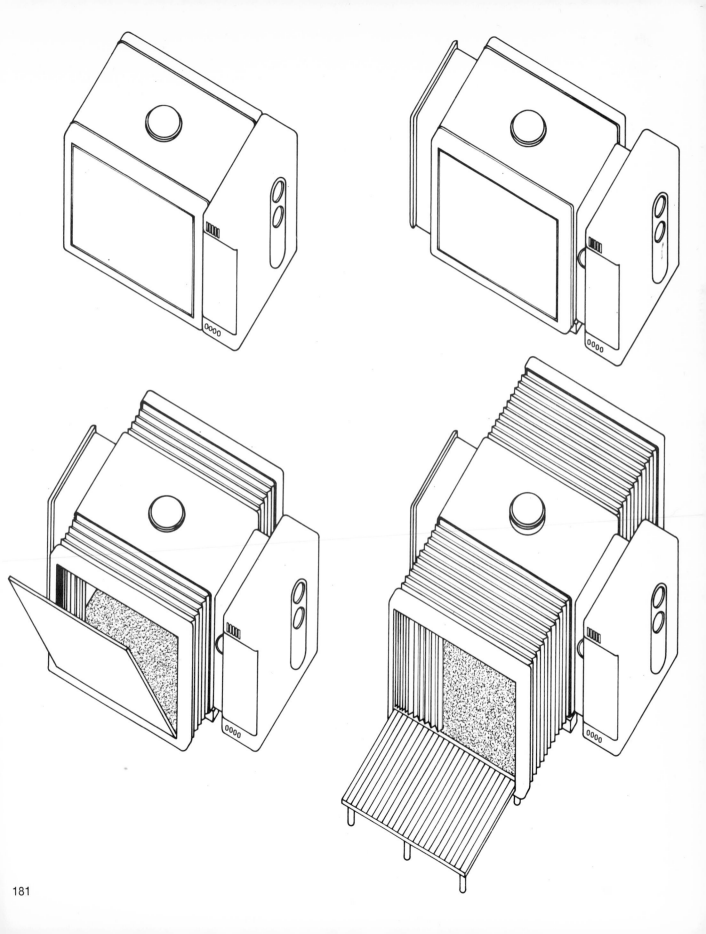

The idea of transforming, or better still, expanding space is closely linked to the idea of the mobile house. The intrinsic mobility of the house-object that is transported from one place to another suggests that the object in fact depends on two conditions, movement and repose, with their differing requirements. Since movement is governed by the circumstances intrinsic to transport, such as road conditions and safety, it demands a small, compact form. Repose means living, and thus a maximum expansion and extension of the potential space available for life and technological requirements. The strictly habitable parts of vehicles designed for the road, therefore, often turn out as a miniature form of a real dwelling, with all living functions reduced to the very meager scale demanded by the road.

But surely we can overcome the limitations of the mobile house by giving it a new form of expression, discovering in it the concept of the mobility of interior space, and of its transformation and connection with other spaces.

Contemporary technology permits us to extend mobility and expansion through the use of lightweight materials and more highly developed mechanisms, for various types of land or air transport.

I wish to propose, first, the possibility of transforming the house-object through an organization of space and technology that would permit an increase not only in its dimensions, but also in its quality. Together with this is a proposal for a new scheme of land use, introducing into the landscape more suitable living receptacles that can be set up or removed with a greater margin of safety: a proposal for the aesthetics of mobility and transformation, as an alternative to a solid block, either stationary or on wheels.

It is indeed possible to envisage a house that conforms to the psychological requirements of life, an object that can be transformed according to the various uses to which it will be put, and that after a certain time can be completely reassembled. This means giving the house not only mobility, but also an interior life of its own, thus offering the user a psychological dimension of space responsive to his own will, with the added pleasure of living in it in different ways, according to the climate and its situation in the landscape.

This mobile environment can be transported by a motor vehicle with a load capacity of between two to two and three-quarter tons (1,500 to 2,000 kilograms). The total size of the vehicle, together with the object to be transported, have been kept within the limits allowed by current European road regulations. During the journey, the object is firmly fixed on the top of an open truck, from which it can be unloaded on arrival and placed on independent supports on the ground. The object has been developed to provide space for use by five or six people, together with modern conveniences and furnishings. Its special characteristic is that it can be transformed from over 1,000 square feet (100 square meters) when in transit, to a maximum of 3,230 square feet (300 square meters) when totally expanded. This expansion takes place on all four sides, by means of a simple system whereby the sides run along telescoping tracks.

Along one axis, the expansion takes place by letting down two platforms and lengthening two folding walls, made of plastic. Along the other axis, two metal capsules containing services and a storage box extend outward.

All necessary furnishings are stored in the interior space during transit; thus, when the object is closed, it is a general container for furnishings and modern conveniences. Two people can easily effect the expansion without the use of any special equipment.

When the area is open, it offers various possibilities for arranging and subdividing the space for both night and day use. It can be divided into three basic areas:

1. A central area, illuminated from above, with service and closet capsules opening off it. It is chiefly intended for use during the day, as a general or dining area.

2. A rear area, with two folding beds and closets attached to the outside walls. It can be shut off by drawing a curtain.

3. A front area, connecting with the central area and a terrace. It is flexible, and can be used as a living area during the day, or as a space for two or three beds at night. The terrace platform can be drawn up to seal off the area. Light plastic curtains allow a simple subdivision of this area according to the wishes of the users, or to allow the furniture to be arranged in different ways.

The capsule is of aluminum throughout and is mounted on a steel frame, to which are attached the tracks for the moving parts. The light materials used for both mobile and fixed structures are the result of technological research on automobile bodies and aeronautics.

Following are other possible ways in which the basic sections might be arranged:

1. As a house with only one expandable side wall.

2. With windowed walls as an alternative to the terrace.

3. With two side walls for beds, and a central or living area.

4. Several capsules might be joined together to provide more space and a wider range of uses.

This mobile house could serve as a model for either individual or group living. In urban and holiday areas, the houses could be so disposed as to make use of common service and utility systems and could be scattered over a stretch of land unencumbered with vehicles.

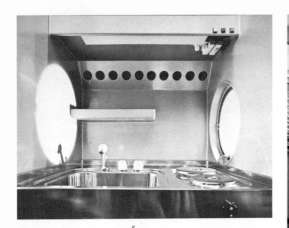

MARCO ZANUSO
RICHARD SAPPER

Marco Zanuso was born in Milan in 1916, and since 1945 has been active in that city in the fields of architecture, design, and city planning. He served as editor of *Domus* from 1947 to 1949 and of *Casabella* from 1952 to 1956. In 1964, he was appointed by the Council for Cultural Cooperation of the Council of Europe to compile and edit a comprehensive study, 'Creation and Industrial Production for Everyday Life — 20 Years of European Experience.' From 1966 to 1969, he was president of the ADI (Association for Industrial Design). He has won numerous prizes for products he has designed and in 1966 received the award of the INA (National Architectural Institute) for Lombardy and Venetia. Since 1970, he has been Director of the Institute of Technology of the Faculty of Architecture at the Politecnico, Milan, as well as holding a professorship there in industrial design and technology.

Richard Sapper, born in Munich in 1932, worked in the styling department of Mercedes-Benz before coming to Italy in 1958. After working first with Gio Ponti and the Rinascente, he became associated with Marco Zanuso in designing products for Brionvega, Kartell, and Siemens. Since 1970, he has been consultant to FIAT for its experimental safety-vehicle programs, and to Pirelli for advanced pneumatic structures.

Designers: Studio Zanuso, Marco Zanuso and Richard Sapper
Patrons: ANIC-Lanerossi, FIAT, Kartell, Boffi
Producers: FIAT, with the participation of Boffi; Kartell
Film: directed by Giacomo Battiato

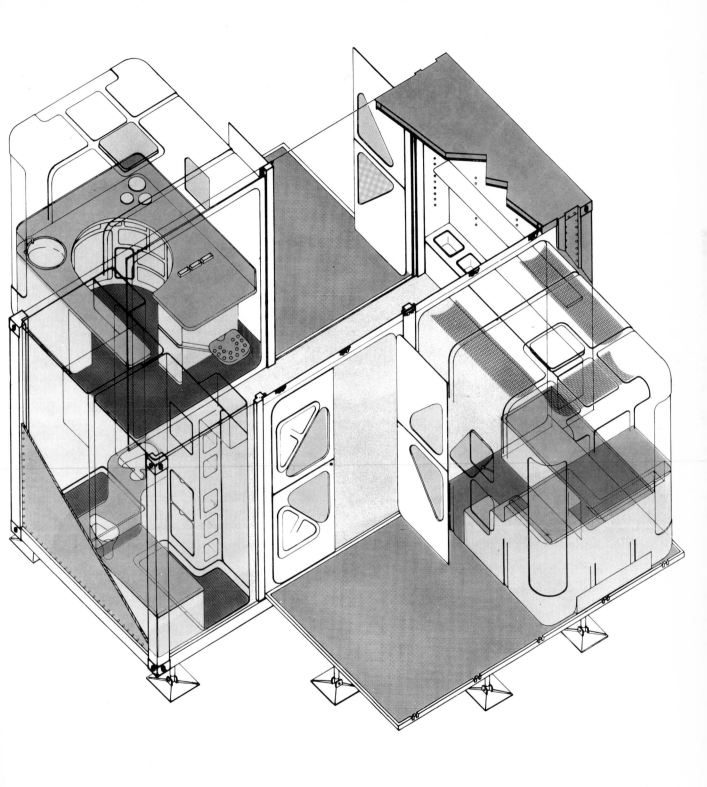

Complete and fully equipped habitations, easily transportable and ready for immediate use. This is the theme.

The theme suggested not so much single habitations to provide city-bound families with a place of occasional retreat, as living quarters for entire communities, transported far from metropolises and urban areas:

a. Working communities engaged in large-scale public works — road and dam construction, land reclamation — for which quarters of a provisional and highly mobile character are needed.

b. Communities of rescue workers carrying out first-aid operations in areas struck by catastrophe, where ready-made, fully equipped living quarters should be available if the workers are not to be diverted from the job in hand.

c. Tourist colonies, where it is necessary to respect the natural surroundings, and where living quarters must be strictly temporary, without permanent structures.

The units are designed for all situations that require immediately available, easily transportable living quarters that do not spoil the natural environment.

This proposal relates less to mobility at the family level than to mobility at the urban level: the immediate transport of communities and living quarters to any part of the world by conventional means of transport.

The prototype is one of a number of similar elements that can be assembled and coordinated to provide living quarters for communities of varying sizes. The units are constructed entirely in the workshop; their outer shell also serves as protective packaging during storage and transport. Like containers, from which they are derived, the units can be stacked, reducing storage and transport space to the minimum.

Eight hundred units — living quarters for sixteen hundred to two thousand people — can be stacked in an area of about 27,000 square feet (2,500 square meters), the size of a normal courtyard. The same number of units can be transported by a 10,000-ton ship; a train could transport up to two hundred and fifty, a large airplane twelve, a truck two, a helicopter one.

The size of the units and their structural characteristics, which are the same as those of a 20-foot container, enable them to be transported by all the above-mentioned means of transport, which means that they can be set up without loss of time anywhere on the face of the earth. Once the unit has been placed on the desired site, it takes only a few minutes to make it ready for use: the time needed to open the two lateral doors and slide the two alcoves out along them horizontally.

The unit is equipped with a water tank, a waste-disposal tank, and an electrical system. With the aid of water distribution and refuse removal by tank trucks, and with electrical power supplied by a generator, the unit is independent of permanent installations for distribution and drainage. The limited weight of the unit (about 3 tons) and its adjustable supports make it easy to eect on any terrain without having to construct foundations.

A 10,000-ton ship would suffice to transport a community of, for example, fifteen hundred persons to any coastal point on any continent in the world. Unloading and placing the units in position could be done by helicopter; the units would then be ready for immediate use.

The prototype is designed to provide living accomodations for two persons, but it can be combined with similar units to furnish living quarters for four, six, or more persons.

The units can be combined either horizontally, or vertically up to a height of three or four stories, making It possible to achieve groups of considerable complexity. The units are to be regarded as provisional living quarters, and therefore their presence on the site will be only temporary, respecting the natural surroundings to the maximum. Once removed from the site, the units are immediately available for further use. A particular advantage of these units for aid and rehabilitation in emergency situations due to natural disasters (earthquakes, cyclones, floods, fires) is that they can be stockpiled in suitably distributed storage depots, making it possible to provide completely equipped living quarters in a very short time, without diverting the work force from their most urgent tasks.

A study of the prototype and its derivatives points the way to the development of similar units suitable for the construction of such community facilities as schools, hospitals, assembly halls, etc., which may be needed to complement living quarters in isolated communities.

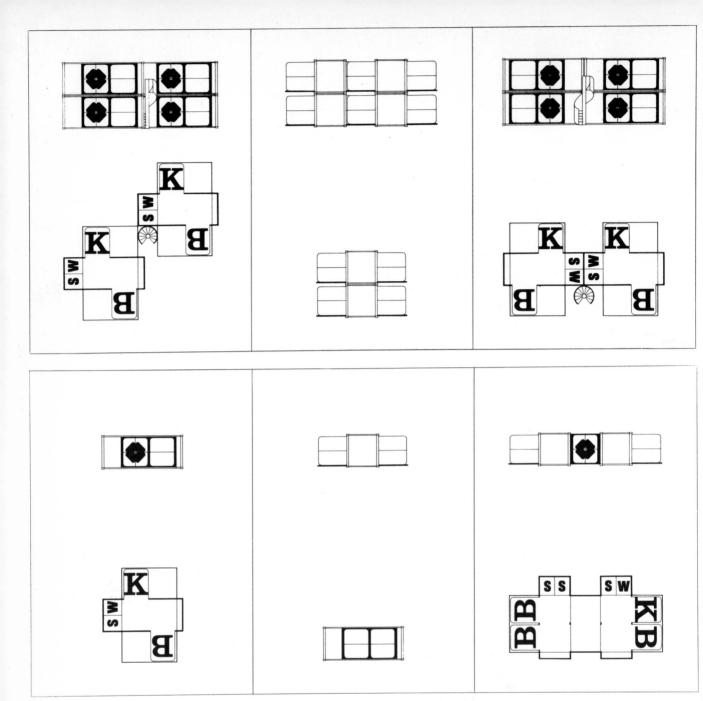

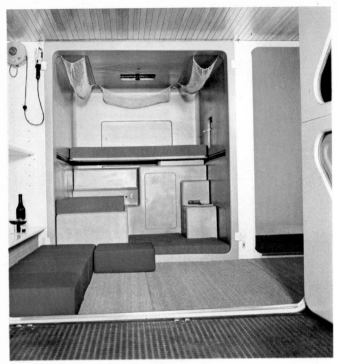

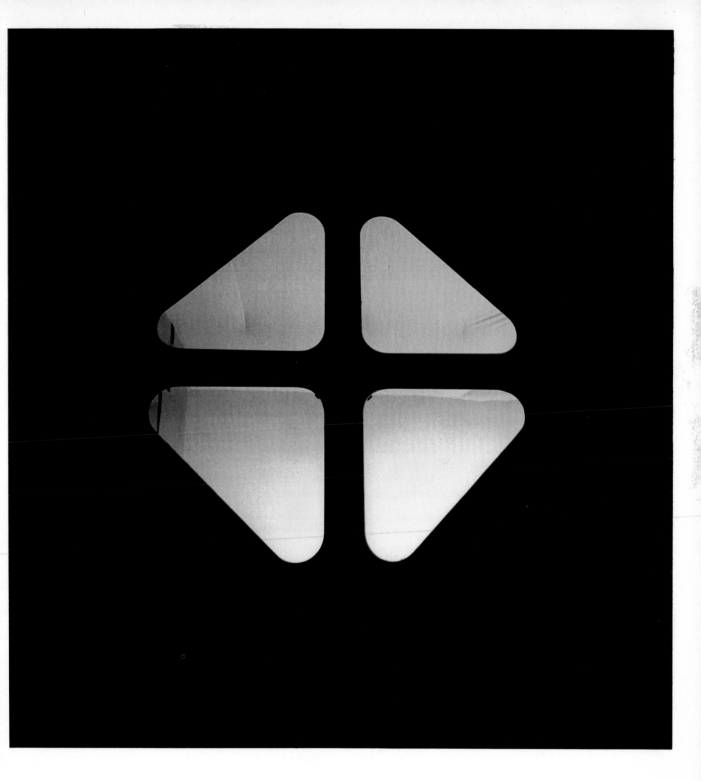

MARIO BELLINI

Mario Bellini received his degree in architecture in 1959. He maintains his studio for architecture, industrial design, and city planning in Milan. He is at present consultant responsible for industrial design in the microcomputer, calculator, typewriter, and copying-machine sectors of Olivetti at Ivrea, besides producing exclusive designs for Brionvega, Cassina, Poggi, Flos, C & B, Bras, and Bacci. His designs have received many prizes, including three 'Compasso d'Oro' awards from Rinascente-UPIM. He is vice-president of the ADI (Association for Industrial Design).

Designer: Mario Bellini; collaborators Dario Bellini, Francesco Binfaré, Giorgio Origlia; collaborators for technical development: Centro Cassina
Patron/producers: Cassina, C & B Italia; with the contributions of Citroën, Pirelli .
Film: directed by Davide Mosconi; visual ideas by Mario Bellini, Francesco Binfaré, Davide Mosconi, Giorgio Origlia

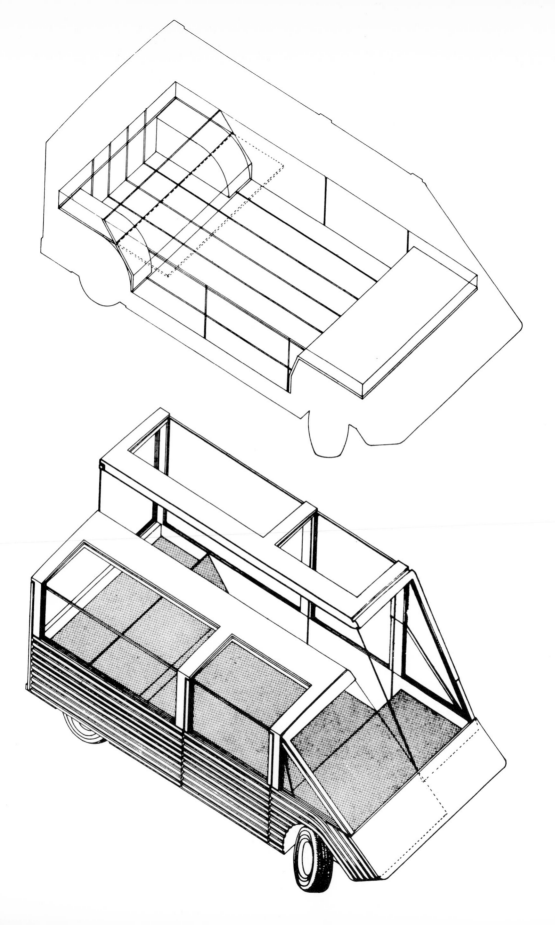

The automobile will die out, and will kill us in the process. It chokes our cities and poisons the air; it stuns us with its noise. It devours disproportionate amounts of expenditures. City planners, ecologists, and sociologists join in decrying it. Some envision utopian cities made to the automobile's own measure, while others propose such ingenuous consumer remedies as special city cars or — more drastically — recommend either abolishing the automobile or abandoning the megalopolis. In urban areas, the crisis can be overcome only by eliminating the automobile as a means of internal communication. But in less densely populated areas, there seems to be no effective substitute for the automobile, now or in the future, as a means of nonpublic transport.

In present-day automobiles, however, we have no option other than to enter, sit — either alone or, less comfortably, in motionless groups of five or six; smoke, think, read a bit, talk to the passenger next to us or others behind us, switch on the radio, watch the landscape out of the corner of our eye, and finally get out. More important, we can speed along, accelerate, roar down the road like real sports heroes, loving the automobile itself and hating the people in the cars we overtake; we can permit others to admire our virility and economic power, of which our car is a symbol; we can implicate the car in obscene attempts at lovemaking; we can ruin ourselves for the car, kill others with it, or die in it ourselves.

Despite all this — and precisely because of all this — I believe that it makes sense today to give more thought to the automobile, to rethink the automobile, the automobile that can still stay with us.
While it goes without saying that, sooner or later, motors will be cleaner and quieter, it certainly won't be the automatic navigational devices or systems of propulsion dreamed of in science-fiction flights of fancy, nor the macabre anxiety that would make the automobile into a swifter, surer bier, that will initiate the redemption of this fascinating mechanical monster.

Now, immediately, we must do away with the parameters of the AUTOMOBILE-MAN system and instead make the car a MOBILE HUMAN SPACE, intended for human and not automotive rites: a mobile space into which one may enter and sit down, be seated even more comfortably, stretch out, sleep, smile, converse face-to-face, observe the outside world and breathe in its essence, enjoy the sun, stand up, take films while underway, change places, sit with one's back to the driver or sideways, play cards, eat a sandwich and drink at a little table, consult a map, put away and pick up all kinds of objects, carry children and play with them, make love in a manner not conditioned by the automobile, transport baggage and things — many things and fewer people, a load of apples plus the driver, 'completely empty, with only the two of us and some cushions; on the way we'll buy a horse or a piano.'

Why not all this, and heaven knows what else besides? Why not in millions of automobiles, why not for millions of dollars a day? Why exchange so many thousands of millions of hours of life for the squalid pagan rites of the KAR, when we still have years and years of gas, oxygen, and life to burn in the automobile? Why let ourselves be taken in by the strategy of marketing the coupé, stunned by motion, like astronauts or the drivers of racing cars? Or is this perhaps just what we are unconsciously seeking, accepting it as compensation for other voids and for the tyranny of the car itself, which we are forced to drive in exhausting, interminable suburban journeys to and from work each day, or during weekends, when we feed both our illusions and the industry of escape from the city?

I don't know if there's any way out for this human race composed of ex-car owners, car owners, and would-be car owners, classified hierarchically according to cubic inches, the number of cylinders, horsepower, rapidity of pickup, and maximum speed. The only answer may be in the prospect of becoming less of a motorist and giving to the automobile, as its chief function, the role of providing human space in motion — space for more significant events; making it an effective instrument serving our need to communicate and get to know the country, or also, thanks to its capacity for mixed transport, making it also serve as a real working tool. These roles could be carried out in combination with, or as an alternative to, mass transport, saving us from becoming merely frustrated chauffeurs enslaved by our own need to get about.

It is principally in this perspective of human space in motion that the automobile must discover its own proper role, the reasons for its own survival as a positive force. The dream of the automobile salesman — an automobile each year for every year of every human being — cannot possibly last much longer. In any case, as a means of locomotion serving almost exclusively individual ends, and at very high cost, the automobile would in any case reach some critical limit in the extent of its indeterminate multiplication. One interesting solution to the problem might be to organize a large number of vehicles of this kind through a capillary network of international rent-a-car companies serving the aims both of tourism and work; this would allow investments in time and space to be more rationally and fully utilized.

The prototype created in response to the invitation of The Museum of Modern Art is intended only as an indication, a proposal for the present, not as a borrowing from science fiction. It seems to us much more 'revolutionary' at this stage in the 'civilization of the automobile' — at a time when more than five million automobiles are being produced and rabidly consumed each year — to propose a car that is subtly different. This proposal, however, has no reference whatsoever to the concept of the trailer home, which is a faithful and often grotesque miniature of the mythical vacation house, a transportable space rather than an environment in motion, a totem that is a substitute for the urban way of life, conceived as a way of reproducing, anyhow and everywhere, the same impenetrable domestic rites.

KAR-A

EVERYTHING SEEMS NORMAL, ALMOST AS IF IN A CAR, IF IT WEREN'T FOR WHAT IS ABOUT TO HAPPEN...

SOMEONE IS LOOKING BACK; AND THUS THE DISCUSSION GAINS INTEREST....

LOOKING AHEAD IS LIKE ENTERING, LOOKING BEHIND IS LIKE COMING OUT, LOOKING SIDEWAYS... AND WHY NOT PICK UP THOSE THREE WHO SPOKE ONLY INDIAN?

WHILE ON THE ROAD, ONE CAN SLEEP, MEDITATE ON INSIDE AND OUTSIDE, LOOKING AND NOT SEE OR LOOK IN THE EYES SHE WHO DRIVES, DEPRIVED OF THESE IMAGINARY LIBERTIES

PEEPING TOM OR SAINT ANTHONY, CAN ONE FIND A DRIVER ABLE TO KEEP IN MOVEMENT THIS MAGIC UNDER THE SKY?

I REMEMBER AN UNUSUAL TRIP IN TWELVE, HOLDING THE BAGGAGE RATHER THAN DIVIDING OURSELVES INTO TWO CARS, WITH ONE MORE DRIVER BUT SO MUCH LESS

UTRA

UNTIL EVERYONE IS OVERWHELMED BY THE ANXIETY OF SUDDENLY CONFRONTING, AS IF IN A MIRROR, THE IMAGE OF THEIR IMPRESSIONS

THE FORWARD POST IS CLEAR, THE TOP OPEN; STANDING, TENSE, NOTHING SEPARATES THEM FROM THE WORLD OUTSIDE; EXCEPT EYES, LENSES, GUNS

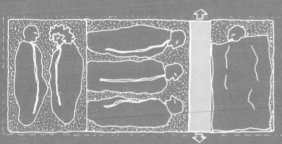

AT DUSK THEY STOPPED, TIRED, FAR AWAY FROM EVERYTHING; THEY PISSED IN THE WOODS AND SLEEP NOW: TO TRAVEL IN THE DARK IS LIKE FALLING THROUGH EMPTYNESS

FEW WORDS, FEW LOOKS, AND NOW, IMMERSED IN THEIR EXPERIENCE, THEY SEE ONLY SAND, SAND-CASTLES, TUAREGH ON THE HIGHWAY HOME

OPEN OR CLOSED, RAISED OR LOWERED, ONE COULD GO AROUND THE WORLD IN TWO, CARRYING EVERYTHING BUT THE TENT

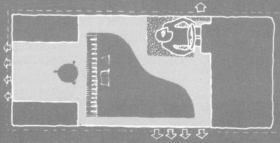

HOW MUCH ROOM FOR DREAMS... A FLYING HAREM... A GRANDPIANO FOR ITINERANT CONCERTS.... BUT NOW IT'S ALL EMPTY, TWO HUNDREDS CUBIC FEET OF WORKSPACE

An essential feature of the new automobile, which is little more than sixteen feet long and six and a half feet wide, is that it has the same dimensions as a normal sedan and an extraordinary flexibility in its load capacity. It can transport a load of something upwards of twenty-six cubic feet, besides the driver; or it can comfortably carry twelve persons without baggage. This is made possible because the loading surface is evenly distributed over sixty-four square feet, and the interior fittings are completely movable. But the most important innovation is the car's ability to carry more people with their baggage — just what is most lacking in the interior fiittings of the traditional automobile, which tends to restrict drastically the possibility of moving about, conversing, or remaining together comfortably as the circumstances of the trip may require. The entire interior fittings of the car consist of a series of cushions, ten inches high by two feet square, which can be variously arranged as seats, backrests, armrests, forming padded ensemble that can assume different shapes: ranging from the traditional double row of three seats to a series of six beds, from a three-sided sofa that allows one to sit facing the direction of the route, or sideways, and can form either separate containers or a continuous soft platform. The special characteristic of these cushion units is their 'plastic inertia,' which allows them to be positioned and kept in place indefinitely, always renewing their shape after receiving the imprints and deformations caused by the pressure of force or of bodies, without losing their special property of elasticity. Thus, the interior of the automobile completely covered by such cushions becomes a 'plastic field,' available indefinitely for any new kind of use and allowing people or things to leave on them their imprint, since they are well adapted both to support and to contain.

Two flat surfaces cover the luggage compartment and the motor compartment; they may be used either to put things on, or as beds. Passengers enter the automobile by the two side doors. The car is loaded through the double rear doors; by removing the platform over the luggage compartment, the whole interior area of the car can be opened to its full extent. To permit even more flexibility, the roof of the car may be raised almost two feet by means of a pneumatic device. This does not affect the car's driving potential but increases its load capacity and also allows passengers to stand up, change places, shift objects around in the car, and get in and out more easily.

A second important characteristic of the car is that the entire top section may be opened completely by retracting the windows, leaving only three roll-bars, a central cross bar, and the windshield. Thus, the potentials of utilizing the mobile space are greatly increased, not only for those who like to travel in an open car, but also for those who wish to take photographs, make moving pictures or, if they must, shoot; as well as for those who simply don't want any filters or partitions between themselves and the environment. When the car is completely open, one may stand and move about freely; one could transport a giraffe or stretch out and take a sunbath while traveling. Considering the automobile is to be used on all kinds of terrain, we have proposed the most extraordinary 'vehicular philosophy' in all the history of the automobile: the use of hydraulic suspension devices that offer more heights above the ground and exceptional safety and comfort en route. The passengers' traveling comfort is enhanced by a number of devices within the car. Two continuous trays of different widths run along the sides at different levels. They hold in easy reach not only maps, guidebooks, and sunglasses but also bottles, drinking glasses, cameras, purses, etc. Next to the dashboard are a small refrigerator and a container for cumbersome objects. The central crossbar in the roof houses lighting panels, air ducts, and two rows of handrails.

Design as Commentary

GAETANO PESCE

Born in La Spezia in 1939, Gaetano Pesce studied at the School of Architecture and the Institute of Industrial Design in Venice. He was a cofounder, in 1949, of Group N in Padua, devoted to the study of programmed art, and at that time began his first investigations in visual communication. Besides his activity as an interior designer, a field he first entered in 1962, he has worked in many mediums, including silk-screen, programmed art, kinetic art, serial art, assemblage and bricolage, film, audio-visual presentations, and multimedia events involving light, movement, and sound. He has traveled, studied, exhibited, and lectured in many countries and has participated in several international conferences, in Finland and Japan; in 1971, he was Italian representative at the International Congress of Design in Aspen, Colorado.

Designer: Gaetano Pesce
Patrons: Cassina, C & B Italia, and Sleeping International System Italia
Producers: Centro Cassina, with the assistance of Sleeping International System Italia
Film: directed by Klaus Zaugg

The following have also collaborated:
Beatrice Bianco, Piergiorgio Brusegan, Enzo Capelletti, Alessandro Carraro, Gianni De Luigi, Philippe Duboy, Carla Genziani, Pico Lazzarini, Ulderico Manani, Antonio Mantellato, Grafica Mariano, Roy Meneghetti, Luigi Pozzoli, Gianni Predieri, Mino Prini, Nino Scolari, Alessandro Zen

I am indebted to Cesare Cassina for the generosity with which, with the customary graciousness that he has so frequently demonstrated, he agreed to execute this work; and to Francesco Binfaré who participated in the formulation of the hypothesis from the outset and contributed his outstanding experience and valuable enthusiasm. I am particularly grateful to my wife, Francesca Lucco, for her constant and attentive presence throughout the progress of the work, which she frequently influenced by her criticisms. — G. P.

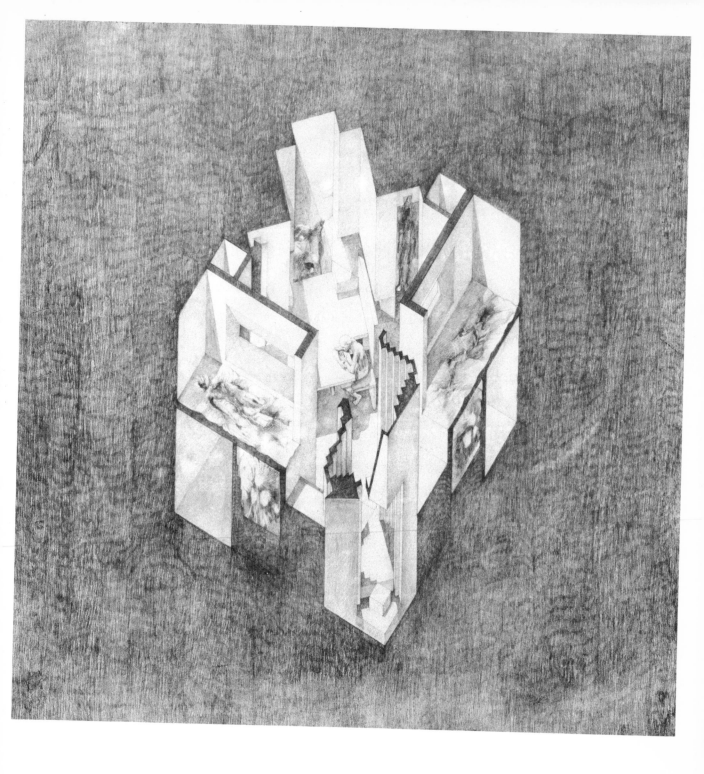

Subject: The discovery of a small subterranean city, belonging to the epoch known as 'The Period of the Great Contaminations'; location: Southern Europe (Alps).

The basic living unit of the urban settlement was found to be in a reasonable state of preservation, and with the salvaging of additional documents, it has been possible to reconstruct models of a probable 'Commune for twelve people,' measuring 13x6,5x5 meters in height, and for a doorway, probably the entrance to the city or some important building, measuring 6x2,5 by 6 meters in height. An original fragment of the latter was discovered, together with a seat of the same epoch. The model, showing a longitudinal section of the ground, represents our hypothetical reconstruction of the city; it is the result of deductions drawn from very miscellaneous documents. Soundings of the subsoil produced no important results, partly because many of the passages had collapsed, etc.

It is nevertheless possible to state with a fair degree of certainty that the communities of the 'Great Contaminations' (to give them a name) exploited underground cavities for their settlements, having first drained off mineral oils, water, etc. The immense hole made in the course of excavations was then closed by a huge stone, hermetically sealing off the interior from the outside world. Once inside, the men of the 'Great Contaminations' began to spread out, looking for further possible spaces in which to settle.

The model is intended to illustrate this situation. The two pyramidal volumes in the upper part represent what may have been stations from which transport to the outside world left, or they may have been advance defense outposts against attacks from the outside. The square spaces are believed to have been hollowed out by man, while the others are preexisting natural formations. The cylindrical volume at the upper right represents the probable water reservoir, while the natural cavern in the lower right was probably an area for leisure, baths, propitiatory rites, etc., on a small lake. The area in the upper left, divided into various levels, may represent zones for production and general work.

Finally, the discovery of a short film, a few minutes in length, was a particularly happy find. It documents some scenes of family life and was probably shot by some member of the family concerned. These documents together give us an indication, even if vague, of the probable conditions of life around the end of the second millennium A.D. and beginning of the third. Further study might throw additional light on the psychological effect that the term 'the year 2000' had on those living both before and after that date, and the consequences that such a 'status' had on their behavior, etc.

In order to be able to understand the origin and development of the architecture of the period of the 'Great Contaminations,' and the geographical, climatic, environmental, technological, and religious laws that influenced its art, we must give a general explanation of the bases underlying the execution of one of its means of expression. Such an explanation is indispensable, in view of the fact that the infrastructure of those inter-European civilizations has almost no points in common with that of our own day.

Within that context, the evolution took a completely different course from what we habitually regard as normal; furthermore, the setting in which this culture developed bears no similarity to that which gave rise to the great agrarian empires of remote antiquity. There is actually no affinity between the plains that were periodically flooded and irrigated by such great rivers as the Nile, the Tigris, the Euphrates, the Indus, or the Yellow River when in spate, and the great excavations of the 'Contaminated' era (their cities lay underground, several tens of meters below sea level). It is difficult to determine the conditions for survival, though we can easily imagine that they were based on the filtering of the many layers of subsoil, its internal resources, heat, etc.

It is natural, therefore, to begin with this somewhat superficial preliminary description of the characteristics of that environment, for they are indicative and will help us to understand the origins of its architecture and art. In this connection, the innumerable cities of the 'Contaminated' era that spread over the territory in question offer many rich and varied examples of architectural material of considerable interest. Unfortunately, however, only the stone objects have survived; and though the stone outer shells have come down to us more or less intact, all the structures in wood, ABS plastic, melamine, polyurethane, etc., have been irreparably lost or damaged by heat and humidity. But although most of the dwellings and public buildings leave insufficient evidence for interpretation, the habitat in question, presumably a dwelling for two people, is surprising well preserved. Its basic form is characterized by several distinctive features, which we shall discuss immediately, as they determined the structure of all buildings of the 'Great Contaminations.'

The habitat was almost always placed on a base, with steps leading up to the entrance. It was made of dry mortised blocks of rigid plastic and was never more than 60 centimeters in height (in the case of our example, 40 centimeters). The function of this base was to serve as a thorough insulation, protecting the inhabitants from the infiltration of residual dampness during the period of great condensation. The dwelling itself was then fitted onto a stone masonry floor. The plan was square, with the principal axis running diagonally, and the whole measuring 4,80x4,80 meters, by 3,60 meters in height. There was only one door (about 60 cm wide and 220 cm high), placed in the corner of the parallelepiped, perpendicular to the principal axis. There was normally no other opening, either in the walls or the roof, and except for the drains, there were no other openings or windows. The walls, fixtures, furnishings, etc., were of mortised blocks of rigid polyurethane, while the seats were of soft polyurethane.

However distant and strange the architecture of the 'Contaminations' may seem, and however difficult it may be to establish links with the buildings of our day, the reticular module on which this architecture was based is nevertheless enlightening and is still of great significance for us.

With regard to the interior space of this habitat, I consider it appropriate and interesting at this point to quote some fragments from a volume that the present author discovered at the site in question. Though apparently obscure, these fragments, taken together with the pictorial documents, will undoubtedly help to make clear the contents of our habitat:

'...in the architecture of the "Great Contaminations," the square and the rectangle are the absolutely fundamental forms, permitting no deviations and not allowing any variations in construction. These basic forms must not be covered nor hidden by any decoration, though this may exist to a limited extent and has an emblematic character...'
'...It can only be deduced that the module as such has no architectural value. It is the meaning with which the architect invests it that determines its actions and reflexes in architecture... It is not difficult to find the geometric key to a plan, but when it is a question of explaining the esoteric content of the constructions and the formulae encountered therein, one is forced into the realms of conjecture...'
'...The house in itself is not only a mass of blocks arranged for a precise

purpose, but it nurtures and conceals an interior pregnant with symbols that are unequivocally related to the mode of representation of that time....'

Thus, having arrived at an understanding of the spirit of these spaces, with the aid of the foregoing passages, we can draw up a list of the motivations that, in our opinion, led to the establishment of this architectural typology. These motivations will be discussed point by point in the following chapter.

1. Urban hypothesis
   a. importance of space
   b. exploitation of the planet's interior
   c. incompatibility between human environment and the atmosphere
   d. urban spaces in depths of the subsoil
   e. vertical mechanical transport routes
   f. special equipment for defense of surface routes
   g. exploitation of sites suited by nature for strategic
   defense (e.g., mountains, etc.)

2. Effect on man's behavior
   a. need for isolation
   b. rejection of human contact
   c. non-communication as a characteristic of life
   d. human contacts considered necessary only in the context
   of work relationships
   e. concept of productivity replaced by a need for transcendental
   symbols
   f. reinstatement of taboos
   g. state of mind similar to that of the period before 1000 A. D.
   h. decline of the technological dream
   i. insecurity as the prospect of the future

3. Habitat
   a. habitat as a refuge of isolation from everything else
   b. 'domestic landscape' and the function of anxiety
   c. spaces for events of ritual character
   d. tendency for the architecture to express 'safe' and
   meaningful spaces for the user
   e. segregation as the hypothesis of an objective choice
   f. habitat as a house of eternal life
   g. tendency to overcome fear through inflating the idea of death
   h. symbolism as a refuge
   i. user not the master of physical space but perhaps an anxious
   protagonist of ritual events within the extremely small arc of his life
   (yearning for a flexible and elastic habitat)
   j. the architect as adversary
   k. architecture as a means of negation (end of collaboration in
   architecture)

Discussion of the introductory statements:
Etc.   .   .   .   .
       .   .   .   .
       .   .   .   .
       .   .   .   .
       .   .   .   .

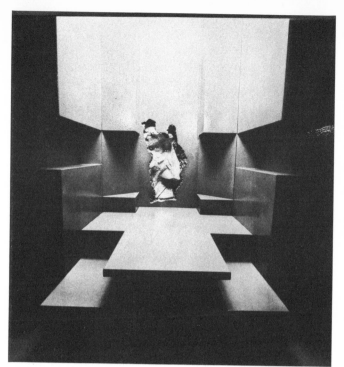 

 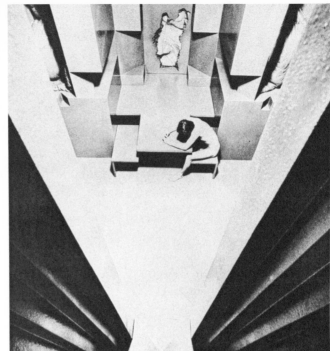

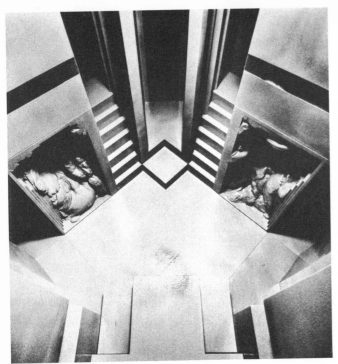
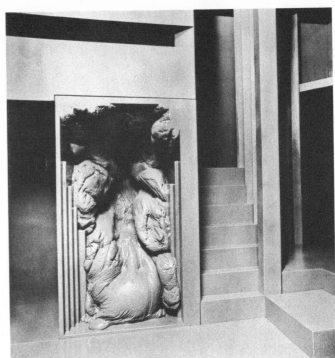
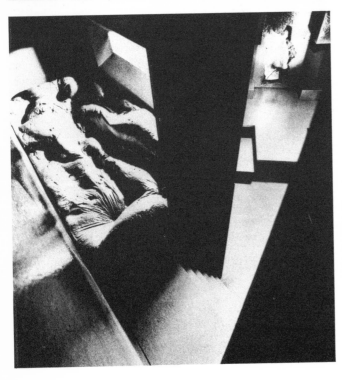

221

Counterdesign as Postulation

UGO LA PIETRA

Ugo La Pietra lives and works in Milan, where he carries on extensive activities in design, research, teaching, and writing. His researches have focused not only on practical problems of mass production and the uses of new materials but also on theoretical problems regarding the morphology and social role of design. His work has been exhibited internationally in one-man and group shows. The author of numerous articles, he is at present editor-in-chief of the periodical *IN: Argomenti e immagini di design.* His theory on 'Unbalancing Systems,' which he elaborates here, has been propounded in two books: *Sistema disequilibrante* (Milan: Toselli, 1970) and *Sistema disequilibrante, II* (Genoa: Masnata, 1971).

Designer: Ugo La Pietra
Patron: ABET-Print, with the collaboration of Silcon and Moro
Audio-visual program: Ugo La Pietra with Piero Castiglioni

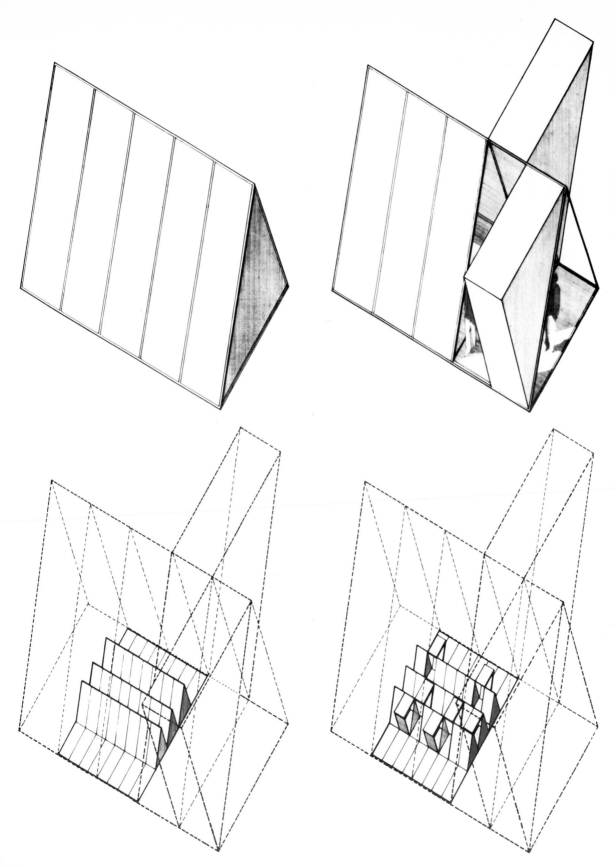

The Domicile Cell: A Microstructure
within the Information and Communications Systems

The proposal for the space that I have designed for the 'New Domestic Landscape' exhibition is not meant to be one of the many contributions to the development of the 'domicile cell' through individual environmental solutions; still less is it an attempt to rediscover through the invention of 'new objects' a possible solution to the knotty social problems that find one of their chief outlets of expression in the domestic environment.

'Object' reality must be sought by analyzing a social and environmental situation in which objects and reified persons interact in a mutually exploitative relationship. The objects' loss of reality corresponds to that moment in which they went beyond their elementary message to become the bearers of structural realities that can no longer be explained by the simple relationships among labor, purchasing power, and use value.

As soon as a process of acculturalization develops among classes, the yardstick for measuring to which class one belongs is the manifestation of one's power to own and possess objects that historically are suited to this role. It is precisely in this respect that the reality of objects can be analyzed from a sociological point of view. Status-symbol objects can, in fact, either be quite useless and overloaded with messages through a semantic operation on some work that fulfils consumer needs — thus, a pure expression of the expenditure of labor; or else, while having a real use value, they can be indirectly subjected to superexpressive mediating materials and thus be merely simulacra of real objects.

The formalizers — the architects — have played a predominant role in this regard, at the service of that class which asked them only for adornment of their ownership, without seeking an aesthetic accomplishment.

Today, the difficulties in which any formalizer who dissents from this position finds himself must, however, drive him toward a logical rejection of utopia, since utopia implies a flight into a reality that is not actual but accessory. Denial of utopia means getting involved in a head-on collision with the logics of production; and trying to find

Fig. 1, left, above: Comprehension model A: 'Immersion'

Immersions are an invitation to a behavior that departs from reality to discover a kind of 'privacy' that is a separation and a means of testing the possibilities for intervention by way of disruptive elements that can displace codified, traditional terms. In this way, one sets in motion a dynamics of relationship, with the free behavior of the individual giving meaning to the potentialities inherent in the spatial presence.

While the containers push toward a certain behavior, they define a space in which the individual believes that he can rediscover an environment for independent decision. In actuality, by having chosen to enter into this enclosure, he has become shut off from interaction with the surrounding environment and thus becomes the object of a formal intention that he is powerless to affect.

The result is a 'crisis' between the user's desire to isolate himself from the context, and his aspiration for an unbalancing inclusion in the system. But this very ambiguity, which is a clash between the aspiration for freedom and the limitation that every choice imposes upon freedom itself, is seen as an 'awareness' that liberation from the social and psychological conditioning by the context proceeds through personal immersion in a space that offers itself as a point for critical and imaginative reflection on the context itself.

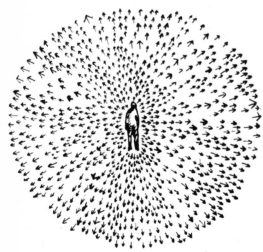

Fig. 1. Comprehension model A: 'Immersion.' Progressive flooding of a container. Metallic container with protective methacrylate coating
Fig. 2. Individual communication with outside world

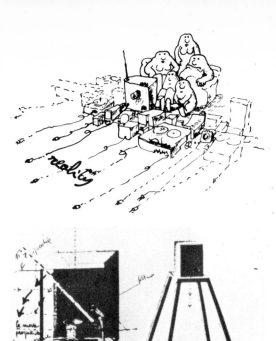

within oneself forces that can be used to give to the forms one creates a superstructural quality in view of a structural transformation.

The operative attitude chosen is that of identifying the degrees of liberty that still exist within the 'organized' social structure. The strategy that the formalizer should adopt is to identify those forms that correspond to the subversive logic of the productivity-oriented system. I have therefore decided to give this complex of forms and behaviors the name UNBALANCING SYSTEM.

Political groups perform unbalancing actions in terms of the political and economic order. The formalizers must operate through environmental phenomena, that is, the relations that grow out of the presumed categories of person, environment, and object. One must therefore select strategies that can, in a professional manner, lead toward the awareness of a phenomenon of uneasiness with respect to behavior and form. (Fig. 1: Comprehension model A).

In the extensive case-history of 'objects' that are accumulated in the domestic landscape, the ones that I maintain should have priority as the subject for analysis and new proposals for use (degrees of freedom) are the audio-visual objects for information and communication. At this particular moment in time, I believe that the most characteristic elements for the transformation and complete overturning of all formal principles, acquired typologies, and rites developed within the domestic landscape are to be found precisely in the use of these instrument-objects. Their typology, too, comes under the heading of objects for decoration, in which one may discern the kind of fetishism, semantic redundancy, and 'social' and 'authoritarian communications' that we are accustomed to observe in the symbols of middle-class models. We must therefore free ourselves from the object that is still regarded as a 'message,' whose characteristics are always those that relate it to a specifically identifiable class of users.

Furthermore, the audio-visual devices that technology has been perfecting for some time guarantee that every one of us (in his own privacy) can communicate with the outside world, giving us the possibility of expanding our physical structure and enlarging the space that we physically use and are aware of (Fig. 2). The relationship established between the individual and the outside world is, nevertheless, increasingly achieved through information (processed by others) that we all accept, or rather submit to, 'like a pig in a poke.'

We must therefore rid ourselves of these intermediary tools, i.e., eliminate the filter that the apparatus interposes between us and reality (Fig. 3); we must transcend them and gear our own minds and our own behavior to a subjective vision of reality, if need be, thus eliminating every risk (beyond the naive dreams of the technocrats) that our society can become perfectly organized and managed solely by means of such 'instruments.'

Figs. 4 and 5, left: Comprehension model B: 'The New Perspective'

In executing this particular object, my aim has been to point out how necessary it is, at this moment in history, to repropose the problem of direct awareness of reality (overcoming the 'barrier' of the 'instrument' that acts as a mediating filter between us and reality).

In this case, the object (recovered from history; cf. the camera obscura used by seventeenth- and eighteenth-century landscape painters for their perspective renderings of landscape) has been reproduced without the mirror tilted at a 45° angle and without the lens — elements that represented the indispensable technical 'filter' required to transpose the real image into the fictive one, projected onto the horizontal plane (the plane of the drawing). Once this filter has been eliminated, it is possible to see the landscape (reality) directly.

Fig. 3. The barriers interposed by communications media between us and reality
Figs. 4 and 5. Comprehension model B: 'The New Perspective.' Perspective chamber with instruments: mirror and lens. Perspective chamber without instruments: 'the new perspective.'

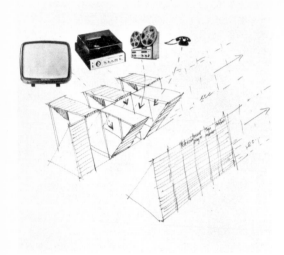

My proposal is intended to express:

The desire to use the information and communications media, while keeping them 'under control' (that is, never submitting to their presence either as 'objects' or as 'instruments'); destroying their 'design' and postulating their development only with respect to their technical, mechanical, and electronic characteristics; using them in a way that will overcome the 'barrier' that they create between us and reality (Figs. 4 and 5; Comprehension model B).

In the project, all this is expressed through concealment of the mechanisms (Fig. 6); this concealment is not total (inasmuch as the containers rise out of the floor with inscriptions that indicate the position of the apparatus). Thus, their presence can give a 'sense of security,' on one hand, and on the other demonstrate their availability for use (Page 225: Comprehension model C).

The domicile cell can thus also be considered as a place in which (by means of the communications and information media) information from outside can be recovered and collected (either directly or indirectly); after having been processed, these data can be put back into circulation and then compared with others. This proposal seeks to raise the possibility of bringing about direct participation in, and awareness of, the physical, behavioral, and mental characteristics that develop within the texture of the city.

All this implies charging the domicile cell with an indispensable role in the evolution of 'organized society.' The domicile nucleus thus assumes a further role; it becomes a center for gathering, processing, and communicating information; a microstructure that can intervene in the information system by enlarging and multiplying exchanges among people, with everyone participating in the dynamics of communication.

Page 225: Comprehension model C: 'Microenvironment'

This microenvironment is meant to stress the problem outlined above, regarding the relationship between the domicile cell and the urban structure by means of instruments of information and communication.

The domicile cell is symbolized by an elementary volume (triangular in section), which can open up, thus placing its interior into contact with the exterior. Within this 'model' one can discern a series of containers (at floor level), in which are hidden all the instruments of communication and information. These containers, too, can be opened so that the instruments they contain become accessible.

Figs. 7, 8, 9: Comprehension models D, E, F:

Within these three models one may perceive — always expressed in a symbolic way — the possibility of considering the private domicile structure as one among many points in which information and communications are gathered, processed, and put into circulation.
The first model shows the possibility of placing individual 'apparatus' (within the urban structure) to record all the telephone messages dispatched from private units, so that they can subsequently be listened to by anyone, as a progressive sum of participations.
The second model is again based on the same apparatus, which is used, however, to transmit messages recorded by anyone in the urban area, and then available for listening to within the private domicile area.
The third model, on the other hand, by utilizing television equipment (television camera, recorder, video screen), is meant to indicate the possibility of being able to record, and then transmit to a central collecting point, audio-visual information and communications, processed by every one of us in our own privacy, and then projected by enlarged television screens throughout the urban scene. (This model can also be utilized for the inverse process.)

Fig. 6. Proposed concealment of mechanisms

Fig. 7. Comprehension model D:
Audio-microevent, urban privacy-system

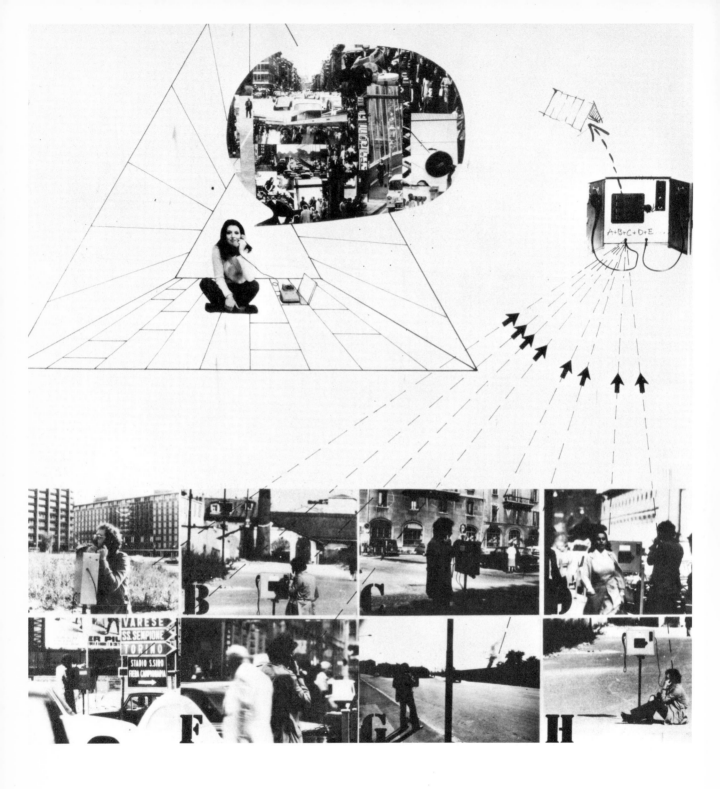

Fig. 8. Comprehension model E:
Audio-microevent, urban-privacy sistem

230

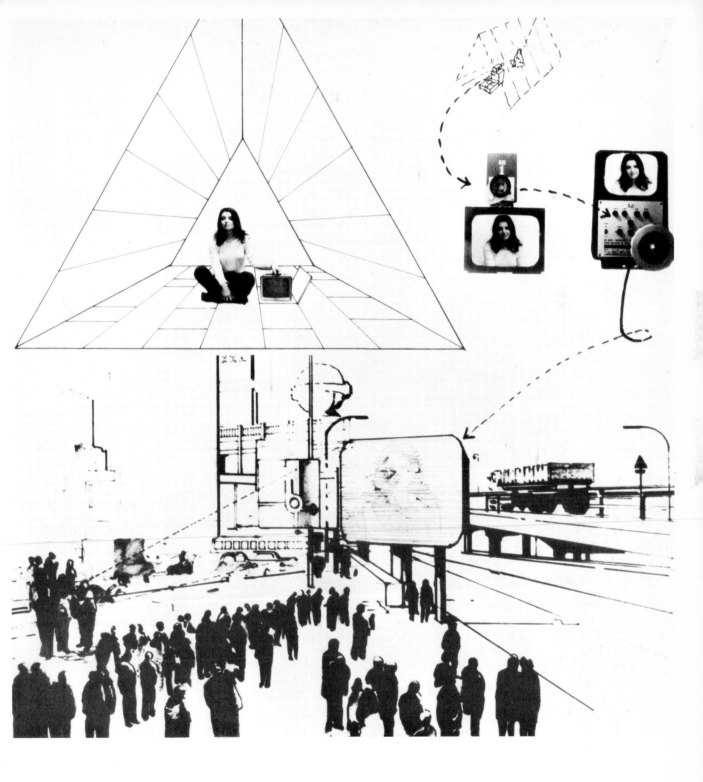

Fig. 9. Comprehension model F:
Audio-visual microevent, urban privacy-system

231

ARCHIZOOM

Archizoom Associati have been active as a group since 1966, maintaining a studio in Florence for product design, architecture, interior design, and the installation of exhibitions. In 1966, they organized the first exhibition of 'Superarchitecture' at Pistoia, followed by a second one in Modena in 1967, and by the 'Center for Eclectic Conspiracy' stand which they presented at the XIV Triennale, Milan, 1968. Since then, they have carried on their polemical activities through such exhibitions as 'No-stop City (Residential Parking),' 1970; participation in the VII Biennale, Paris, 1971; and conferences and publications, including the preparation of a special issue of the periodical *IN* (June 1971) devoted to 'the destruction of the object.' In 1970, in collaboration with Superstudio, they undertook a research project in airport design under the auspices of the Cesare Cassina Center of Studies, Meda (Milan) and in the same year entered two national competitions, at Catanzaro and Genoa, for the designing of airports.

Designers: Archizoom (Andrea Branzi, Gilberto Corretti, Paolo Deganello, Dario Bartolini, Lucia Bartolini, Massimo Morozzi)
Patron: ABET-Print
Audio-score: Giuseppe Chiari

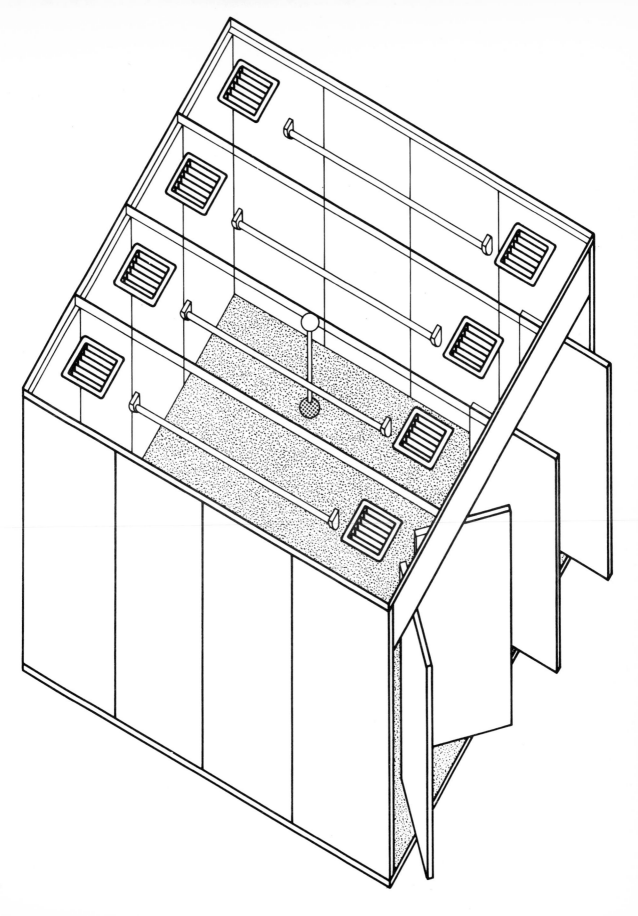

'Listen, I really think it's going to be something quite extraordinary. Very spacious, bright, really well arranged, with no hidden corners, you know. There will be fine lighting, really brilliant, that will clearly show up all those disordered objects.

'The fact is, everything will be simple, with no mysteries and nothing soul-disturbing, you know. Wonderful! Really very beautiful — very beautiful, and very large. Quite extraordinary! It will be cool there too, with an immense silence.

'My God, how can I describe to you the wonderful colors! You see, many things are really quite hard to describe, especially because they'll be used in such a new way. And then, there'll be glass, wood, linoleum, water, plants, vases, and many of those boxes they used to use, in wood or plastic, and all empty....

'What's really extraordinary about all this is that many of these things will be handmade, especially the largest ones. Of course, others will obviously be machinemade. The household equipment will be just perfect, in wonderful colors, neutral colors, I should say... All the rest will be bright, and there'll be a big swing with room for two.

' You see, there'll be a lot of marvelous things, and yet it will look almost empty, it will be so big and so beautiful... How fine it will be... just spending the whole day doing nothing, without working or anything... You know, just great...' (And so on, starting all over again at the beginning.)

What we use, then, in creating our environment is the least physical thing in the world, namely, words. Of course, that doesn't at all mean that in postponing the physical realization of this environment, we have avoided picturing it. On the contrary, we have refused to complete a *single* image, our own, preferring instead that as many should be created as there are people listening to this tale, who will imagine this environment for themselves, quite beyond our control.

Not a single utopia, then, but an infinity of utopias, as many as there are listeners. Not just a single culture, but one for each individual. We have given up making an environmental model, partly because, in general, we think it's time to begin learning how to do without models; and partly because, if we were to present just a single model — the one we think the best — we should have to eliminate an infinite number of others, and that would be quite a waste. Creating culture today is no longer — at least, it shouldn't be — the privilege of a few intellectuals, who provide users with the critical apparatus with which to explain the world and organize the form of their environment, too. The right to go against a reality that lacks 'meaning' (because it is a reality produced by a system that is 'meaningless' in itself) is the right to act, modify, form, and destroy the surrounding environment. This is an inalienable right, and a capability each one of us possesses.

Culture, and the making of culture, the formal act of doing so, doesn't mean expressing oneself through allegory or metaphor. It is a political right, not a subject for linguistic criticism. It is the task of rediscovering and asserting those physiological capacities that are linked to the body's material substance and the electrical energy of the nervous system. Thanks to these, the making of art or the making of culture take on a meaning quite complete in themselves, like a kind of liberating psycho-physiotherapy.

The self-production and self-consumption of culture imply the ability to free oneself from all those repressive systems that 'official culture' has woven around us, by attributing an infinite variety of 'values' and 'meanings' to the reality around us and thus, in fact, taking away our freedom to modify that very environment at will. Our task, then, is to reduce to zero the moral weight of things, methodically questioning all the patterns of religious, aesthetic, cultural, and even environmental behavior.

In putting aside for a moment the formulation of our ideas as designers, we also interrupt that one-way circuit by which things become a means of communication, with the users remaining passive and unconscious recipients. We can't let anyone else plan our private models of behavior, nor limit ourselves to hanging up inside some reproductions of French Impressionists.

It isn't much use to go on planning 'different' houses, if the way in which they are used is always the same — that is, the relationship that society imposes with respect to everything not directly connected with work: a limbo in which to regain that minimum of energy and minimum of balance needed to let us produce again the next day. And so our home becomes primarily the place in which we try to conserve ourselves most, where we try to establish all our contacts with a motionless landscape that doesn't stimulate us, but instead soothes us with its immobility, while we contemplate our own 'status.' It is clear, then, that the form of the house is not a political problem; as Engels once said, the problem is rather that of commandeering the houses already built. And the only effective way to do this is to take possession of them.

To become their masters, that's the problem; since first of all, we must be able to become 'masters' of our own existence, our own time, our own health, and our own actions.

To become the master of one's own life, one must first of all *free oneself from work*. This is the only way in which to recover all the untapped creative faculty that man has available within himself, which has become atrophied throughout the centuries because of frustrating work.

The home — that small, functional organization at the service of the tiny production company that is the family — imitates in its forms and furnishings that culture and those choices no user has ever made, nor has ever had time enough to test.

Only by rejecting work as an extraneous presence in one's life can one picture a new use for the home: a perpetual laboratory for one's own creative faculties, which are continually being tried out and continually being surpassed.

So, then, it's no longer of any importance to imagine the form of this home, because the only thing that matters is the use made of it. Its image is manifold, never final, and has no codified meaning nor spatial hierarchy. A kind of 'furnished parking lot,' from which all antecedent types have vanished, giving way to a spontaneous shaping of the environment; a completely accessible enclosure within which to exercise one's regained freedom of action and judgment.

For us, the problem is no longer that of trying to understand what kind of freedom man is seeking, or perhaps even trying to foresee it in terms of current reality. The problem, instead, is to give man the kind of freedom that will enable him to obtain it for himself.

What we have been trying to do since 1969 is to make a scientific study of the problem of the house and of the city. Through this scientific work, we seek at least to remove all the usual qualitative criteria for present concepts of architecture and city planning.

The streets, squares, boulevards, monuments, theaters, churches, public buildings, and façades of a traditional city are the stage setting for the great spectacle of itself that the city offers, using architecture as medium for presenting thousands upon thousands of different episodes. Every corner of the city, street, or square, every square foot of asphalt, is thus very different from all others, because it occupies a different position in relation to the city's landmarks. The history of individual cities, and their geographical location, make each urban fact seem 'unique' of its kind, not comparable to other urban phenomena but isolated in its own singularity. Each city appears to have its own intrinsic 'quality,' a synthetic expression of its virtues and defects. Up to now, the study of the city has been limited to surveying these separate histories, without any general laws that could be applied to them all, and that could be useful because based on previous experience.

Even modern town planning, despite the profound changes that have taken place in modes of life and in technology, continues to make the same proposals, which are like series of scenic episodes, brought up to date in their forms, but still based on a kind of memory that the city retains within itself — the memory of the 'village.'

The 'village,' in fact, is that experience which, although he has never lived through it, still accompanies the man who dwells in a metropolis, helping him to encounter phenomena of gigantic proportions by means of a lens adjusted to small things and to small scale. The metropolis continues to be a 'village,' to the extent that it presents itself, and develops through, traditional scenic events — the actual streets, squares, buildings, churches, fountains, trees, etc. Thus, the metropolis seems to be the logical successor of some ancient former existence that survives directly, greatly enlarged and confused, but nevertheless retaining those structural characteristics that constitute its deepest nature. When we confront the problem of the modern city, its nonfunctioning, etc., we often continue to accept tacitly the backward model that the city has taken as its point of departure, seeking to introduce technical standards that regularly continue to fail, and just as regularly are reproposed for each new urban experiment.

The ideology of the middle-class city derives from the seventeenth-century discovery of the city as a 'natural object,' that is, a reality homogeneous with the surrounding countryside, in that it has been created in accordance with the same rational laws that regulate the entire world of nature; these give the city a universal significance as a 'naturally' existing reality, rather than as an artificial creation with an autonomous logic of its own. Moreover, since the city is the common product of man's labor, it represents in middle-class ideology the civilized bridge between man and nature, and between man and society. Social balance and ecological balance must be achieved simultaneously. Thus, in the City Plan, what is sought is a not-impossible harmony between the Public, the Private, Nature, and Buildings; these diverse entities are regarded as mutually incompatible elements. The end result, in fact, need not be a 'unity,' but rather a 'harmonious succession' of different and contrasting logics. Having become a 'citizen,' man enjoys an equilibrium achieved through forms, celebrating his 'natural' integration into society. The city always seems to him something far more complex and more spiritual than any practical use that he can make of it in his daily life; to be a citizen means to adopt a mode of behavior that is fully conscious of the existence of that cultural 'unity' upon which society is based. Thus, the middle-class city becomes an ideological superstructure, a screen between the individual and the hierarchical systems of society.

The middle-class city is that spontaneous mechanism in which goods circulate freely and acquire 'value.' Production is hidden away in the remote outskirts, without ever attaining the urban dignity of a recognized function. In this way, the city — completely organized as a means of realizing the ideology of Consumption — inverts the reality of its systems by emphasizing the act of exchange, while concealing the far more real one of production.

Violent conflicts, uncontrolled disorder, and the spontaneous growth of means of communication, are the shock tactics that the city adopts to compel the citizen to integrate himself within consumer society. Urban chaos is the most common mechanism for accomplishing this process of integration and induction, but it is also the least easy to control; it is still the outcome of a system of free competition. Nowadays, the use of programmed electronic media has replaced direct urban procedures; they ensure a far deeper penetration of patterns of consumption into social reality than the fragile channels of information that the city affords.

By arranging city events as if they were different theatrical scenes, the metropolis maintains a visual relationship with its own inhabitants. For the individual, this visual relationship becomes the fundamental system whereby he organizes his own experiences and his own memory of the city. The stratum of the city that he uses has a depth of two meters of ground; as Kevin Lynch has said, he builds memory-tunnels, which are completely personal, and which allow him to arrange his own system of images, composed of illuminated advertising signs and other isolated elements of the town's furnishings. In practice, the city as a unified entity of images and culture is no longer conceivable.

Architecture is no longer a means for representing a system, because the city no longer offers an enjoyable image. The attempt to make the *maquiniste* logic of industrial civilization coincide with a *maquiniste* architecture is an historical blunder, the result of a simple linguistic updating. Actually, the logic of a single factory, or a single machine, cannot be brought, like the part to the whole, into relation with the reality of Planning. The rules of montage permit combining different parts of an object or a building, but the ideology of Planning permits combining different parts of the whole of society, making it a homogeneous production without any functional contradictions or any 'other' kinds of reality.

The modern city, as an intensive concentration, has its origins in the territorial distinction between town and country. Capitalism, born of the city, was identified at the outset with urban management of the land; but as a result of Planning, it now extends its own rational power throughout the whole country, which it organizes for the sake of productivity. The 'urban condition' — being a citizen — does not imply being more integrated than a non-citizen, since there are no longer any territorial regions not organized within the system. The city, therefore, has ceased to be a 'place' and has become a 'condition'; and Consumption disseminates this condition homogeneously throughout the country and throughout society. The town, therefore, corresponds to the size of the market, and not to the size of any particular locale.

Avant-garde architecture no longer seeks to design a 'better' city in opposition to the present one. It performs another function: it challenges the ideology of the middle-class city, denouncing its fraudulent role, and establishes city planning anew as a system to be put to a different use, other than an instrument for social manifestation and induced values. Besides, as Engels said, the problem of a different city does not arise for the worker; his problem is, rather, to take possession of the city that already exists. The new town, in fact, will be born from an inversion of the town as it is at present.

It is to this end that our efforts are directed: to see and understand the city, no longer as a cultural unit, but rather as a structure to be used;

a homogeneous ensemble of services, upon which is superimposed a mesh of scenic happenings, of spatial episodes, that give this ensemble of functions a unified cultural meaning. The street, which divides and serves this compact mass of facilities, becomes a dynamic sequence, in which the flat surfaces of the façades of individual buildings allow the growth of an architectural language, whose forms serve to verify the various functional organisms.

Within this make-believe unity, the citizen must achieve his own civic integration, to the extent that he adopts the patterns of social behavior that the city, as a cultural unit, demands. We are trying, therefore, to overthrow this relationship by defining the city as 'a latrine every hundred square meters,' rejecting the problem of its 'form,' and defining its space, not as an ensemble of architectural volumes, but rather as a 'hollow space.'

A workers' city doesn't exist — because there exists no workers' culture. The belief that the worker, in the course of his struggle to work less and earn more, develops a culture of his own, replete with new meanings and values, is typical middle-class humbug. As a matter of fact, not only does the working class fail to produce a new culture, but it even uses the cultural patterns of the middle class as a strategy to mark the stages of its own economic advance. As it uses them, it also destroys them, precisely because in attaining these stages it destroys the economic balance on which the patterns were based.

The places in which the industrial system has completely realized its own ideology of Planning are the Factory and the Supermarket. Consumption and Production are not, in fact, contrasting logics, for both predicate a social and material reality that is entirely continuous and undifferentiated; both the Factory and the Supermarket are optimal urban systems, potentially unlimited, in which the functions of production and of merchandising information are freely organized according to a continuous plan and are made homogeneous by a system of artificial ventilation and lighting, without any interferences.

An outer image of these organisms 'does not exist,' since the plane of their façades does not indicate the linguistic structure of the organism but is simply the surface where two situations in different degrees of integration encounter one another.

If we were to apply the technological level and functional organization that has already been attained in these sections of the city to housing as well, we should see a complete transformation of the city. In fact, the metropolis today, like the traditional city, still adheres to certain standards of natural lighting and ventilation; no factory may exceed an established depth, in order that light and air may permeate its interior. This has resulted in a continuous 'formation' of architectural blocks made up of inner courtyards, façades, and interruptions. II we were to introduce on an urban scale the principle of artificial lighting and ventilation, we should see that it is no longer necessary to follow the procedure of a continual breaking-up into apartments, in the typical fashion of a traditional city; the city would become a continuous residential structure, without empty spaces, and hence without architectural images. Traffic would no longer divide the city into sections but would be arranged in an autonomous, optimal fashion, uniformly distributed throughout the land.

The city as a technical artifact and the territory that it occupies would no longer coincide. Nature and the city would run on two parallel tracks, without interfering with one another; while in the middle-class city, nature seems to the citizen to be the perfect reconciliation of the industrial system with natural laws, something to be used as a means of consolation, in our hypothetical 'homogeneous city,' nature is no longer an urban episode, but recovers its own complete autonomy. No longer would the individual have his contacts with nature contaminated by architectural elements that tend to attribute to it some cultural significance. Nature would, instead, remain a neutral area, lacking any meaning and therefore entirely open to physical awareness, without any intermediary.

These large microclimatized plans make it possible to overcome the limitations of present-day residential typology and the prevailing concept of architecture. Actually, architecture today, seeking to provide the greatest possible number of various kinds of freedom to the user, is bound to recognize in the urban phenomenon its true destiny, and, in the private phenomenon, its true nature. And so, in a contradictory fashion, on every occasion it sets forth the general state of affairs, while simultaneously coming to the defense of the partial, individual experience as against the collective experience. It thus serves symbolically as mediator in the conflict between public and private life.

The limitations of this mediating role cannot be overcome by the processes of design, but only by a change in the use made of architecture itself. It must be regarded as a neutral system, available for undifferentiated use, and not as an instrumentality for the organization of society; as a free, equipped area in which it may be possible to perform spontaneous actions of experimentation in individual or collective dwelling.

SUPERSTUDIO

Superstudio began their group activity in Florence in December, 1966, working in the fields of architecture and interior and industrial design. Since then, they have carried on research on architectural theory and system design, participated in national and international competitions, exhibited in Italy and abroad, and delivered lectures, as well as publishing articles and issuing a series of booklets, posters, and prints under their own imprint. In 1970, together with Gruppo 9999, they began a Separate School for Expanded Conceptual Architecture (or Sine Space School), engaged in experimental teaching and the exchange of information. They participated in two 'Superarchitecture' exhibitions at Pistoia and Modena in 1966 and 1967, respectively, and in the VII Biennale, Paris, 1971; in the latter year, they also exhibited in Rome their 'Twelve Ideal Cities.'

Designers: Superstudio (Piero Frassinelli, Alessandro Magris, Roberto Magris, Adolfo Natalini, Alessandro Poli, Cristiano Toraldo di Francia)
Patron: ANIC-Lanerossi
Film: Superstudio

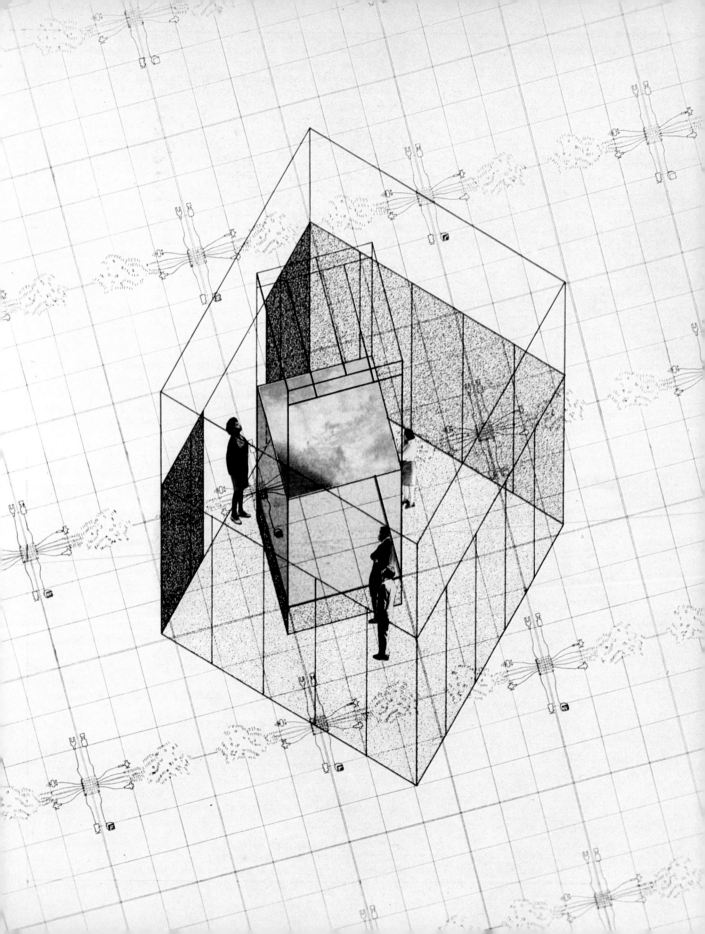

The proposed microevent is a critical reappraisal of the possibilities of life without objects. It is a reconsideration of the relations between the process of design and the environment through an alternative model of existence, rendered visible by a series of symbolic images. The microenvironment is like a room with walls; the floor and ceiling are covered with black felt; thin luminescent lines make the corner angles stand out clearly.

A cube about six feet wide is placed in the center on a platform about sixteen inches high. All the walls of the cube, except the one facing the entrance, are made of polarized mirrors, so that the model inside becomes clearer and clearer as we move to the end of the room. This model, repeated to infinity by the mirrors, is a square plate of chequered laminated plastic, with a little 'machine' out of which come various terminals. One of the terminals is connected to a TV screen, which transmits a three-minute movie, a documentary on the model seen in various natural and work situations. The sound-track gives information about the original concepts for the model. Meteorological events will be projected on the ceiling: sunrise, sun, clouds, storm, sunset, night. The lighting of the cube varies according to the phenomena projected. The rest of the room is permanently plunged in darkness.

Specific Considerations

In this exhibition, we present the model of a mental attitude. This is not a three-dimensional model of a reality that can be given concrete form by a mere transposition of scale, but a visual rendition of a critical attitude toward (or a hope for) the activity of designing, understood as philosophical speculation, as a means to knowledge, as critical existence.

Design should be considered as a 'cross-discipline,' for it no longer has the function of rendering our requirements more complex through creating a new artificial panorama between man and environment. By finding a connection between data taken from the various humanistic and scientific disciplines (from the technique of body control to philosophy, the disciplines of logic and medicine, to bionomics, geography, etc.), we can visualize an image-guide: the final attempt of design to act as the 'projection' of a society no longer based on work (and on power and violence, which are connected with this), but an unalienated human relationship.

In this exhibition, we present an alternative model for life on earth.

We can imagine a network of energy and information extending to every properly inhabitable area. Life without work and a new 'potentialized' humanity are made possible by such a network. (In the model, this network is represented by a Cartesian 'squared' surface, which is of course to be understood not only in the physical sense, but as a visual-verbal metaphor for an ordered and rational distribution of resources.) The network of energy can assume different forms. The first is a linear development. The others include different planimetrical developments, with the possibility of covering different, and gradually increasing, parts of the habitable areas. The configuration (typology) of the environment depends solely on the percentage of area covered, analogous to the way in which we distinguish a street from a town, a town from a city.

Some of the types:
10 percent covered: The network is developed like a continuous ribbon extending over the territory.
50 percent covered: The network is developed like a checkerboard, with areas measuring one square kilometer alternating with squares of

open land.
100 percent covered: The network is transformed into a continuous development, the natural confines of which are formed by mountains, coasts, rivers.

It is an image of humanity wandering, playing, sleeping, etc., on this platform. Naked humanity, walking along the highway with banners, magic objects, archeological objects, in fancy dress...
The distances between man and man (modified); these generate the ways in which people gather, and therefore 'the places': if a person is

alone, the place is a small room; if there are two together, it is a larger room; if there are ten, it is a school; if a hundred, a theater; if a thousand, an assembly hall; if ten thousand, a city; if a million, a metropolis...

Nomadism becomes the permanent condition: the movements of individuals interact, thereby creating continual currents. The movements and migrations of the individual can be considered as regulated by precise norms, the distances between man and man, attractions/reactions — love/hate. As with fluids, the movement of one part affects the movements of the whole.

The diminished possibility of physical movement results in an increase in conceptual activities (communications). The model constitutes the logical selection of these developing tendencies: the elimination of all formal structures, the transfer of all designing activity to the conceptual sphere. In substance, the rejection of production and consumption, the rejection of work, are visualized as an aphysical metaphor: the whole city as a network of energy and communications.

The places where humanity is concentrated in great numbers have always been based on the city network of energy and information, with three-dimensional structures representing the values of the system. In their free time, large crowds on the beaches or in the country are in fact a concentrated mass of people 'served' by mechanical, mobile miniservices (car, radio, portable refrigerator). Concentrations such as the Isle of Wight or Woodstock indicate the possibility of an 'urban' life without the emergence of three-dimensional structures as a basis. The tendency to the spontaneous gathering and dispersing of large crowds becomes more and more detached from the existence of three-dimensional structures.

Free gathering and dispersal, permanent nomadism, the choice of interpersonal relationships beyond any preestablished hierarchy, are characteristics that become increasingly evident in a work-free society.

The types of movement can be considered as the manifestations of the intellectual processes: the logical structure of thought continually compared (or contrasted) to our unconscious motivations.

Our elementary requirements can be satisfied by highly sophisticated (miniaturized) techniques. A greater ability to think, and the integral use of our psychic potential, will then be the foundations and reasons for a life free from want.

Bidonvilles, drop-out city, camping sites, slums, tendopoles, or geodetic domes are all different expressions of an analogous desire to attempt to control the environment by the most economical means.

The membrane dividing exterior and interior becomes increasingly tenuous: the next step will be the disappearance of this membrane and the control of the environment through energy (air-cushions, artificial air currents, barriers of hot or cold air, heat-radiating plates, radiation surfaces, etc.).

Through an examination of the statistics of population growth, an analysis of the relationship between·population and the territory that can be exploited for living purposes, new techniques for agricultural production, and ecological theories, we can arrive at a formulation of various hypotheses for survival strategies:

a) hypothesis for the creation and development of servoskin: personal control of the environment through thermoregulation, techniques for breathing, cyborgs... mental expansion, full development of senses, techniques of body control (and initially, chemistry and medicine).

b) hypothesis for total system of communications, software, central memories, personal terminals, etc.

c) hypothesis for network of energy distribution, acclimatization without protective walls.

d) mathematical models of the cyclic use of territory, shifting of the population, functioning and non-functioning of the networks.

General Considerations

If we look closely, we can see how all the changes in society and culture in this century (or since 1920) have been generated by one force only — the elimination of formal structures as a tendency toward a state of nature free from work.

The destruction of objects, the elimination of the city, and the disappearance of work are closely connected events. By the destruction of objects, we mean the destruction of their attributes of 'status' and the connotations imposed by those in power, so that we live *with* objects (reduced to the condition of neutral and disposable elements) and not *for* objects.

By the elimination of the city, we mean the elimination of the accumulation of the formal structures of power, the elimination of the city as hierarchy and social model, in search of a new free egalitarian state in which everyone can reach different levels in the development of his possibilities, beginning from equal starting points.

By the end of work, we mean the end of specialized and repetitive work, seen as an alienating activity, foreign to the nature of man; the logical consequence will be a new, revolutionary society in which everyone should find the full development of his possibilities, and in which the principle of 'from everyone according to his capacities, to everyone according to his needs' should be put into practice. The construction of a revolutionary society is passing through the phase of radical, concrete criticism of present society, of its way of producing, consuming, living.

Merchandise, according to Guy Debord, in bourgeois society (which acts and perpetuates itself through its products — including political parties and trade unions, which are essential parts of the spectacle) becomes the contemplation of itself.
The production machine produces a second poverty (Galbraith), perpetuating itself even after the fulfilment of its goals, or beyond its essential ends (the satisfaction of primary needs), constantly inducing new needs.

Once clarified that:
a) design is merely an inducement to consume;
b) objects are status symbols, the expressions of models proposed by the ruling class. Their progressive accessibility to the proletariat is part of a 'leveling' strategy intended to avoid the conflagration of the class struggle;
c) the possession of objects is the expression of unconscious motivation: through analysis, the removal of the motivation underlying their desirability may be reached;
...then it becomes urgent to proceed to destroy them... or does it?

Metamorphoses become frequent when a culture does not have sufficient courage to commit suicide (to eliminate itself) and has no clear alternatives to offer, either.
The theory of intermediate states is the book of changes?

Thus, while the merchandise-form continues on toward its absolute realization, we reduce operations to a minimum. Reducing operations to a minimum, in all fields, is part of a general process of 'reduction.' Only through this reduction process can the field be cleared of false problems and induced needs. Through reduction, we proceed toward a mental state of concentration and knowledge, a condition essential for a truly human existence.

Earlier, we defined the destruction of the syntactical ties that bind the object to the system, the destruction of its significance as

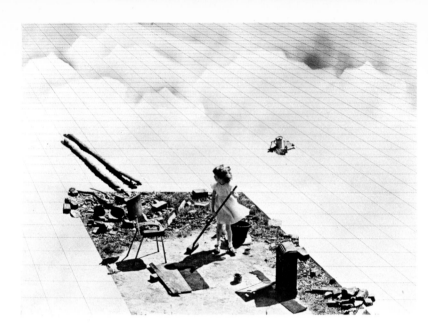

superimposed by the ruling classes, as 'destruction of the object.' We have formulated an hypothesis of the reduction of objects to neutral, disposable elements. To this, we can add the hypothesis of the construction of the object through its metamorphosis. The present process of 'overloading' meanings onto an object is part of that strategy of disgust to which we have already referred.
Through the psychological rethinking of an object, we can try for its 'reconstruction.' And this through discontinuous and alogical action, refusing guarantees of value (licenses issued by the system), aspiring to identify with life and total reality.

Objects thus cease to be the vehicles of social communication to become a form of reality and the direct experience of reality.

The metamorphoses which the object has to go through are those during which it is reloaded with the values of myth, of sacredness, of magic, through the reconstruction of relationships between production and use, beyond the abolition of the fictitious ties of production-consumption.

When design as an inducement to consume ceases to exist, an empty area is created, in which, slowly, as on the surface of a mirror, such things as the need to act, mold, transform, give, conserve, modify, come to light.

The alternative image (which is, really, the hope of an image) is a more serene, distended world, in which actions can find their complete sense and life is possible with few, more or less magical, utensils.

Objects, that is, such as mirrors — reflection and measure.

The objects we will need will be only flags or talismans, signals for an existence that continues, or simple utensils for simple operations. Thus, on the one hand, there will remain utensils (with less chrome and decorations); on the other, such symbolic objects as monuments or badges. Objects perhaps created for eternity from marble and mirrors, or for the present from paper and flowers — objects made to die at their appointed hours, and which even have this sense of death among their characteristics. Objects that can easily be carried about, if we should decide to become nomads, or heavy and immovable, il we decide to stay in one place forever.

A Journey from A to B

There will be no further need for cities or castles.
There will be no further reason for roads or squares.
Every point will be the same as any other
(excluding a few deserts or mountains which are in no wise
inhabitable).
So, having chosen a random point on the map,
we'll be able to say my house will be here
for three days two months or ten years.
And we'll set off that way (let's call it B)
without provisions, carrying only objects we're fond of.
The journey from A to B can be long or short,
in any case it will be a constant migration,
with the actions of living at every point along the ideal line
between A (departure) and B (arrival).
It won't, you see, be just the transportation of matter.

These are the objects we'll carry with us:
some strange pressed flowers,
a few videotapes, some family photos,
a drawing on crumpled paper,
an enormous banner of grass and reeds interwoven with
old pieces of material which once were clothes,
a fine suit, a bad book...
These will be the objects.
Someone will take with him
only a herd of animals for friends. For instance:
a quartet of Bremermusikanten,
or a horse, two dogs and two doves
or twelve cats, five dogs and a goat.
Yet others will take with them only memory,
become so sharp and bright as to be a visible object.
Others will hold one arm raised, fist clenched.
Someone will have learnt a magic word and will take it with him
as a suitcase or a standard: CALM, COMPREHENSION, CONFIDENCE,

COURAGE, ENERGY, ENTHUSIASM, GOODNESS, GRATITUDE,
HARMONY, JOY, LOVE, PATIENCE, SERENITY, SIMPLICITY,
WILL, WISDOM (dark blue).
(This is the complete set of cards in the 'Technique of Evocative
Words' by Roberto Assagioli, M.D.)
But almost everybody will take only himself from A to B,
a single visible object, like a complete catalogue
as an enormous Mail Order Catalogue

What we'll do

We'll keep silence to listen to our own bodies,
we'll hear the sound of blood in our ears,
the slight crackings of our joints or teeth,
we'll examine the texture of our skins, the patterns made by the hairs
on our bodies and heads.
We'll listen to our hearts and our breathing.

We'll watch ourselves living.
We'll do very complicated muscular acrobatics.
We'll do very complicated mental acrobatics.

The mind will fall back on itself to read its own history.
We'll carry out astonishing mental operations.
Perhaps we'll be able to transmit thoughts and images,
then one happy day our minds will be in communication with
that of the whole world.

That which was called philosophy will be the natural physical activity of
our minds, and will at the same time be philosophy, religion,
love, politics, science....
Perhaps we'll lose the names of these disciplines (and it will be no
great loss)
when everybody will be present in essence in our minds.
We'll be able to create and transmit visions and images,
perhaps even make little objects move for fun.

We'll play wonderful games, games of ability and love.
We'll talk a lot, to ourselves and to everybody.
We'll look at the sun, the clouds, the stars.
We'll go to faraway places, just to look at them and hear them.
Some people will become great story-tellers: many will move
to go and listen to them.
Some will sing and play.
Stories, songs, music, dancing will be the words we speak and tell
ourselves.
Life will be the only environmental art.

The happy island

A lady of our acquaintance
became hysterical at hearing all this story
and said: I certainly have no intention of doing without
my vacuum-cleaner and the mowing machine, and the electric iron and
the washing machine and refrigerator, and the vase full of flowers,
the books, my costume jewellery, doll and clothes!
Whatever you say madam!
Just take whatever you like, or rather equip a happy island for yourself
with all your goods.
The only problem is that the sea has receded all round and
the island is sticking up in the middle of a plain without any messages
in bottles.

The distant mountain

Look at that distant mountain.... what can you see?

is that the place to go to? or is it only the limit of the habitable?
It's the one and the other, since contradiction no longer exists,
it's only a case of being complementary.
Thus thought a fairly adult Alice skipping over her rope, very
slowly, though without feeling either heat or effort.

The encampment

You can be where you like, taking with you the tribe or family.
There's no need for shelters, since the climatic conditions
and the body mechanisms of thermoregulation have been modified
to guarantee total comfort.
At the most we can play at making a shelter, or rather at the home,
at architecture.

The invisible dome

All you have to do is stop and connect a plug: the desired microclimate
is immediately created (temperature, humidity, etc.); you plug in to the
network of information, you switch on the food and water blenders....

A short moral tale on design, which is disappearing

Design, become perfect and rational, proceeds to synthesize different realities by syncretism and finally transforms itself, not coming out of itself, but rather withdrawing into itself, in its final essence of natural philosophy.
Thus designing coincides more and more with existence: no longer existence under the protection of design objects, but existence as a design.
The times being over when utensils generated ideas, and when ideas generated utensils, now ideas are utensils. It is with these new utensils that life forms freely in a cosmic consciousness.

If the instruments of design have become as sharp as lancets and as sensitive as sounding lines, we can use them for a delicate lobotomy.

Thus beyond the convulsions of overproduction a state
can be born of calm in which a world takes shape without
products and refuse, a zone in which the mind is energy and raw
material and is also the final product, the only intangible object for
consumption.
The designing of a region free from the pollution of design
is very similar to a design for a terrestrial paradise....
This is the definitive product — this is only one of the projects for a
marvelous metamorphosis.

GRUPPO STRUM

Gruppo Strum (Group for Instrumental Architecture), centered in Turin, is a loosely knit association of architects and designers, whose individual members have won numerous national and international competitions and have frequently had their works published in Italian and foreign periodicals. Their common interest is in the use of architecture as an active instrument for political propaganda, by means of activities and theories connected with the physical organization of space. The group had its origin in a series of experiments conducted by its participants in the fields of architecture, design, scientific research, and teaching. Most members of the group are on the faculties of architectural schools in Italian universities and share a concern with the problems of teaching in such schools, as well as with the organization of education at all levels and the policies governing scientific research in universities. The political control of housing, particularly as regards the struggles of the proletariat now in progress throughout Italy, is another of their major concerns. Gruppo Strum defines its field of action as 'The Mediatory City,' a theme clarified in the photo-stories prepared for the present exhibition.

Designers: Gruppo Strum (Piero Derossi, Giorgio Ceretti, Carlo Giammarco, Riccardo Rosso, Maurizio Vogliazzo)
Consultant for photography: Paolo Mussat Sartor
Patrons: Gufram, Casabella

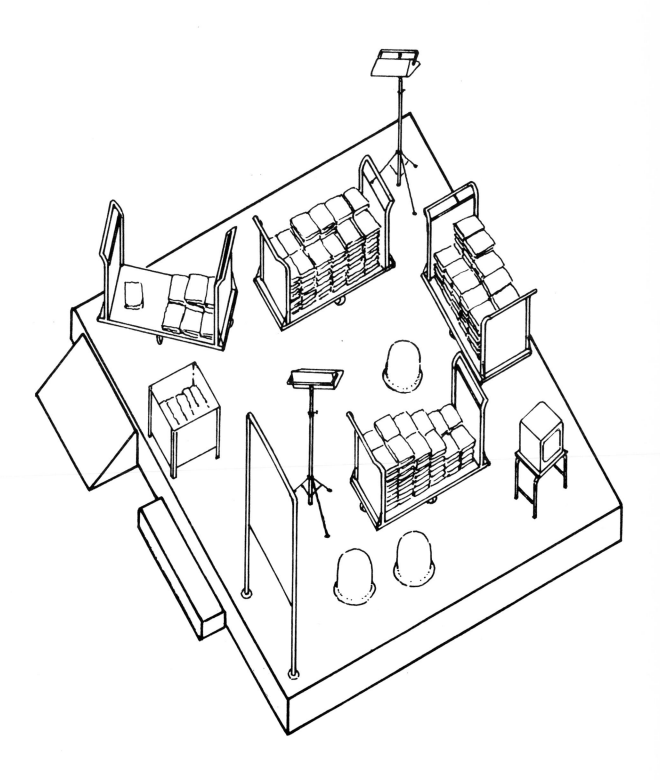

We should have liked to respond in a simple way to the commission by The Museum of Modern Art of New York for the exhibition of Italian design, by 'designing' the space entrusted to us. But, after discussing the directives given for our argument and examining the possible 'reference groups' to aim it at (the good people who will come to the Museum on Sundays to inform themselves about Italian design, the American manufacturers frightened by foreign competition, the Italian manufacturers who need to boost their exports, our designer friends or enemies who want to understand and categorize us, the critics who will have to pass judgment, or all these together, etc.), we were unable to invent any 'physical forms' capable of communicating our thought — or even part of our thought — on the problems of the 'New Domestic Landscape' in Italy. Not the softest of armchairs, not the descending flights carpeted with thick pile, not the colored 'perspex,' not the gleaming stainless steel surfaces, not the complex and multipurpose devices, not the plain, functional furniture; and not even empty space as a negation, mirrors with their magic, disturbing psychophysical sensations, nor all-encompassing displays.

And yet, we had plenty of things to say. We wanted this important exhibition at The Museum of Modern Art to be an opportunity for saying them, and saying them in a direct way, really to provide information with precise data and documents.

As the instrument for developing our approach, we chose a sort of newspaper, or rather, 'photo-stories with documents' — an odd kind of symbiosis between an illustrated adventure story and the descriptive catalogue of a department store: a stratagem that allowed us to tackle big topics in a sufficiently synthetic way.

The area assigned to us by the Museum has become a simply equipped space for distributing free of charge to the public three series of 'photo-stories'; each series illustrates an aspect of design that we believe to be of particular importance. If someone has doubts about how to decorate his house, or wants to know if design is art, or more exactly, if Italian design is good, he will certainly find an answer in the information that we shall give him, though the answer will be indirect and will deal with broader and more involved topics than normally expected.

The problems we have examined are:

The struggle for housing (white paper): Many people in Italy do not have a decent home to live in, and others have no home at all. If they are not given one (and for the time being, no one is likely to give it to them), they must get homes for themselves by organizing themselves into a political movement capable of overturning the trend of the current system, of which their fringe existence and exploitation are functions.

And not only must they get hold of homes of some kind, in order to have a place to sleep after many hours of alienating work, but they must get themselves homes that are really comfortable for their individual lives, and that will form the initial elements of a different collective life. What they must procure, therefore, is not merely a home to live in, but a city, so as to ensure for themselves a freer social life, and one more in keeping with their needs.

There have been many struggles for these goals in Italy. The photo-story on white paper recounts, and illustrates, these facts, showing also how these struggles for housing continually reshape cities, by attacking and defeating the capitalist organization of the territory together with the symbolic values that formalize it.

Utopia (green paper): It is impossible not to consider a world organized in a different way, where people would be happier and all the current shortcomings eliminated. Imagining a heavy development of technology (or a return to country life), one can predict flexible, adaptable, mobile, provisional urban settlements, controlled by computers, with extremely high levels of communications, information, etc...

Refusing to attribute autonomous values (the fruits of a neutral conception of science) to these alternative models, we are concerned with rediscovering UTOPIA as an act of provocation, and as a negation of the objectivity of the present-day system of production; in short, we want to try to use UTOPIA as a means of intervention, directly linked with the organization of the struggles against the programmed reorganization of capital.

The photo-story on green paper shows diagramatically the development of UTOPIA at the present moment, to demonstrate some possible connection with the themes emerging from the organization of political vanguards, and to counter, by unveiling their objectives, the ideological use that the owners of the realm are trying to make of UTOPIA. Each design of the city, or of a situation within the city, can accept or reject UTOPIA; or it can implement this alternative by intervening directly in the actual political situation today, and taking specific mediatory action.

The mediatory city (red paper): Workers, students, technicians, and soldiers fighting against the capitalist system and its bosses are organizing themselves into groups to discuss policy, coordinate action, inform militants, and defend themselves against repression. Taking part in this political activity means changing individual and collective behavior from the patterns that the bourgeois city tries to impose. With their directed rigidities, the physical layouts of traditional urban centers become obstacles to the expansion of destructive activities.

Economic blackmail, repressive laws, police controls, whitewashing lies, and the bait of illusory reforms are the constant enemies. The development of a revolutionary organization depends also on the capacity of groups to procure the physical space required for their work, and to find new tools that can be used for the struggle without being undermined by the bosses. This objective requires a reconsideration of the city to counteract the various ideological interpretations made by the bosses' 'watchdogs,' and also to identify the priorities and modes of action best suited to an overall strategy; moreover, they entail learning to use a physical reality that is consistently available for the possibility of rendering the action feasible.

The city becomes a complex set of old and new tools for use, places to be conquered, and objects to be altered; a great storehouse available to proletarian creativity, enabling those who have rejected the capitalist city, and who are struggling to destroy it, to survive. These are mediatory actions that take place in a continuous process, and they form a mediatory city every day.

The Struggle for Housing
Photo-story. Contents:

1. The Board Room of a large company. The Chairman makes a speech on the necessity for raising production at any cost. He illustrates the means at his disposal to control the functioning of the city.
2. Four realistic examples of the living standards of the Italian proletariat, from the standpoint of the relation between housing and factory labor, with particular reference to the problem of migration.
3. At a general meeting of workers and students, a politically conscious worker explains the workings (and shortcomings) of public building in Italy.
4. Four realistic examples of struggle that end up with an equal number of occupied houses. The police defend the landlords.
5. The struggle for homes induces the proletariat to organize collective services that they themselves manage, thereby producing red nursery schools, red health clinics, red markets, people's housing developments, etc., etc.
6. The position that the political vanguard should take on the housing issue is debated during the general meeting. The struggles in the various districts become general and link up with the struggles in the factories, affecting the whole city.

THE STRUGGLE FOR HOUSING

fotoromanzo

capitalist

barracks

students

with:

GesCal.

PRENDIAMOCI LA CITTÀ

with data and documents

free distribution of »fotoromanzi« to the visitors

Utopia
Photo-story. Contents:

1. Theoretical writings on the values and limits of utopian city proposals. Diagrammatic scheme of the institutions, agencies, persons, texts, and journals concerned with utopia in Italy.

2. Description, with theoretical explanations and visual presentations, of the principal utopias (the utopia of the social-democratic, or comfort, city; the utopia of the communication city; the utopia of the flexible city; the utopia of the mobile city; the utopia of the continuous city; etc., etc.).

3. Diagrams for counterpositions of possible ways by which the implications within the utopias described are manipulated and deformed by the bosses and find a place in the attitudes of the fighting political vanguards.

4. Theoretical discourse on how utopia is to be brought about. Utopia as a means of intervention. List of the most significant positions.

5. Catalogue of conditions and situations in which utopia can play an immediately positive role in backing up the organization of political vanguards.

6. The mediatory city as a means of overcoming utopias.

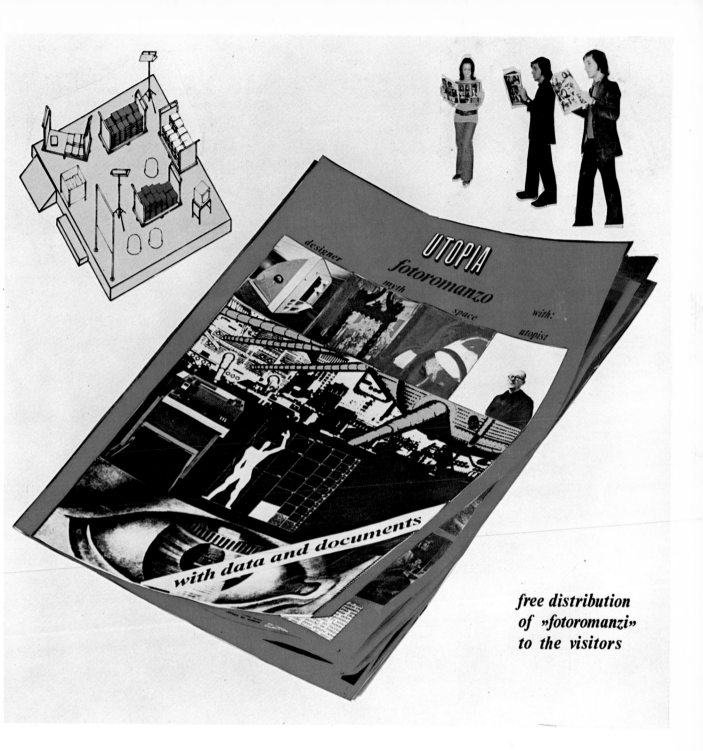

UTOPIA
fotoromanzo

designer

myth

space

with:

utopist

with data and documents

free distribution
of »fotoromanzi»
to the visitors

The Mediatory City
Photo-story. Contents:

1. Five realistic examples of the precarious living conditions and the struggles on the margins of ruling-class legitimacy. The common denominator is rejection of the established order as limiting creativity.

2. General meetings of workers and students on the theme 'Extend the struggle from the factories to the whole of social life.' Catalogue of political papers on the relation between the factory and the city. A resolution is passed to start specific activities designed as phases in organizing the struggle.

3. Catalogue of actions and objects that can be used in organizing the proposed activities. Specialists supply the requisite technical and economic data.

4. The construction of red bases must become a mass action. All the workers in the factories and districts throughout the city must be informed by means of appropriate campaigns, and their active participation must be requested.

5. The reformists try to direct the needs born of the workers' struggles back into the framework of institutional enterprises. They propose projects for functional and modern social centers, alternative cities, and programs for serious cultural activities.

6. Mass action rejects the compromise and carries on with organizing the cause. Red bases spring up throughout the city; the proletariat defends them from police attacks.

Organization of the cause redefines the functions and form of the capitalist city.

THE MEDIATORY CITY

fotoromanzo

militant

architect

workman

with:

student

SLEEP DRI

SLEEP DRI

with data and documents

free distribution of "fotoromanzi" to the visitors

ENZO MARI

Enzo Mari was born in 1932 and works in Milan. Beside his extensive activity as a designer, since 1952 he has devoted himself intensively to theoretical research, especially on the psychology of vision, systems of perception, and the methodology of design. He has taught design both at the Umanitaria School in Milan and the Experimental Film Center in Rome. His research has centered especially on two aspects of perception, which he has defined respectively as 'the ambiguity of interior tridimensional space' and 'the analogy between serial systems of natural phenomena and the programming of the phenomena of perception.' He has recently summarized his research, theories on design, and work as a designer in a monograph, *Funzione della ricerca estetica* ('Function of Aesthetic Research'; Milan: Comunità, 1970). In recent years, he has been especially concerned with the role of the designer in relation to contemporary society.

Knowing Mr. Mari's position, the Museum extended him a formal invitation *not* to design an environment. He consented and produced the following essay, in which he attempts to reconcile the fact that, although he is the designer of many beautiful objects, including a number presented in this exhibition, he nevertheless does not believe that the task of designing objects, as physical articles to be executed and sold, has any significance today. He proposes, instead, that the only valid sphere of action for the designer is that of communications, and that the only honorable strategy open to him is that of renewing language — the alphabet included. — E. A.

Foreword

All human activity is, first and foremost, communication. We must clarify whether the 'communications' inherent in the two parts into which the exhibition 'Italy: The New Domestic Landscape' is divided are the same, or different. The Museum has used two different criteria in organizing these sections. In one, it has made a critical selection of objects produced by Italian designers in the past ten years. For the other, it has invited individual designers or groups to present their 'philosophies' in the form of environments.

But, if these 'philosophies' are identical with those implicit in the other section of the exhibition, their mise-en-scène is redundant. If, on the other hand, they differ, it is illusory to think that the differences concerning the socioeconomic aspects of design (since this is all that can be in question) can be communicated scenographically. In fact, within the context of an exhibition of this kind, this scenography can find backing for its execution only within the limits of communicating superstructural aspects to the visitor, or at any rate by using an experimental, and hence obscure, language.

That is why I have refused to participate directly in this section. (I have, however, allowed some objects designed by me, and chosen by the Museum, to be included in the second part of the exhibition; for, in today's competitive climate, the profession to which I belong must rely solely on formal quality for its patronage, and hence for its survival.)

My 'philosophy,' however, can be more clearly and usefully explained in the following

Proposal for Behavior
directed to my colleagues:

Producing art, and the philosophy related to it, are aspects of communication among mankind. More specifically, they concern research into the appropriateness of modes of language in their historical development. The usefulness of this research lies in the fact that, by means of new models, it proclaims (or should proclaim) the sclerosis of current communications.

Communication is the most important factor in social relations and their evolution. Social evolution today can be determined solely by the class struggle. Communication is the determining element in the class struggle. (If there is a ruling class and a subordinate class, it is not only because the former has more guns than the latter, but because a large part of the subordinate class is still not clearly aware of the implications of its own conditioning; in short, there has been no communication.) Any revolutionary activity, therefore, is above all a matter of communication. But while, on the one hand, it is clear in what direction the collective force is moving to achieve a new order, it is less clear what strategic choices are required if truly revolutionary results are to be brought about quickly.

It is just this lack of clarity that accounts for the diversification of current ideological and political research, and hence of related communications.

Political research must inevitably be recognized as being far and away more urgent than any other kind of investigation, precisely because it serves to determine the conditions under which it might be possible — for everyone — to achieve everything else.

But, inasmuch as collective political maturity is itself closely associated with a refusal to delegate, the priority of political research does not entail a refusal to acknowledge other branches of research. It does mean, however, that while such researches (including communications research) must be free to define their own scope (since this cannot be evaluated by researchers in other fields, including the political), they must nevertheless recognize that their rationale can be found only within the political framework.

Accordingly, in the case of language research, precisely because the ways in which political communications can be conveyed must be defined and analyzed, this research — although at liberty to prefer one angle to another — must necessarily respect its own field of study: communications (whose destination, whether we like it or not, is always useful to any political faction). Those 'artists' who profess to agree with the cause of the subjugated class must, therefore, be aware that their search for an idiom should be a valuable ingredient in the class struggle. But we know very well that nowadays their research, simply because it is conducted on the individual plane (and is therefore unavoidably conditioned by that fact) is ultimately still a useful instrument for maintaining the privileges of the ruling class. In fact, the only use which that class makes of the artistic manufactured object is for its own cultural vainglory; and it certainly does not avail itself of the implications of research in design. On the contrary, it favors the irrational, inexpressible, and sacral aspects of art, precisely because this sacral quality is the means best suited for the ostentation in question. The most serious thing about all this is that the subjugated class is so strongly conditioned by the 'mystic' aura that even those belonging to its most politically progressive fringes still look upon art and culture in general as 'places' wherein the individual finds his gratification (and alienation).

On the other hand, those 'artists' who want to make a different contribution to the moment of destruction, beyond that of their own technical ability, fail to realize not so much the willfulness of their self-gratification (the only way in which we can describe it) as the fact that the effectiveness of this different contribution is in any case irrelevant, in comparison with the political effectiveness of their day-to-day expressions as tools of the ruling class.

Ultimately, the only correct undertaking for 'artists' is that of language research — that is, critical examination of the communications systems now in use, and critical acts affecting the ways in which man's primary needs (rather than ideologies as such) are conveyed — and almost always manipulated. For this reason, 'artists,' and those connected with their work, must not confine themselves to experimenting and devising new modes of expression but must show a fundamental concern for the manner in which the substance and implications of their research are communicated and received; and especially they must question who the interlocutors are. Only a constant and diligent informational campaign can succeed in narrowing the margin of manipulation to which the researcher's work is now liable, and consequently help to destroy the cultural myth-making of the ruling class.

Since all the 'artists,' although in varying degrees, profess to share the revolutionary will of the proletariat, it would seem easy for them to operate in this way. But as they are, in fact, so badly conditioned that they actually connive with the dominant class, they end up by cloaking their connivance in a kind of formal jargon and, worse still, in arguments that are often 'justified' as Marxian.

Perhaps it may be possible to formulate a theory of behavior capable of overcoming this situation.

Those who maintain: a) that social evolution can be resolved only by the class struggle; b) that 'artistic' activity today has no alternative other than to be used as an instrument; and c) that they desire to carry out collective action — must impose on themselves a code of behavior that can liberate research dialectics from all its mystifying superstructures. At the same time, they must consent to carrying on a dialogue with those who are not in agreement with them on points a), b), and c), only on condition that the latter are willing to 'communicate' the nature and extent of their dissent -- and in terms that are not politically ambiguous.

It is proposed, therefore, that all communication of the artist's own 'artistic' or critical activity should take place according to the following scheme:

I. Enunciation of his own utopian vision of the development of society.
II. Definition of the strategy deemed fitting for the attainment of this ideal.
III. Statement of what tactical moment of this strategy he has now reached.
IV. Synchronization of his research with that tactical moment.
V. Communication of the work of research in question (being at pains to remember that this should be with special reference to the foregoing points).

Furthermore:
The progression proposed should always be followed, so that the different communications can be compared with one another.

The scheme must be followed long enough to allow its results to be checked. For the same reason, it should be followed regardless of the importance, scope, or frequency of the communications — that is, without examining each time the pertinence of this method of procedure.

In short, it is not a question of simply making abstract pronouncements, detached from the daily practice of one's profession, but of constantly bringing one's work (especially one's critical work) into relation with one's contingent reality, one's own will to make statements and clarify them, and one's own free, ideological choice, which alone can explain the motivations.

It might be objected:
That this attempt, too, can easily be manipulated. Quite so. But perhaps it will allow such manipulation to become more clearly apparent.

That this proposal will be ignored by many of those to whom it is directed. This may well be suspected; but precisely in the sense that what is 'suspected' is not so much their avowed adherence to the dominating class, as their feigned adherence to the class that is dominated.

GIANANTONIO MARI

Gianantonio Mari was born in Varese in 1946 and received his degree in architecture at the Politecnico, Milan, in which city he now has his studio. For several years, he was associated in architectural and design projects with Joe Colombo, until the latter's death in 1971.

Designers: Gianantonio Mari and Studio Tecnico G. Mari
Planning: Gianantonio Mari
Project line and texts: Ezio Mari
Graphics: Ornella Selvafolta
Collaborators: S. Ando, M. Matsukaze
Photography: S. Pazzi

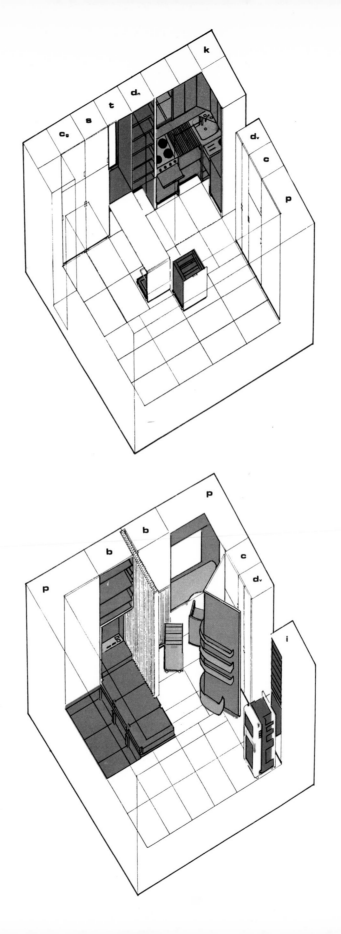

1. Living and Housing Poetry. The program proposed for the competition was considered by Studio Tecnico; G. Mari remarked that the first consideration should be the poetry of living in and inhabiting a house, namely the combined feelings of warmth, possession, protection, and security that are normally regarded as the attributes of a traditional home. These should not be ignored nor diminished but, on the contrary, should be carefully considered, revaluated, and if possible, improved by a plan that would emphasize the human factors, eliminating any extreme or utopian techniques.

2. Environment. The designers' fundamental premise was that the concept of environment in itself exceeds the limits of physical space, and therefore the design should take into consideration the totality of relations between man and space, the space itself being regarded as the sum of physical dimensions and the equipment included within those dimensions — that is, space should be regarded as dynamically organized.

3. Rituals and Ceremonies. Rituals and ceremonies represent different ways of using a place, and the relations between man and live space, seen in relation to man's own physical, biological, psychological, and social needs. An analysis of the various needs and functions (rituals and ceremonies) of man within his dwelling, and a grouping of them according to the points they have in common, led to dividing the various rites and ceremonies into five major categories:

Privacy: including study, work, hobbies, meditation, and man's need for reflection and being by himself.

Sleeping: including the psychological and physiological activities of man when he goes to bed, such as dreaming, erotic intercourse, resting.

Dining: Although divided into two states — the preparation and the eating of food — these can be directly and quickly related.

Leisure: including relaxation in its widest sense, information (usually through mass media, such as radio, TV, newspapers, magazines, etc.); and also living in terms of relations with other people and with those outside the home environment (public relations, social contacts outside the family, friends, relations, etc.).

Sensory: including everything in the environment that stimulates the sensory and perceptive faculties, such as light, music, air, real images or those created by fancy.

4. Modularity. The final consideration was to regard the spatial limitations — 15 feet 8 inches (4.80 meters) square by 10 1/2 feet (3.20 meters) high — not as a final size but as a basic dimension. This led to the decision to use modular and repeatable elements in the design, which would allow the environment to be expanded or contracted according to circumstances, and to provide suitable modular elements, starting with the primary unit of life (man), and progressing to two (man-woman), to three, four, and so on, thus overcoming the spatial limitations of the terms proposed for the competition.

5. Method Used in the Design. The foregoing analysis made evident the need for a synthesis that would take into consideration the definition of the environment and examine all the possibilities for superimposition. The catalytic element of this synthesis was the factor of time — namely, the fourth dimension seen in the context of the *timing* of the rituals, and consequently the use of the environment. The final phenomena that were taken into consideration in determining the method used were:

a. The existence and almost simultaneous enactment of the rituals

Leisure, Living, Privacy, and *Dining.*

b. The existence within a well-defined time-span (practically, only at night) of the ritual *Sleeping.*

c. The necessity of foreseeing occasional situations or modes of use (rituals) requiring independent action, even in the time shared with others, and therefore studying ways of providing for partial isolation.

Examining sleeping and its environment (normally during the nocturnal hours), the designers decided to consider this as the basic activity on which all other environments should be superimposed. This led to the need to design objects that would be available when they must function or be used, and that are not present (therefore not encumbering) when no ritual requires their use.

This conclusion, taken in connection with the values originally ascribed to the environments, led to a process of designing types of equipment that could be distributed throughout the entire available space, including those areas traditionally known as the ceiling and the floor. Subsequent analysis of the spatial characteristics of the areas, in accordance with prosemic considerations, led to determination of the areas respectively assigned for the functioning of the equipment.

Each element, consequently, has been planned to be contained (or incorporated, when not in use) within the walls, floor, and ceiling of the habitat unit, so that in the absence of each ritual, when its equipment is accordingly not functioning, it is symbolically represented by the continuity of the neutral white walls; while the ritual that is alive (that is, whose equipment is at work) is indicated by giving the various elements different colors, consistent with the respective functions that they perform.

Having agreed upon this method, the planners went through all the considerations entailed in creating product design, intentionally basing their plan on four axioms:

a. The non-sense of designing a piece of equipment or a product that is not an instrument for the total solution of the habitat unit (i.e., avoiding designing objects limited to furnishing a space).

b. The necessity of being able to participate in the technical and structural decisions regarding buildings themselves, by offering design products that can be modularly multiplied and assembled according to precise systems, the purpose being to develop (starting at a basic typological, microurbanistic level) a different townscape, more consistent with the dynamism of modern-day development.

c. Scaling for execution, basing the project on actual structures and techniques of construction that can be carried out by modern industrial systems (hence taking into consideration criteria related to economics and production). This might allow a partial and gradual application of the design, as well as one that could be used Immediately in collateral branches of building, such as trailers, houseboats, small prefabricated habitats (cottages), as well as motels, tourist colonies, and even temporary quarters for migratory workers.

d. The requirement, derived from the scaling, of a design that would permit rational, modern, and therefore economically valid mass production; this meant that all the equipment must be completely modular, enabling simplification in assembling and installation, so that the unit can be constructed and completed in the shortest possible time.

The designers therefore decided to adopt in their project the principle of designing all the equipment as completely modular and combinable units, which, together with a simple grill-support structure, are totally repeatable and can be used for structures of larger size by adapting the module, practically without any limitations of dimension. The units constitute elements of separation from the outside environment, while, within, they provide the equipment that allows enactment of all the functions of living — the rituals.

Intentionally, the purely formal aspects of the equipment as designed has not been taken into consideration, because following a planning method meant to create a unit that should more perfectly serve man's needs has by implication led to going beyond the concept of a product with purely formal qualities.

The resulting design, however, has not been the outcome of technical considerations alone, carried to extremes and hence utopian. It involves a creative technique that looks to the future and is based on human and psychological considerations of 'living and dwelling'; therefore, special attention has been given to those human values peculiar to family life, with the view of providing technical solutions that could increase flexibility and enjoyment of the environment — privacy, leisure, living, sensory functions — all too rarely to be enjoyed in traditional habitats today.

The planned modular system permits either horizontal or vertical extensions, allowing coordination with standard environments, and also other dimensions of the habitat units of which this is the prototype.

Two persons can sleep together, or each alone privately; and if desired each one can be isolated from the rest of the habitat.

Materials Proposed for Modular Elements:

Starting from the premise that traditional materials are foreseen for the traditional furnishings, all the modular parts are to be executed in plastic resins. Anticipating that, when supported by an appropriate metal grill system, the external part of each module could also form the element of separation from the exterior (as the ceiling and floor modules form the element of separation from other superimposed elements), the use of plastic materials strengthened with fiber glass is proposed for these external parts.

The internal parts can be executed with any type of plastic resin, provided it can ensure the necessary mechanical resistance. Between the internal and external parts (that is, between the plastic resin strengthened by fiber glass and the normal plastic resin) there should be a hollow space of expanded polyurethane to provide thermal and acoustical insulation. This hollow space can be of different types of polyurethane, so that the internal part of the element can be made from the same material.

It is proposed that all floor modules be covered with mohair carpeting of the usual type, except the kitchen-cooking area, for which water-repellent material (synthetic rubber or the like) is suggested; however, if desired, all floors, including that in the cooking area, can be covered with mohair, provided it is waterproof.

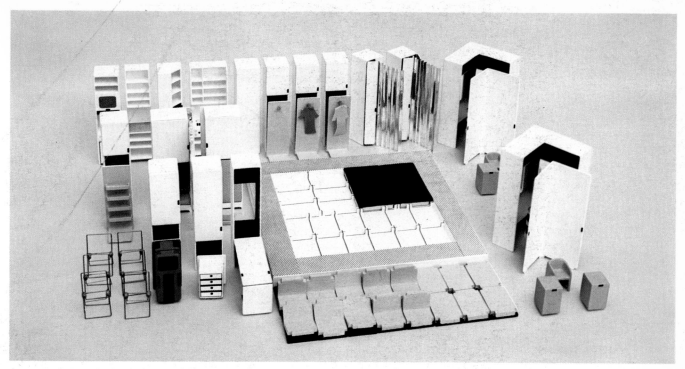

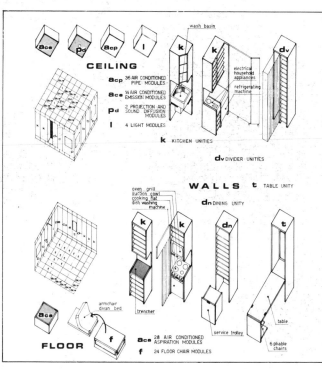

CEILING

acp 36 AIR CONDITIONED
PIPE MODULES

ace 14 AIR CONDITIONED
EMISSION MODULES

pd 2 PROJECTION AND
SOUND DIFFUSION
MODULES

l 4 LIGHT MODULES

wash basin

k KITCHEN UNITIES

electrical
household
appliances

refrigerating
machine

dv DIVIDER UNITIES

WALLS **t** TABLE UNITY

oven grill
suction cowl
cooking flat
dish washing
machine

dn DINING UNITY

armchair
divan bed

trencher

service trolley

table

6 pliable
chairs

ace 28 AIR CONDITIONED
ASPIRATION MODULES

FLOOR **f** 24 FLOOR CHAIR MODULES

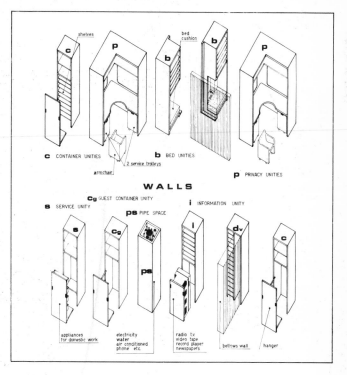

shelves

bed cushion

c CONTAINER UNITIES

2 service trolleys

armchair

b BED UNITIES

p PRIVACY UNITIES

WALLS

Cg GUEST CONTAINER UNITY

s SERVICE UNITY **i** INFORMATION UNITY

ps PIPE SPACE

appliances
for domestic work

electricity
water
air conditioned
phone etc.

radio t.v.
video tape
record player
newspapers

bellows wall

hanger

273

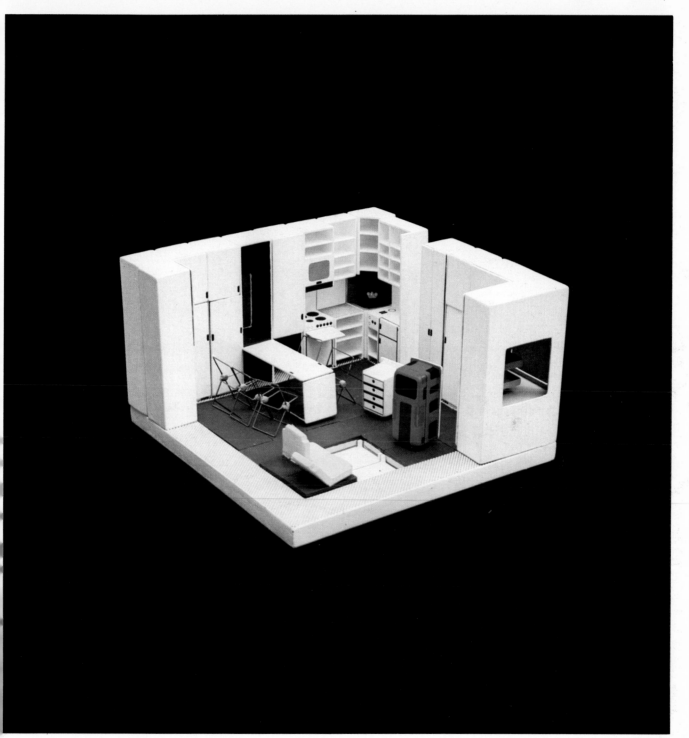

The four Florentine artists, now in their early thirties, who comprise Group 9999 have been especially interested in research on the theater as applied to architecture and the other arts. In 1968, they projected a 'Design Happening' on the Ponte Vecchio, and in the following year executed a multimedia environment for the discothèque 'Space Electronic' in Florence. They were co-founders, with Superstudio, of the Separate School for Expanded Conceptual Architecture in 1971, besides participating, in the same year, in the International Institute of Design summer school in London and conducting lectures and seminars on 'Life, Death, and Miracles of Architecture' at Florence. Their work has been exhibited in Graz and Salzburg, as well as throughout Italy.

Designers: Group 9999 (Giorgio Birelli, Carlo Caldini, Fabrizio Fiumi, Paolo Galli)

Up until now, technology has followed a completely autonomous path, one, we might say, in conflict with nature. Through its power, it has influenced every mode of thought and every kind of design operation, whether for art or for industry. It has produced dross and unwanted waste, and architecture has reflected some of the most typical aspects of this general phenomenon.

Our project is an attempt to pursue a direction entirely different from the one followed up to now. It parallels the investigations of recent researchers in diverse fields who have been studying the cyclical utilization of energy. Their work allows us to formulate hypotheses and new spatial models that no longer follow the one-way route of present-day systems of technology and production. We want to avoid falling into that aesthetic and formal environmental chaos which today has resulted in a crisis of global proportions.

Our project seeks to offer a solution based on new cyclical relations among man, nature, and technology. We want to propose returning once more to elements that have long been lost and are by now forgotten: ancient and primordial things like food and water, side by side with technological inventions. It is our attempt to bring man back into relation with nature, even in this modern and hectic life, without returning to mythical and impossible conditions. Our project must be understood, therefore, as the model of a real object, which must find its place in the home. It is an eco-survival device, to be reproduced on a global scale. It is itself a habitable and consumable place in accordance with the principles of the recycling of resources. Intentionally, it makes use of very simple elements: a garden, water, and an air bed.

Man is in direct contact with nature; he follows its growth and development; he cultivates and uses its products. He establishes a symbiotic relationship. Man, himself a product of nature, participates in the cycle of the seasons, in the variations of the stars.

If technology keeps on destroying nature, the possibility of having contact with the vegetable kingdom in its integral cycle will assume even greater significance. The vegetable garden will become the sacred place of a new religion.

Our project is completed by the air bed, which has both real and symbolic functions. It is a place for meditation and thought. Its abstract and religious symbolism is that of a resting place suspended above the ground but still near to it. Air jets scented with natural elements help to maintain this contact.

The technology used is simple and tested. But its physical aspect is hidden from sight, and this is essential to the project.

We should like to apply to the situation we have created a description written two thousand years ago by Vergil in the 'Georgics,' V. 125-45:

. . .   pauca relicti
Iugera ruris erant, nec fertilis illa iuvencis
nec pecori opportuna seges nec commoda Baccho:
hic rarum tamen in dumis olus albaque circum
lilia verbenasque premens vescumque papaver
regum aequabat opes animis, seraque revertens
nocte domum dapibus mensas onerabat inemptis.
primus vere rosam atque autumno carpere poma,
et cum tristis hiems etiamnum frigore saxa
rumperet et glacie cursus frenaret aquarum,
ille comam mollis iam tondebat hyacinthi
aestatem increpitans seram Zephyrosque morantis.
ergo apibus fetis idem atque examine multo
primus abundare et spumantia cogere pressis
mella favis; illi tiliae atque uberrima tinus,
quotque in flore novo pomis se fertilis arbos
induerat, totidem autumno matura tenebat.
ille etiam seras in versum distulit ulmos
eduramque pirum et spinos iam pruna ferentis
iamque ministrantem platanum potantibus umbras.

...a few acres of unclaimed land, and this a soil not rich enough for bullocks' ploughing, unfitted for the flock, and unkindly to the vine. Yet, as he planted herbs here and there among the bushes, with white lilies about, and vervain, and slender poppy, he matched in contentment the wealth of kings, and, returning home in the late evening, would load his board with unbought dainties. He was first to pluck roses in spring and apples in autumn; and when sullen winter was still bursting rocks with the cold, and curbing running waters with ice, he was already culling the soft hyacinth's bloom, chiding laggard summer and the loitering zephyrs. So he, too, was first to be enriched with mother-bees and a plenteous swarm, the first to gather frothing honey from the squeezed comb. Luxuriant were his limes and laurestines; and all the fruits his bounteous tree donned in its early bloom, full as many it kept in the ripeness of autumn. He, too, planted out in rows elms far-grown, pear-trees when quite hard, thorns even now bearing plums, and the plane already yielding to drinkers the service of its shade.

TAV
4

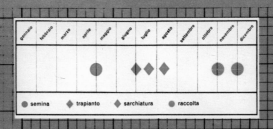

semina   trapianto   sarchiatura   raccolta

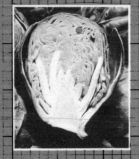

Cavolino di Bruxelles
Famiglia: *Cruciferae* – Specie: *Brassica oleracea*

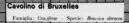

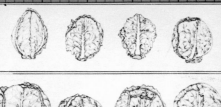

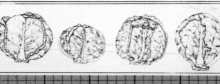

Cavolo
Famiglia: *Cruciferae* – Specie: *Brassica oleracea*

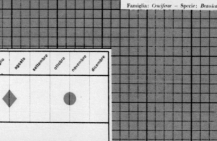

semina   trapianto   raccolta

Cavolo

199

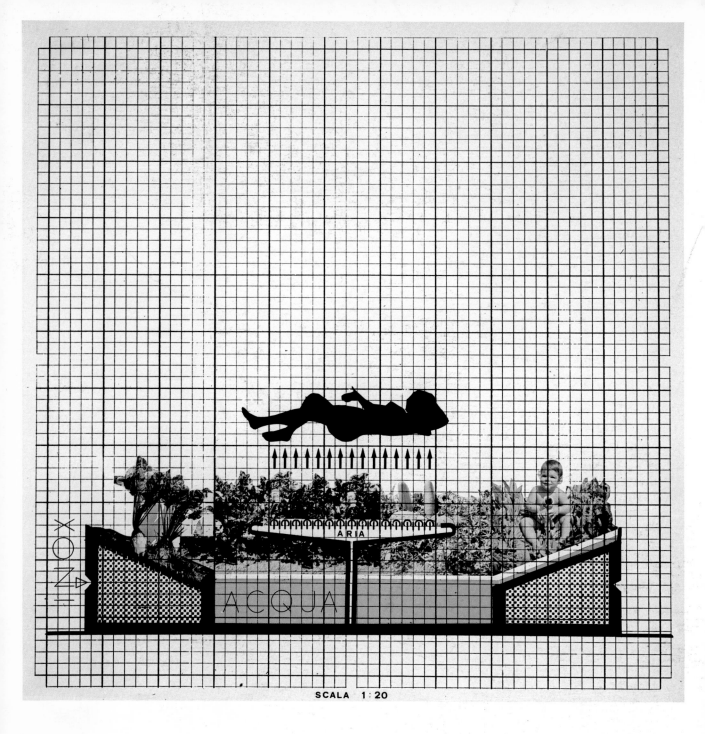

SCALA 1:20

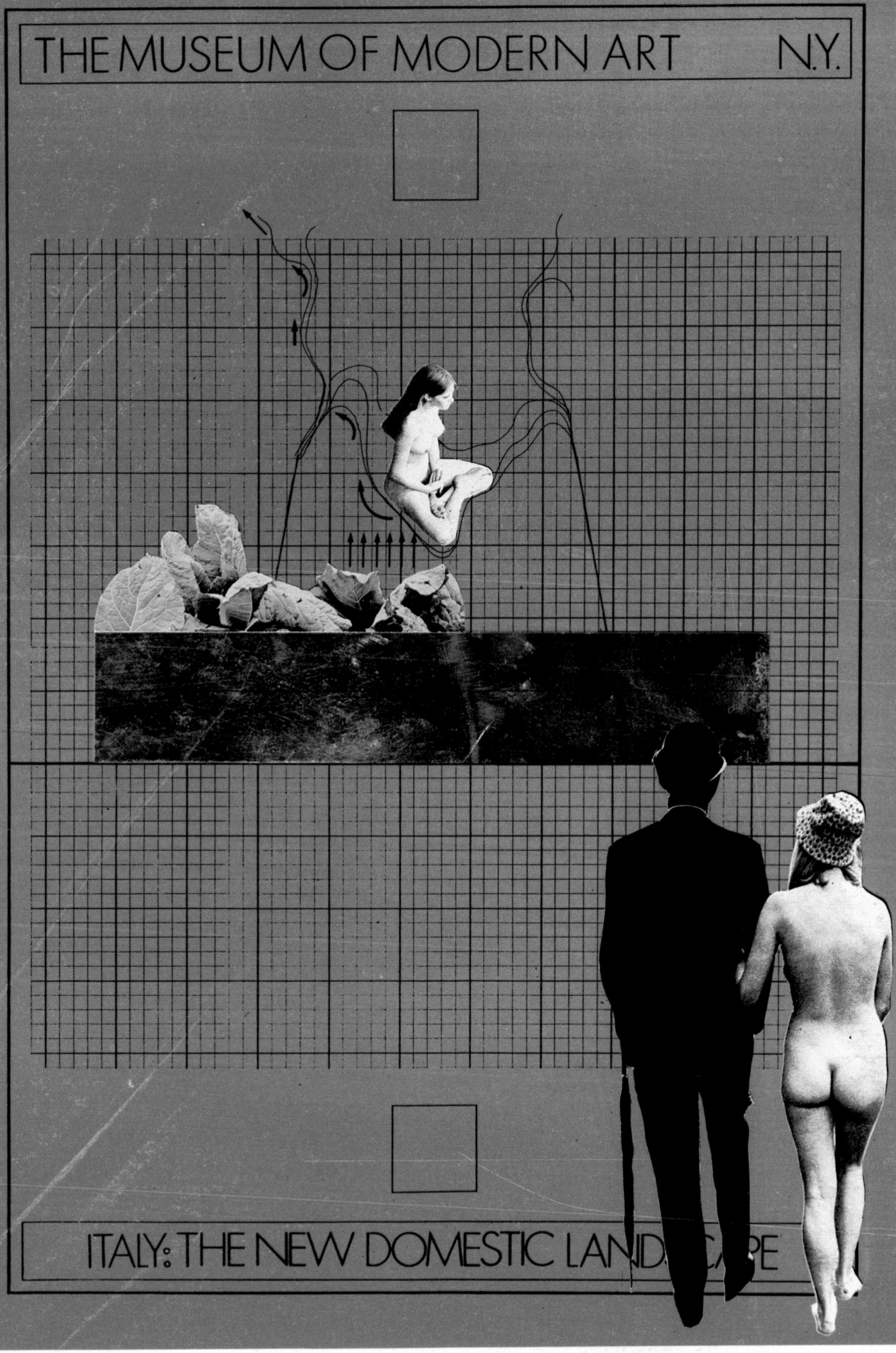
THE MUSEUM OF MODERN ART N.Y.
ITALY: THE NEW DOMESTIC LAND

INTRODUCTION

For an American audience, especially, it has seemed indispensable to provide on this occasion some background on the evolution of Italian design during this century. The four articles in this section are intended to serve as an historical basis not only for the exhibition itself but also for the critical articles that follow. Although the American public is well acquainted with the work of Italy's outstanding modern painters, sculptors, and printmakers, for the most part it is only the postwar achievements of her architects and designers that are familiar to us. Even these we have tended to evaluate for their formal qualities alone, with little understanding of the ideological positions they represent — a situation the present exhibition and publication seek to rectify. Our ignorance in this regard is perhaps understandable in view of the fact that, to a great extent, twentieth-century architecture and design in Italy developed outside the mainstream of European currents. In reading the literature, in fact, one is constantly surprised to note how frequently the Italians compare conditions in their own country with those in 'Europe,' as though they inhabited, not a peninsula, but an island quite separate from the continent.

The essays deal respectively with the *stile Liberty* or *arte floreale*, the Italian version of Art Nouveau (Paolo Portoghesi); Futurist concepts of design (Fagiolo dell'Arco); the period between the two World Wars (Leonardo Benevolo); and the postwar years, during which Italy emerged as a major international force in design (Vittorio Gregotti). These periods are of interest not so much because of the different 'styles' that they generated, but rather because in them we encounter a succession of circumstances and ideas that, for all their contrasts and divergences, still underlie the situation today and frequently persist either as survivals or revivals.

Throughout these articles, one may trace the story of the continuing efforts of Italian architects to rid their profession of backwardness and provincialism. *Aggiornamento*, meaning both 'updating' and 'renewal,' has been an aspiration of Italian architects and designers throughout this century. Those who espoused Art Nouveau did so because that style seemed to them to epitomize 'modernity'; the Futurists were even more zealous, seeking to abolish and destroy every vestige of the past in their ardent embracing of the new. The proponents of both movements met with stubborn resistance from deeply entrenched traditionalist attitudes and institutions. Their struggles were, nevertheless, far less anguished than those of the Italian vanguard of the 1930s, who found themselves in the dilemma of striving to practice and promulgate what we, following Henry-Russell Hitchcock, generally call the 'International Style' (to which in Italy the more polemical term 'Rationalism' is often applied), while simultaneously, over the opposition of their colleagues, presenting it to the Fascist regime as the style best suited to Italy's then-virulent nationalism.

It was only after the Second World War that Italy finally won worldwide recognition for her leadership in product design. Yet, paradoxically, it may well be not the tangible objects, but the searching questions now being raised by many of her designers and critics, and the counterdesign approach of some of her avant-garde, which today firmly establish Italy's position in the center of international thought on design and its role in society.

ART NOUVEAU IN ITALY
Paolo Portoghesi

The penetration into Italy of Art Nouveau theories and style (more frequently known in that country as *stile Liberty*) manifested itself in three ways: the formation in several cities of groups of artists interested in assimilating and disseminating the new artistic principles; the founding of factories and commercial firms seeking to involve artists in the process of production; and the establishment and strategic use of art periodicals and exhibitions that served to acquaint the public at large with objects in the new style.

The last decade of the nineteenth century, which in England, France, Belgium, and Germany was the golden age of Art Nouveau, in Italy was a time of preparation and observation. Cultural circles began to feel the influence of John Ruskin and William Morris, and self-criticism of eclecticism became increasingly harsh and unsparing; but it was only in the final years of the century that architecture and objects reflecting the new style made their appearance. Thus, it was seven years after *The Studio* was founded and Horta's Tassel House built (1893), and six years after the 'Exposition de la Libre Esthétique' in Brussels (1894), that there was any active Italian participation in Art Nouveau. This notable delay was not due to any difficulties in communication or lack of information, but rather to a cultural resistance that was stronger and more widespread than in other countries, and that made the span of the style's development even more brief and ephemeral than elsewhere, restricting it to little more than a decade.

The most significant events marking that development were Italy's participation in the Paris Exposition of 1900; the Promotional Exposition of Fine Arts at Palermo in the same year (Fig. 1); the opening of the International Exposition of Modern Decorative Art at Turin in 1902; the 1906 opening of the exhibition in Milan to celebrate the completion of the Simplon Tunnel; and two exhibitions held in Turin and Rome in 1911. At Paris, which afforded the first major opportunity for a comparison with work being produced elsewhere in Europe, the Italian pavilion revealed the backwardness of the country's art, and the extent to which it was retarded by adherence to the outworn modes of an academic historicism, aggravated by nationalistic rhetoric. A few exceptions, however, testified to the existence of a group of avant-garde artists and craftsmen bent on playing a role in the struggle against imitation of the past. One of the first prizes went to a Milanese craftsman, Eugenio Quarti, for his furniture inlaid with silver and mother-of-pearl, in which the new linear style was rendered with exceptional technical mastery; while Carlo Bugatti's furniture, which blended exoticism with a clear desire to break with the past, aroused interest and controversy.

The Turin Exposition of 1902, organized by a group of enlightened critics, provided an important opportunity for evaluating the achievements of Art Nouveau and gave Italian artists a chance to come into direct contact with the most significant work by current masters. It was a tendentious exhibition that excluded everything that did not conform to the efforts to establish a new style. The Italian participation bore witness to the existence and tireless activity of some ten groups, who, despite opposition and difficulties, had managed to assert themselves in the most important cities of Italy — Turin, Milan, Bologna, Florence, Rome, and Palermo. Official sanction was given to the show by the inaugural address, delivered by the Minister of Education, but actually written by the art critic Angelo Conti, a last-minute convert to the new movement.

With the V Biennale at Venice in 1903, interior decoration made its first appearance in a show devoted to the major arts; the 'regional exhibitions of fine and applied arts' were intended to demonstrate

Paolo Portoghesi received his degree in architecture in 1957 and has since combined the active practice of architecture with the teaching of architectural history. Until recently, he was chairman of the School of Architecture at the Politecnico in Milan; he is now living in his native city, Rome. In 1963, his distinguished criticism was awarded the national prize of the Italian Institute of Architecture (INARCH). He is the author of a number of monographs and has contributed to many periodicals, as well as serving as director of the *Quaderni dell'Istituto di Storia dell'Architettura* and of the architectural section of the review *Marcatré*.

287

the fusion of painting and sculpture with furnishings. Here, the work of the Basile-Ducrot team of Palermo was outstanding for its originality, coherence, and careful execution (Fig. 2), giving the lie to those who would naturally have assumed that artistic circles in Northern Italy would be more open to the new tendencies than those in the South.

Four years after the Turin exhibition, the 1906 exposition celebrating the Simplon Tunnel permitted an estimate of encouraging progress, especially in the minor arts and furniture. Whereas architecture remained dominated by a compromise between innovation and tradition, craftsmen had found in the new style a source for renewed impetus and self-expression; they took it up enthusiastically, often achieving original effects. Yet, the seeds of self-destruction were already manifest, which even before the second decade of the century would threaten the continuing development of the 'Liberty' style. The most complicating factor was a more or less explicit tendency, already widespread, to follow the lead of the Viennese, and Otto Wagner in particular, in transforming the broad, diversified repertory of Art Nouveau into a single idiom based on closed spatial patterns, with rounded edges at the intersection of planes, and projecting elements and fascias that sought to reproduce mechanically, in simplified form, the compositional system of the classical architectural orders. The encounter with middle-class Italian taste, essential for the creation of a market, had already made its negative effects felt, depriving the resulting works of their artistic force. As a result, the young generation of intellectuals could only express their severe condemnation, thus perpetuating a new dialectical round of self-criticism and leading to the new tangential departures so vital to the survival of the myth of innovation.

The two exhibitions of 1911 marked the inglorious end of Art Nouveau in Italy. On the one hand, the traditionalists reacted by initiating new archaisms and new historical revivals; while on the other hand, Futurism proposed renovation of a kind quite different from that patterned on the achievements of Morris and the Arts and Crafts Movement; in its destructive aims, Futurism drew no distinction between eclecticism and Art Nouveau, even though it drew much of its vital impetus from the latter.

Cultural Centers

Turin: As early as the beginning of the 1890s, Leonardo Bistolfi's predilection for plant forms and linearism predetermined his infatuation with the flowery style of Art Nouveau. It is significant that out of his circle came two sculptors, Giacomo Cometti and Celestino Fumagalli, who devoted themselve enthusiastically to the applied arts. But while Bistolfi has, rightly or wrongly, been regarded as the pioneer in the art of Turin, this title should more correctly be given to Enrico Thovez, one of the most gifted and courageous literary critics of his generation. His contribution lay in inveighing against the prevailing traditions of Carducci and D'Annunzio, and proclaiming Leopardi's modernity and his decisive role in initiating the progressive developments in Italian lyric poetry. Thovez proved himself equally prophetic with respect to the visual arts and took up the cause of the new style with apostolic zeal. Together with Bistolfi, Calandra, the architect Reycend, and the decorator Giorgio Ceragioli, he organized the 1902 exhibition, as well as founding the periodical *L'Arte decorativa moderna*, which in Italy was to occupy the place held elsewhere in Europe by such polemical magazines as *The Studio*, *Jugend*, and *L'Art décoratif*.

In the field of architecture, apart from the structures that Raimondo D'Aronco created for the exposition, Turin was the first city to

1. Ernesto Basile. Building for Promotional Exposition of Fine Arts, Palermo, 1900
2. Basile and Ducrot. Screen shown at V Biennale, Venice, 1903
3. Celestino Fumagalli. Silver tea service. c. 1902

288

experiment with the new style (as an important project of 1899 evidences), even though it could claim no outstanding works; the effort to give a higher value to the work of the mediocre decorator Pietro Fenoglio is unconvincing. The new style nevertheless influenced much building activity for a decade and left enduring traces in the Piedmontese provinces.

But undoubtedly the objects produced at the time were of a far higher level of quality than the architecture. Giacomo Cometti had already acquired considerable experience as a decorator when Thovez reproduced his work in January, 1902 in the first issue of *L'Arte decorativa moderna.* Cometti's furniture, which he produced in a small workshop adjoining his sculpture studio, was of a high level of crafsmanship, showing influences derived from, but not slavishly copying, the work of Horta, Gaillant, and Guimard. In particular, his use of fretwork was inspired by a series of furnishings for the Solvay House in Brussels, while a glass-fronted cabinet showed the beaked motif so recurrent in Guimard's architecture (Fig. 4). His most original pieces were some living-room chairs, in which a floreate Art Nouveau motif becomes the logical adjunct of a rational structure. In the furniture published later in *Ambiente moderno* and other magazines, we can trace the progressive deterioration of his work, evident sometimes in the geometric simplification due to Viennese and Scottish influences, and elsewhere in the choice of dark wood and the relatively minor role played by decoration, which is often reduced to superimposed metal plaques in accordance with the taste prevailing at the Turin exposition of 1911. Cometti's specialty became composite pieces of furniture, combining divans, glass-fronted cabinets, bookshelves, and étagères in elongated, asymmetrical structures that reveal a certain opposition to the more rhetorical and monumental achievements of the new styles.

Even more important than Cometti's furniture as indicating the precocious interest of Bistolfi's circle in Art Nouveau is Celestino Fumagalli's work in silver (Fig. 3). In 1902, Thovez stated that it had been produced 'several years ago, when little or nothing was known ... of the work being produced in the same field by foreign artists.' While the basic form of the objects remained traditional, the accessory parts — handles, spouts, bases — frequently assumed the serpentine forms characteristic of Horta and the School of Nancy, but with an accentuated plastic solidity that is typically Italian.

The enthusiastic climate of commitment generated by Thovez inspired even a sculptor of as high quality and renown as Edoardo Rubino to undertake the design of plaques and other objects, among them the very fine lamp published in *L'Arte decorativa moderna*, in which the electric bulbs are arranged in a ring upheld by the hands of a graceful female figure (Fig. 5).

Among the makers of furniture in Turin was also V. Valabrega, whose work, illustrated in the albums *Meubles de style moderne*, were also commercially distributed internationally. He specialized in pieces in which virtuoso decorative effects testify to the expert skill in intaglio of the craftsmen at his disposal.

Milan: As early as 1900, the works shown by Quarti and Bugatti that were exhibited in the Paris Exposition gave evidence of the presence in Milan of a favorable environment, as well as of the innovating influence of Segantini (a relation of Bugatti). The influence of Art Nouveau on architecture was more important in Milan than in Turin, affecting such notable masters as Sommaruga, Moretti, and Stacchini. These men became the proponents of an Italian Secession, which was to misunderstand and betray almost all the aesthetic

4. Giacomo Cometti. Glass-fronted cabinet and chair. c. 1902
5. Edoardo Rubino. Bronze lamp. c. 1902

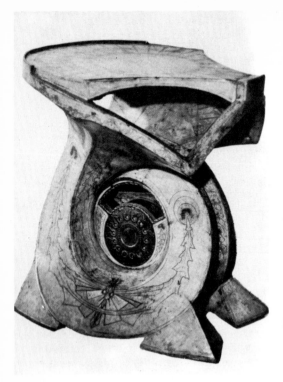

principles that gave validity to the work of Horta, Van de Velde, and Wagner.

Carlo Bugatti, whose theatrical furniture had caused a sensation when exhibited in Turin in 1898, remained an outstandingly bold and original master, worthy of further research. His rebellion against every previously existing formal type aligned him with Futurism. He differs from that movement, however, because of his archaism, at once refined and brutal, which has led some to speak of Oriental or Islamic influences in his work — without specifying, however, what those hypothetical sources might be. Apart from the many-sidedness of his work, one of its most interesting aspects is its extraordinarily forceful plasticity, attained by huge, dense architectonic masses that expand and distort Art Nouveau motifs and presuppose a kind of brutal rhetoric (Fig. 6), well adapted to certain reactionary inclinations of the Italian middle class, which would later find an outlet in Fascism.

Eugenio Quarti, a member of Bugatti's workshop, was influenced thereby only to the extent of acquiring a commitment to the new style. In contrast to the violent manner of Bugatti, Quarti's mode of expression, at the height of his career, was a rather anemic delicacy and an inclination toward elegance, rather simplistically satisfied by overlaying his elongated structures with silver and mother-of-pearl intarsia (Fig. 7). Drawing inspiration from very diverse sources — Horta, Olbrich, Mackintosh, and Majorelle — he attained a cotherent style by finding in everything he did a common denominator of clarity and gracefulness, which in only a few of the worst examples deteriorates to weakness.

Even earlier than Quarti, the furniture designed by F. Tesio for his Dormeletto villa in 1898 is of interest as showing an amateur's liking for the Art Nouveau style. In these designs, simple, traditional structures are brought up to date by the inclusion of curvilinear panels devoid of any naturalistic elements.

By about 1910, when the first impetus toward an international style was succeeded by neoacademic reaction, the number of Milanese makers of furniture who were to some extent involved with the new fashion had grown immensely. It included Gianbattista Giannotti, who, together with Alfredo Melani, founded the periodical *Per l'arte.* Confining his use of flowing lines to details in his furniture, Giannotti conceived of these pieces as blocky volumes with corners enclosed by broad, curved surfaces. In some examples, such as the decorations for a banker's house published in *L'Ambiente moderno,* he achieved a fine funereal opulence, a misunderstood rendering of the Wagnerian mode so in vogue at that date.

Other furniture makers who achieved renown between 1900 and 1910 were Carlo and Pietro Zen, who were influenced by Gallé and Majorelle. Among the architects who concerned themselves with interior decoration, mention should be made of E. Monti, A. Campanini, Moretti, and G. G. Arata — the latter serving to demonstrate the extent to which pomposity and melodrama were rooted in the architectural standards of the time.

But it was not only in architecture that Milan was an imporant center of Art Nouveau. Critics such as Melani and Vittorio Pica, who were vitally interested in the problem of industrial arts, helped to improve the cultural climate. Whereas Pica, who had directed the magazine *Emporium* since 1897, took a detached view of the vicissitudes of the style, Melani threw himelf into the fray to champion every sort of undertaking that might promulgate the modern movement, associating himself first with Thovez in editing *L'Arte decorativa moderna* and subsequently founding *Per l'arte* and *L'Ambiente moderno.*

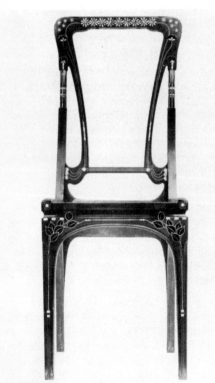

6. Carlo Bugatti. Wood and parchment table with metal plaques and painted dragon-fly motifs. 1902. Private collection, Pavia
7. Eugenio Quarti. Mahogany chair with silver and mother-of-pearl intarsia and gilt-bronze appliqué. c. 1900. Private collection, Milan

In the field of crafts connected with building, the outstanding figure was Alessandro Mazzucotelli, whose technique in wrought iron was entirely different from that of other Art Nouveau masters. Unlike Horta, Van de Velde, and Mackintosh, who used uniform sections of iron to express the purely graphic quality inherent in their material, Mazzucotelli, a stubborn metalworker eager to prove himself as an artist craftsman, forged and hammered out the iron, beating and transforming it into concave and convex surfaces in constant metamorphosis. Reviving a very old technical tradition, he made it serve the purposes of his own active and exuberant imagination, always seeking out rhythms and allusive forms (Fig. 8).

Thanks to the Società Ceramica di Laveno, and above all to Giulio Richard, who in 1873 founded a firm that in 1896 incorporated the old Ginori factory of Doccia, ceramics were preeminent among the applied arts produced in Art Nouveau style in Lombardy. Under the direction of the Lombard Luigi Tazzini, Richard-Ginori manufactured porcelain and ceramics in the *stile Liberty* that often showed considerable originality by comparison with contemporary Bavarian and French production (Fig. 9). Some figures in white porcelain, particularly, made as lamp bases or handles for large trays, introduced into the Jugendstil repertory and that of the School of Nancy a certain restraint in design, and an Impressionist technique borrowed from Medardo Rosso.

Venice, Florence, Bologna, and Rome: The cities that had the oldest cultural traditions offered the greatest resistance to the infiltration of the new style. In Venice, it was only in the glass produced at Murano that Art Nouveau style had any influence, which was revealed in a sinuous and showy plasticity filtered through a kind of exuberant eclecticism. In Florence, the sad fate of Giovanni Michelazzi, who committed suicide after having erected three polemical structures that challenged the 'Florentine manner,' indicates the extent of local reluctance to adopt the style, only partly overcome by Galileo Chini, who, together with Domenico Trentacoste and Count Giustiniani, founded the Arte della Ceramica factory. This gained international fame through its delicate use of metallic glazes. The style of this workshop was rooted in local tradition, from which it slowly freed itself, without, however, succeeding in abandoning a certain archaistic propensity that wavered among Mycenean, Renaissance, and medieval prototypes (Fig. 10).

In Bologna, the architect Alfonso Rubbiani, who was responsible for some restorations that are somewhat debatable from the point of view of style, joined with a group of artists and ladies of title in founding the Aemilia Ars society. Their rather ambiguous version of Art Nouveau style was essentially different from that elsewhere in Europe, being somewhat the outcome of a sterile and provincial overrefinement, exemplified in some stiff pieces of jewelry reproduced by Clementi.

In Rome, although the battle for the new style was completely lost in the field of architecture, it won some ephemeral and numerically insignificant victories in decoration and taste. The interior of the Caffé Faraglia, decorated by Basile, together with numerous shops and a few well-done interiors, opened a door that was promptly slammed shut by conservative reactionaries. The most interesting development was the activity of the magazine *La Casa*, which sponsored an organization to produce furniture and useful objects for home decoration. Vittorio Grassi and Duilio Cambellotti, the most notable artsts working for *La Casa*, were typical exponents of the most progressive elements in Roman society, animated by humanitarian and socialist political ideals. The plastic exuberance of the French and Belgian modes of expression, and the luxuriousness of the

8. Mazzucotelli and Sommaruga. Wrought-iron sconce for the Palazzo Castiglioni, Milan. 1903
9. Richard-Ginori. Nereid vase in white porcelain. c. 1901. Private collection, Milan
10. Arte della Ceramica, Florence, Vases

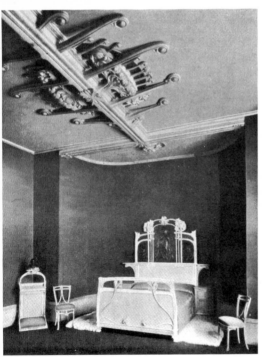

Viennese Secession, were both spurned in favor of the austerity of certain English and Dutch models.

Palermo: While a lack of reliable sources makes it difficult to ascertain to what extent the new style may have influenced popular taste in Naples (and it was probably not negligible in the case of ceramics and objets d'art), its notable development in Palermo can be traced in its entirety. For all its contradictions and unevenness, the work of Ernesto Basile represents Italy's most outstanding contribution to Art Nouveau. It should be noted that, just as in the case of Horta and Gaudi, it was an industrial patron who gave Basile the opportunity for concrete expression of his creative abilities. Although to a lesser extent than Güell or Solvay, Florio had both the economic means and the theoretical interest for an undertaking suited to his ambition — typical of upper-class Sicilian society at the turn of the century — to establish direct contact with European centers, bypassing, if necessary, the Italian mainland. If the collaboration of Florio and Basile was essential to inspire the European outlook manifest in the decor of the Villa Florio at Ulivella (Fig. 11), another factor was necessary to bring success to Basile as designer. That was the enthusiastic initiative of the industrialist Vittorio Ducrot, who radically transformed his equipment in order to produce the furniture that he presented with such notable success at Turin in 1902 (Fig. 12).

But if it is easy to account for the circumstances that led to the success and quality of Ernesto Basile's work, it is not so easy to explain its cultural origins. Alone among the Italians who contributed to Art Nouveau, Basile devoted himself to interior decoration with the same enthusiasm that he brought to his large-scale works, and achieved in this field even more striking results. Whereas D'Aronco, Moretti, and Sommaruga made use of artisans whose taste harmonized with theirs, Basile — like Horta, Guimard, Hoffmann, and Mackintosh — had an all-encompassing creativity that he applied equally to objects, spaces, and mass. The bedroom he exhibited at Turin in 1902, with its light-colored furniture and plastic divisions of the ceiling, was one of the few entries that could hold their own by comparison with the best European productions of those years. Naturalistic and Art Nouveau motifs were reduced to their essentials and rendered with faultless delicacy, and with a sense of balance and control that makes Basile outstanding even with respect to the European masters. The composition of the framing structure, while consistently dynamic, suggests simple movements of orbit and rotation, eliminating all oblique or helicoidal trajectories and regaining certain rhythmic qualities of Greek decoration. Unlike D'Aronco, Basile did not merely reproduce Art Nouveau themes, but rather he instinctively embellished them with the luster of a rational control consistent with the quest for a linear syle; and unlike Sommaruga, this embellishment did not constitute either a revision or a deviation.

Once his work in architecture and in design had reached its zenith, however, a regressive tendency toward classicism and academicism set in. This coincided with his abandonment of a commitment to the industrial arts, which had led him from the isolation of being a mere observer in far-off Sicily to direct creative participation in the mainstream of European culture.

11. Ernesto Basile. Villa Florio, Ulivella. c. 1902
12. Ernesto Basile. Bedroom for the Vittorio Ducrot establishment, Palermo. c. 1902

THE FUTURIST CONSTRUCTION OF THE UNIVERSE
Maurizio Fagiolo dell'Arco

Maurizio Fagiolo dell'Arco teaches the history of art at the University of Rome and the Academy of Art in that city. His researches range from sixteenth-century Mannerism to contemporary art; currently, he is editor of the baroque section of the periodical *Storia dell'Arte* and Roman correspondent for *Art International*. He has written extensively on Futurism, publishing an important monograph on Balla in 1968 and, this year, *Il Silenzio dell'avanguardia: Futurismo, Dada, Pop art.*
293

'The hands of the traditionalist artist suffered for the lost Object; our hands agonized for a new Object to create. And that is why the new Object (the plastic construction) appears miraculously in your hands.' — Marinetti to Depero and Balla

1. From Art to Environment to Life

On February 20, 1909, with the publication of Filippo Tommaso Marinetti's manifesto, the Futurist adventure began. It was a howl of protest against the past, a hymn of praise to the present, and a prophecy for the future. The painters were Umberto Boccioni, Carlo Carrà, Luigi Russolo, Giacomo Balla, Gino Severini; within a few short years, they succeeded in putting into practice some of the theories announced in their manifestos, and they attracted other followers, such as Anton Giulio Bragaglia, Fortunato Depero, and Enrico Prampolini. The concept of simultaneity apparent in their paintings was the plastic expression of an absolute idea (speed), of a new spectacle (modern life), and a new passion (science). They wished to get in step with their own times, with the hectic pace of progress. Shortly before his death in 1916, Boccioni could sum up the Futurist experience in these words: 'We have always acted with a savage delight in destruction but above all with a zeal for reconstruction.'

It was a fundamental movement because of this very notion of the avant-garde; thanks to the impetus given by Marinetti, it sought an outlet both in the theater and in real life. The 'Futurist gala' was in this sense a precursor of the Dada Cabaret and the 'happenings' of the more recent avant-garde. We may recall the invention of *parole in libertà* ('free words' — not literature but phonetic writing and typography), of the 'synthetic theater,' a forerunner of the theater of the absurd, the new music (Russolo's art of noise), the visionary architecture (Sant'Elia), the experiments in film (Bragaglia), and also the search for 'the unreal' (Marinetti foresaw the 'automatic writing' that was to preoccupy the Surrealists). Nor should it be forgotten that the manifestos were as important as the actual works, because of their new concept of art seen as theory and propaganda. In the light of present-day experimentation in art, it is evident that Futurism was not a movement confined only to Italy and to a single decade but was a veritable storehouse of plans for a permanent artistic revolution.

It was logical that a movement so eager to escape restraint and extend its range from painting to the world of real life should have been interested in extraartistic projects and models; and logical also that in 1918 Benedetto Croce could write an article entitled 'Futurism as Something Extraneous to Art.' The Futurists looked forward to a future dominated by the machine, but the present still belonged to the craftsman; Italy had not yet been affected to the slightest degree by a true technological modernization. Just as Boccioni's pictures could not have for their background the skyline of New York instead of the outskirts of Milan, so Balla's furniture, while reflecting some Middle European innovations, remained bound to the Italian handcraft tradition ('modernistic' decoration embellishing old forms).

The aspect of Futurism that in fact had the most future was its 'typographical revolution,' which in many respects was related to design. At the Cabaret Voltaire, during the heroic days of Dada, some of Marinetti's 'free-word' compositions were on view (with others by Paolo Buzzi, Francesco Cangiullo, and Corrado Govoni), and several were reproduced in the Cabaret Voltaire pamphlet. Recently, some of the originals reappeared in the sale of the Tristan Tzara collection, among them all that might be considered proto-Dadaist.

A preliminary chronological statement is necessary: the true boundaries of the Futurist movement were spatial (first Milan, then Rome) and temporal (from 1909 to 1919). In the manifesto that

RICOSTRUZIONE FUTURISTA DELL'UNIVERSO

Leggete LA BALZA
GIORNALE FUTURISTA
MESSINA

Col Manifesto tecnico della Pittura futurista e colla prefazione al Catalogo dell'Esposizione futurista di Parigi (firmati Boccioni, Carrà, Russolo, Balla, Severini), col Manifesto della Scultura futurista (firmato Boccioni), col Manifesto La Pittura dei suoni rumori e odori (firmato Carrà), col volume *Pittura e scultura futuriste*, di Boccioni, e col volume *Guerrapittura*, di Carrà, il futurismo pittorico si è svolto, in 6 anni, quale superamento e solidificazione dell'impressionismo, dinamismo plastico e plasmazione dell'atmosfera, compenetrazione di piani e stati d'animo. La valutazione lirica dell'universo, mediante le Parole in libertà di Marinetti, e l'Arte dei Rumori di Russolo, si fondono col dinamismo plastico per dare l'espressione dinamica, simultanea, plastica, rumoristica della vibrazione universale.

Noi futuristi, Balla e Depero, vogliamo realizzare questa fusione totale per ricostruire l'universo rallegrandolo, cioè ricreandolo integralmente. Daremo scheletro e carne all'invisibile, all'impalpabile, all'imponderabile, all'impercettibile. Troveremo degli equivalenti astratti di tutte le forme e di tutti gli elementi dell'universo, poi li combineremo insieme, secondo i capricci della nostra ispirazione, per formare dei complessi plastici che metteremo in moto.

Balla cominciò collo studiare la velocità delle automobili, ne scoprì le leggi e le linee-forze essenziali. Dopo più di 20 quadri sulla medesima ricerca, comprese che il piano unico della tela non permetteva di dare in profondità il volume dinamico della velocità. Balla sentì la necessità di costruire con fili di ferro, piani di cartone, stoffe e carte veline, ecc., il primo complesso plastico dinamico.

1. **Astratto.** — **2. Dinamico.** Moto relativo (cinematografo) + moto assoluto. — **3. Trasparentissimo.** Per la velocità e per la volatilità del complesso plastico, che deve apparire e scomparire, leggerissimo e impalpabile. — **4. Coloratissimo e Luminosissimo** (mediante lampade interne). — **5. Autonomo,** cioè somigliante solo a sè stesso. — **6. Trasformabile.** — **7. Drammatico.** — **8. Volatile.** — **9. Odoroso.** — **10. Rumoreggiante.** Rumorismo plastico simultaneo coll'espressione plastica. — **11. Scoppiante,** apparizione e scomparsa simultanee a scoppi.

Il parolibero Marinetti, al quale noi mostrammo i nostri primi complessi plastici ci disse con entusiasmo: « L'arte, prima di noi, fu ricordo, rievocazione angosciosa di un Oggetto perduto (felicità, amore, paesaggio) perciò nostalgia, statica, dolore, lontananza. Col Futurismo invece, l'arte diventa arte-azione, cioè volontà, ottimismo, aggressione, possesso, penetrazione, gioia, realtà brutale nell'arte (Es.: onomatopee. — Es.: intonarumori = motori), splendore geometrico delle forze, proiezione in avanti. Dunque l'arte diventa Presenza, nuovo Oggetto, nuova realtà creata cogli elementi astratti dell'universo. Le mani dell'artista passatista soffrivano per l'Oggetto perduto; le nostre mani spasimavano per un nuovo Oggetto da creare. Ecco perchè il nuovo Oggetto (complesso plastico) appare miracolosamente fra le vostre.

La costruzione materiale del complesso plastico

MEZZI NECESSARI: Fili metallici, di cotone, lana, seta, d'ogni spessore, colorati. Vetri colorati, cartevelline, celluloidi, reti metalliche, trasparenti d'ogni genere, coloratissimi, tessuti,

launched the movement, Marinetti declared: 'The oldest of us are thirty, so we have at least another decade left in which to complete our work. When we are forty, other men, younger than we, may throw us into the wastebasket like useless manuscripts. — That's just what we should like!' Then came the war and the involvement with Fascism; in art, the Futurist 'style' would run its course, and Neo-Futurism (an historical category that has about the same validity as 'Post-Impressionism') would emerge. Although some of its proponents were to remain active during the 1920s and '30s, Futurism was clearly destined to obsolescence and death.

A word should also be said about the former deprecation of Futurism on political grounds. It must be made clear at once that the same 'avant-garde' goal that may be valid in a bourgeois culture is completely ineffective in a time of revolution. In 1920, Lunacharsky cited Marinetti as a revolutionary intellectual, and from a strictly Marxist-Leninist standpoint, this was a mistake. On the other hand, even Gramsci praised the movement enthusiastically: 'The Italian Futurists have carried out this task within the field of bourgeois culture, they have destroyed over and over again, without worrying whether the new creations produced through their activities would on the whole be superior to the works they destroyed ... They have a completely revolutionary, absolutely Marxist, concept. Within their own field, that of culture, the Futurists are revolutionaries; in this field, it will probably be a long time before the working class will succeed in producing any works better than those of the Futurists.' Obviously, in a revolutionary climate, every kind of idea can be invoked for the sake of total regeneration; today, it is clear that one should not expect a revolution from art (the artist being an individualist, or at best an anarchist). That a utopian society can be brought about by means of an 'aesthetic dimension' is a misguided notion that has persisted from Bogdanov, in the days of Lenin, down to Marcuse.

2. Theories on the Future Environment

Briefly, Futurist activity in design manifested itself in eight fields: interior decoration, environmental architecture, furniture, decorative art objects, windows, toys, illustrations, and advertisements. These were the categories proposed by Prampolini's magazine *Noi* in 1923, in an issue devoted to 'Futurist Decorative Art,' which summarized the experience of a decade (and included not only Prampolini, Marchi, Balla, Depero, and Paladini but also Picasso, Huszar, and van Doesburg).

There were many who held theories about the interior decoration of the future. The list is headed by Balla, followed by Depero and Prampolini, but it also includes Arnaldo Ginna, Nicola Galante, and Francesco Cangiullo. Balla left no specific theory on the subject, but it can be asserted that all the research on the new environment and the object of the future received its impetus from the manifesto 'The Futurist Reconstruction of the Universe,' which he wrote, with Depero's help, in 1915 (Fig. 1). It mentions the 'plastic construction,' the 'toy,' the 'artificial landscape,' and the 'mechanical animal.' What emerges primarily is the vision of a cheerful universe; this vision could only have come from having read Aldo Palazzeschi's *Controdolore* ('Painkiller') manifesto, in which God is imagined to have derived amusement from creating 'the joyous spectacle of the universe.' Above all, there is no longer any clear line of demarcation between the beautiful and the ugly, because there is a distinct sense of being 'out.' (As early as 1912, Balla wrote, 'Neither the beautiful nor the ugly any longer have any boundaries when one enters within, or goes beyond, the boundary lines.') It was also Balla who asserted, when speaking of an exhibition of paintings, that a glittering shop front was

more beautiful, that an electric iron was more beautiful than a sculpture, and a typewriter more important than the wrong kind of architecture (1).

Late in 1916, just after he finished shooting the film *Futurist Life*, Ginna wrote an article entitled 'The First Italian Futurist Furniture' (2). He had, among other things, designed furniture for the actress Karenne, who wanted to promote this type of production. The cabinet makers came from Faenza, precious materials were used, and the designs were fantastic. In his article, Ginna referred to Antonio Sant'Elia's ideas ('what we need is original modernity, health, elegance, and synthetic emotion') and gave due importance to the role of the machine, stating that work done by hand inevitably 'involves a certain roughness and coarseness, and flourishes in the carving'; neverthless, 'lyrical emotion' should be preserved. 'The interior decoration of a house, a restaurant, or a hotel must have character, that is, every room should have a physiognomy suited to its function. I am of the opinion that in this way art may find a worthy task to perform. Art would thus become an intimate part of life, our attitudes, and our needs. The task of introducing a high level of refinement and a new elegance seems to me a worthy one.'

Galante also tackled this subject in an article, 'Notes on Decorative Art,' published in the first issue of *Noi* in 1917 (3). 'Two things are of prime importance: the material and the use of the object. By "object," I mean everything from a building to a fork. The material must be used exclusively in accordance with its own properties, with respect to its elasticity. Experience and new requirements will teach us what materials to employ for a given thing. The form of the object must be dictated solely by the criteria of good usage and health. If form is dictated by tradition, the consequence is thousands of objects that are — and will continue to be indefinitely — identical in their general type, just as many of our customs have remained unchanged from ancient civilization right down to the present. Machines, however, in which every part is governed by its own laws, are leading in a new direction and can form novel compositions. Although the two principles just mentioned are already old, they have never really been put into practice. To respect these principles, without worrying at all about the decorative aspect, means to us reverting to a kind of primitivism, beneficial insofar as it makes us forget about "styles." ' Although the text deals with the importance of decoration and the functional use of color, it presents an almost Rationalist approach: form-function, object-behavior, the machine as an agent devoid of the crime of ornamentation.

Prampolini, another proponent of the movement, worked within Balla's circle; there is even a difference of opinion regarding the theory of 'plastic constructions' (4). In the field of decoration, the designs for the interior of an airplane and of a car (1917) have been preserved. But Prampolini's most important work was the furnishing of the House of Italian Art, completed in 1919 (Fig. 5) The idea for this house arose from the need for a center in which all 'Futurist' research being done throughout Europe could be brought together. Its program declared: 'We are of the opinion that the decoration of a house, if taken full advantage of, is one of the most essential expressions of collectivity, affording the opportunity for a widespread, anonymous collaboration among many different people, and for various kinds of interaction between the artist and his craft. It was in accordance with these criteria, and also with a full awareness of the interplay in values between material and object, form and function, considered in their decorative aspect, that the painter E. Prampolini planned the decoration of the House of Italian Art.' Here one encounters fairly rational objects (we recall Prampolini's early contact with De Stijl),

8. Balla. Futurist environment. 1918. Balla collection, Rome
9. Balla. Futurist environment: lampshades, screen, mask, chandelier, and smoking stand. 1922
10. Balla. Design for a stage set, *Tipografia*, 1914. Pencil and ink on paper, 8 7/8x13 inches. Casa Balla, Rome

but still in Balla's decorative and fanciful style. In 1919, the House held an exhibition of applied arts — furniture, cushions, painted materials, curtains, tapestries, carpets, ceramics, and lamps.

Cangiullo was a strange personality. Like Balla, he was interested in the theater, free verse, and typography. He was also interested in interior decoration, as is apparent from the manifesto 'Futurist Furniture. Astonishing Talking and Free-Word Furniture' (5). He mentions the experiments carried out from 1914 on, and an 'interventionist tricolor chair' called ZANG ('It exclaimed ZANG! and was constructed of wooden letters from the famous free-word poem ZANG-TUMB-TUMB by Marinetti, to whom I offered it in homage'). He was also critical of the furniture of the past: 'Up to now, furniture has been sad, funereal, and gloomy (walnut); icy (English mahogany); useless, boorish, and Goldonian (gilded). Moreover, furniture has always been as obstinately silent as the tomb. Your rooms don't talk to you, and if they do, it is in an utterly funereal tone, because they speak of bygone eras, dead things, and memories of the past.' He follows with a Futurist declaration: 'My furniture will speak cheerfully; it won't wear you out with boredom but, on the contrary, you will have the pleasure and enjoyment of letting it wear itself out by breaking its drawers. It will be practical, comfortable, useful, elegant, gaily colored, economical, and above all HEALTHY. THE ASTONISHING FUTURIST ALPHABET FURNITURE WILL HAVE SPECIAL PERSONAL, URGENT MISSIONS TO PERFORM. Some pieces can be easily moved about, for greater convenience in the home, for example the "tapis-roulent" (*sic*) (rolling carpets), but also in order to be more consistent with itself, that is to say, an article of furniture [*mobile*] should be more mobile.' Propelled by the unrestrained tide of the program for 'the Futurist reconstruction of the universe,' Cangiullo sums up past, present, and future. Furniture must be dynamic (*mobile*, in actuality), and rooms should 'talk.'

3. The House as 'Theater'

'VISIT BALLA'S FUTURIST HOUSE. VIA NICOLO PORPORA 2 ROME.' This notice appeared regularly in large capital letters in the magazine *Roma Futurista* (Fig. 6). It is already indicative of the 'public' meaning that the artist wished to give his work-life: Balla's home was not merely something private but was in itself almost a manifesto of the new Futurist trend in decorative art (6).

This interest in interior decoration indicates a desire to break away from the restricting dimension of painting and enter into the ambience of everyday life. It is symptomatic that such a comprehensive attitude should have arisen after Balla's stay in Düsseldorf (1912), for only in the atmosphere of the Secession could he have become aware of the abolition of all barriers between the 'major' and 'minor' arts. Balla used his pictorial discoveries in interior decoration; for example, the motifs for the decor in the Löwenstein House in Düsseldorf (Fig. 4) derived from his painting *Iridescent Interpenetrations*.

One of his most up-to-date schemes was that for a Children's Room, designed for his daughter Elica (and also for Marinetti and Cosmelli); many designs for it are still in existence (Fig. 2), and a wardrobe is preserved in Balla's own house. The various elements were enlivened by a series of stylized children in different poses: they stand flanking the wardrobe, lean over to hold up the bed, and bend double to act as benches. The pieces of furniture thus became characters and could be considered toys (in one design, they were animals from fables; Fig. 3).

Other interesting designs project a fusion of lights and colors acting as a secret, surprising light that itself becomes line-form-color. While the light is functional, at the same time it causes wonderment.

11. Balla. Design for a screen. Tempera on paper. Private collection, Rome
12. Balla. Dining-room cupboard. c. 1918. Painted wood, 73 1/4 x 65 x 21 1/4 inches
13. Fortunato Depero. Costume designs for *Song of the Nightingale*, 1917-18

Another up-to-date idea is for demountable furniture; its colors are bright, its forms always symmetrical, but its most interesting feature is its simplicity and means of assembly (once the joints have been unscrewed, the piece of furniture becomes a series of planks). Decomposition and recomposition, movement and interpenetration.

His most outstanding achievement was the interior decoration of the Balla House (1918-20; Fig. 7); decorating a place in which to live was to become the painter's *idée fixe* (before reconstructing the universe, he began by enlivening his own environment). Here is his self-portrait in the role of craftsman: 'Like a good laborer, he sets to work, and those delicate, expert hands that could have executed a traditionalist picture with four strokes of the brush, earning thousands of lire, preferred to stick glue, saw wood, and cut out paper and cardboard to make lamps, lampshades, screens, and toys, and sell them the next day for a handful of lire.' The large piece of furniture that can be dismantled, the little table, the stools, the bench, and the divan are still in existence. The predominant colors are green and yellow, applied so that from one angle the pieces appear entirely green, and from another, yellow (Fig. 12). One should also mention the brightly colored screens (Fig. 11), one of which is also still in the Balla House, and which were veritable sculptures in space; as well as amusing small pieces, such as the little smoking table with a spiral of silver smoke painted on it.

Depero worked along the same lines. He was a painter who was not afraid of going beyond painting; he was, as well, a designer of stage sets, a sculptor, a poet, an architect, an interior decorator, a graphic artist, and a choreographer. His work was generally similar to Balla's, even though — in spite of his trips to Paris and New York — it tended to lapse into ornamentation and a kind of folk art. As early as his 1916 exhibition, he produced spatial constructions designed for the new environment, and typographical architecture; many objects of decorative art were included in his one-man show in Milan in 1919. Subsequently, Depero returned to his native town of Rovereto, where according to the *Cronache di attualità*, he opened 'a large factory to produce furniture and objects of decorative art' — cushions, tapestries, and furnishings. It is particularly interesting to explain the significance of one of his gayest and most enchanting environments, which he designed for the house of Giuseppe Sprovieri in 1921. There were yellow walls, a gray carpet with two tightrope-walkers, a bed and two chairs with a heart design, two balls of different sizes, and a large colored club. Sprovieri has told me that he wanted 'a very colorful and fantastic house, that of a Futurist acrobat ... I wanted to be someone who juggled with ideas.'

It is logical that such an interest in the show even of everyday life should have had both its point of departure and its goal in the theater (Fig. 10). Significantly enough, Balla carried out his first experiments in environments for several Futurist evenings at Giuseppe Sprovieri's gallery in Rome, in 1914. Special mention should be made of the production of Stravinsky's *Fireworks* for Diaghilev's Ballets Russes at the Costanzi Theater in Rome in 1916; the ballet had no dancers, and the only movement consisted of colored lights playing over a group of geometrical solids. Balla, in the prompter's box, switched the various lights on and off to the rhythm of the music, in fact 'playing' the light (7). Depero, too, created designs for the Ballets Russes, for the production of *Song of the Nightingale* (Fig. 13). It is well documented in surviving records and is noteworthy because it showed a kind of cardboard jungle, a Futurist reconstruction of nature, a toy on the scale of town planning. And it was not by chance that both Balla and Depero undertook the decoration of cabarets, the former completing

IL VESTITO ANTINEUTRALE
Manifesto futurista

Glorifichiamo la guerra,
sola igiene del mondo.
MARINETTI.
(1° Manifesto del Futurismo - 20 Febbraio 1909)

Viva Asinari di Bernezzo!
MARINETTI.
(1° Serata futurista - Teatro Lirico, Milano, Febbraio 1910)

L'umanità si vestì sempre di **quiete**, di **paura**, di **cautela** o d'**indecisione**, portò sempre il lutto, o il piviale, o il mantello. Il corpo dell'uomo fu sempre diminuito da **tinte neutre**, avvilito dal nero, soffocato da cinture, imprigionato da panneggiamenti.

Fino ad oggi gli uomini usarono abiti di colori e forme statiche, cioè drappeggiati, solenni, gravi, incomodi e sacerdotali. Erano espressioni di timidezza, di malinconia e di **schiavitù**, negazione della vita muscolare, che soffocava in un passatismo anti-igienico di stoffe troppo pesanti e di mezze tinte tediose, effeminate o decadenti. Tonalità e ritmi di **pace desolante**, funeraria e deprimente.

OGGI vogliamo abolire:
1. — Tutte le tinte **neutre**, carine e sbiadite, *fantasia*, semioscure e umilianti.
2. — Tutte le tinte e le foggie pedanti, professorali e teutoniche. I disegni a righe, a quadretti, a **puntini diplomatici**.
3. — I vestiti da lutto, nemmeno adatti per i becchini. Le morti eroiche non devono essere compiante, ma ricordate con vestiti rossi.
4. — L'equilibrio **mediocrista**, il cosidetto buon gusto e la cosidetta armonia di tinte e di forme, che frenano gli entusiasmi e rallentano il passo.
5. — La simmetria nel taglio, le linee **statiche**, che stancano, deprimono, contristano, legano i muscoli; l'uniformità di goffi risvolti e tutte le cincischiature. I bottoni inutili. I colletti e i polsini inamidati.

Noi futuristi vogliamo liberare la nostra razza da ogni **neutralità**, dall'indecisione paurosa e quietista, dal pessimismo negatore e dall'inerzia.

Vestito bianco - rosso - verde
portato dal parolibero futurista Cangiullo, nelle dimostrazioni dei Futuristi contro i professori tedescofili e neutralisti dell'Università di Roma (11-12 Dicembre 1914).

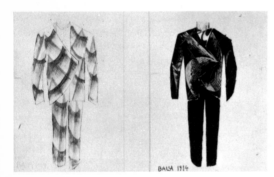

BALLA 1914

Bragaglia's House of Art and the Bal Tic Tac, the latter the Devil's Cabaret, between 1921 and 1922.

4. 'Painkiller' Clothing

Clothing, according to Balla, is a complement to the environment. In Düsseldorf, besides decorating a room, he designed for its owner a dress to match. Before creating a new painting, he designed clothes to wear, to show a new kind of behavior. In 1914, he wrote the Futurist Manifesto on Menswear, later revised by Marinetti and published in September of the same year under the title 'Antineutral Clothes' (Figs. 14-17). Because of its interest, I quote here an extensive excerpt from the original manifesto (8):

'WE MUST DESTROY THE TRADITIONALIST CLOTHING epidermic colorless funereal decadent boring and unhealthy. In materials, abolish: wishy-washy pretty fanciful neutral dull colors and designs with stripes checks dots. *Cut and tailoring.* The abolition of symmetry static lines uniformity cuffs lapels bad cutting, etc. *Put a stop once and for all to those exhumed garments the hypocritical garb of mourning.* The crowded streets assemblies theaters cafés have a desolate funereal atmosphere because the clothes reflect the surly miserable humor of today's traditionalists.

'WE MUST INVENT FUTURIST CLOTHING cheerrrrrrful cheeky brightly colored dynamic simple in line and above all short-lived in order to increase industrial activity and give our bodies the continual delight of novelty. USE for materials MUSCULAR colors brightest violets brightest reds brightest turquoises brightest greens yellows oranges vermilions BONE shades white gray black create dynamic designs expressed, through abstract equivalents: Triangles cones spirals ellipses circles, etc. In the cut use dynamic asymmetrical lines the end of one sleeve and front of one jacket are curved on the left and angular on the right with waistcoats trousers overcoats, etc. This flashing gaiety of clothing moving through the noisy streets transformed by the new FUTURIST *architecture* will glitter with the prismatic splendor of a huge display in a jeweler's shop window...

'We want comfortable and practical Futurist clothes
Dynamic
Aggressive
Amazing
Strong-willed
Violent
Flying (that is giving an idea of flight of rising and running)
Nimble-making
Joyful
Illuminating (to give light in the rain)
Phosphorescent (that is using phosphorescent materials)
Decorated with electric light bulbs.
'*Designs with adaptors* having pneumatic devices that can be used on the spur of the moment; thus everyone can alter his dress according to the needs of his spirit. The *modifications will be:*
Loving
Overbearing
Persuasive
Diplomatic
One-toned
Many-toned
Softly shaded
Polychromed
Perfumed.
'The result will be a needed variety of clothes, even in a city in which the population lacks imagination.

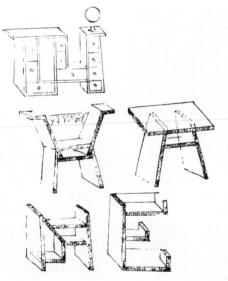

'The gaiety of our *Futurist clothing* will help to bring about the spread of good humor proclaimed by my dear friend Palazzeschi in his *Painkiller* manifesto.'

Balla also promised 'a woman's dress.' In some of his notes, he suggests a new type of psychological badge ('a changeable ribbon falling from the head to the neck') and also real movement in the dress ('woman's dress, a peacocklike mechanism, mobile, changing its shapes'). The key points in the manifesto are: the abolition of all compromises in forms and colors; the search for 'painkiller' gaiety; 'states of mind' clothing. In his still unpublished diary, Bragaglia recalls Balla in the following words: 'When he was in civilian clothes he wore, at least, a colored lamp in a cravat cut like a box, with a celluloid window in it. At the most electrifying moments in the conversation, he would press the button, and the cravat would light up; he made his points in this way.'

5. The Object in the Limelight

Among the achievements listed in the manifesto on Italian Futurism in 1921 appears 'Balla's Futurist decorative art (ceramics, screens, lamps, etc.) (Workshop in Rome); (Bal Tic Tac, imposing night club in Rome, Futuristically decorated by Balla).' By this time, there were countless decorative objects (Figs. 8 and 9). We may mention, among others, a series of trapezoidal lampshades on which the words (orchid, rose, violet) appeared fully written out and painted with colors corresponding to the flowers themselves. There were also lamps, embroidery, boxes, materials, plates, carpets, tapestries, ceramics, and hats. The Bal Tic Tac offered a comprehensive view of this synthesis of color-movement-light.

Interest in the object in relation to the environment accords with the ideas of *pittura metafisica* and anticipates Dada. The question of interior decoration and the human personality of the object interested Marinetti too, especially within the context of the theater (9). In *Musica da Toeletta* ('Music for a Lady's Boudoir'), there was a piano on stage, a pair of elegant little golden shoes that were being dusted by a manservant, while a chambermaid dusted the piano keys with a toothbrush. *Vengono* ('They're Coming') is a forerunner of Ionesco; it is a play about an armchair and eight chairs, which through being moved into various positions by the servants, 'gradually acquire a strange bizarre life. And at the end of the play, the audience, helped by the shadows slowly lengthening toward the door, should feel that the chairs are really alive and moving by themselves to make their exit.'

An example of object invention is Balla's Letter Furniture. The design for Yannelli's living room still exists (Fig. 18); it shows Balla inventing a stool for *Y*, a little table for *A*, two little cabinets for the double *N* and the *E*, a chest of drawers for the group *LLI*, with a lamp representing the dot on the *I*. Another example is The Chair in Which You Are Seated Before You Sit Down, invented for the Devil's Cabaret by Depero (one recalls Sprovieri); it was in the form of a seated man and gave the (metaphysical) impression of a split personality.

The height of the quest for the lost object (adopting Marinetti's definition) was attained by Balla in his Futurist flowers (Fig. 19). Pieces of wood, cut out and joined together, were painted so as to give the impression of an impossible, unreal nature; most of them were huge. The flowers stood as high as five feet (one is still in the Balla House) or else swung from the ceiling — a new kind of house decor. This attempt to revive in a new form the cult of the Nature Goddess in opposition to the Machine Devil is also a legacy from Art Nouveau.

Opposite:
14. Manifesto, 'Antineutral Clothes,' September 11, 1914
15. Balla. Designs for Futurist clothes, 1913-14. Tempera on paper. Rome, Casa Balla
16. Balla. Design for a Futurist woolen sweater. 1917. Tempera on paper, 7 7/8x5 1/2 inches. Casa Balla, Rome

Above:
17. Balla. Futurist clothing, from the original manifesto, 1914
18. Balla. Design for Letter Furniture forming the name of the owner, Yannelli. Casa Balla, Rome

19. Balla with his daughters Luce and Elica and some Futurist wooden flowers
20. Frame from the film *Vita futurista*, 1916: 'Balla marries a chair and a little stool is born'

(It is no mere coincidence that England and Belgium, the two most industrialized countries, should have been the centers of revolution at the turn of the century.) Balla constructed a flower as tall as a tree, which looks like a wardrobe, and expressed an organic phenomenon in geometrical terms. He recreated the Creation, patiently demonstrating that even nature can become artificial — a proposition that has had an extensive following in recent art of our own day.

The new significance given to the object appears most clearly in the film *Vita Futurista* ('Futurist Life') (10). Here is one of its scenes: 'Drama of objects. Marinetti and Settimelli, with utmost caution, approach objects that are strangely jumbled together, in order to view them from new aspects'; and another: 'Conversation between a foot, a hammer, and an umbrella — Taking advantage of the human expressions of objects in order to plunge into new areas of artistic sensibility.' The following scene, of which we fortunately have a photograph, has up to now been interpreted as 'Distorted Caricature of Balla,' but is in reality the scene described as 'Balla falls in love and marries a chair, and a little stool is born' (Fig. 20). At this point, any comment seems unnecessary: this marriage ceremony between the artist and the object is already Dada.

I should like to thank, for their aid and advice, Luce and Elica Balla, Giuseppe Sprovieri, Vittorio Orazi, Francesco Cangiullo, Arnaldo Ginna, Antonella Vigliani Bragaglia, Gaspero del Corso and Cesare Bellici, Mario Bulzoni, and Maurizio di Puolo.

Basic bibliography: For Futurism in general, see the texts containing notes on the environment, in *Archivi del Futurismo*, edited by Maria Drudi Gambillo and Teresa Fiori, Rome: De Luca, vol. 1, 1958, vol. 2, 1962 (included in *Sintesi del Futurismo*, facsimile of the manifestos, Rome, 1968); *Futurism*, edited by Joshua C. Taylor, New York: The Museum of Modern Art, 1961; Maurizio Calvesi, *Dinamismo e simultaneità nella poetica futurista*, Milan, 1967-68 (L'Arte moderna, vol. 5); Marianne W. Martin, *Futurist Art and Theory*, Oxford: Oxford University Press, 1968. On Balla, see: *Giacomo Balla*, edited by Enrico Crispolti and Maria Drudi Gambillo, Turin: Galleria Civica d'Arte Moderna, 1963; Maurizio Fagiolo dell'Arco, *Omaggio a Balla*, Rome, 1967; idem, *Futur-Balla*, Rome, 1968 (2nd ed., 1970). On Depero, see: *Fortunato Depero*, edited by B. Passamani, Bassano del Grappa, 1970. On Prampolini, see: Filiberto Menna, *Prampolini*, Rome, 1967; Enrico Crispolti, 'Dada a Roma, Contributo alla partecipazione italiana al Dadaismo,' *Palatino*, vols. X, nos. 3/4 - XII, 1968. For more specific references, see the following notes.

(1). The passage by Balla referred to dates from 1918 and reads: 'Any shop in a *large modern city* with its elegant windows displaying useful and delightful objects affords more artistic enjoyment than the much-praised exhibitions of the traditionalists. An *electric iron*, metallic white, smooth, shining, highly polished, delights the eye far more than a nude statuette set on a badly made pedestal tinted for the occasion. The *typewriter* is more architectonic than the building projects awarded prizes by academies and competitions. The window of a perfume shop with its boxes, small boxes, and tiny boxes, its jars and little jars, all with Futurist-colored reflections in elegant mirrors. The clever and flirtatious shape of woman's shoes, the bizarre ingenuity of many-colored umbrellas. Furs, valises, dishes — all these are more pleasing to the eye than the dingy, insignificant little pictures of a traditionalist painter hanging on a gray wall.'
(2). For Ginna's article, see *L'Italia futurista*, December 1916 (summarized by Menna, *Prampolini*, pp. 45-46; see also M. Verdone, 'Ginna e Corra. Cinema e letteratura del futurismo,' *Bianco e nero*, 1967, nos. 10-12, pp. 59-60.
(3). Although the article did not appear in *Noi* until June 1917, it was written in 1916.
(4). For Prampolini's designs, see Menna, *Prampolini*, figs. 35-37. For the House of Italian Art, see V. Orazi, 'La Casa d'arte italiana,' *Strenno dei romanisti*, 1968, pp. 277-79; Menna, *op. cit.*, pp. 42-52; its program of activities was reported in *Noi*, nos. 5-7. For the controversy regarding 'spatial constructions,' partially noted in the *Archivi del Futurismo*, the essential document is the last letter Prampolini wrote to Balla, in which he withdrew his accusations and acknowledged the importance of Balla's work (see Fagiolo dell'Arco, *Futur-Balla*, pp. 76-77).
(5). For Cangiullo's various activities, see my *Futur-Balla*, passim; the manifesto, published in *Roma futurista*, February 22, 1920, is reproduced in

vol. III, p. 95. Regarding the furniture, I add what Cangiullo himself wrote me: 'Many years have passed. But I recall, among other things, that I also made a kiss-box (*baciario*); it was a kind of little shelf in the form of a lacquered pink mouth, in which one was supposed to deposit kisses. You — husband, lover, mother, father, children, for example — in leaving the house, would say, "I have left eight kisses in the kiss-box." Whoever wished to go and put his lips to the kiss-box would have the sweet illusion of receiving kisses from that mouth.' For Balla's circle, see also the article by Volt, 'La Casa futurista Independente, Mobile, Smontabile, Meccanica, Esilarante,' *L'Italia futurista*, 1920.

(6). For Balla and his house, see the catalogue of the Turin exhibition 'Giacomo Balla,' and my *Futur-Balla*, vol. III, pp. 94-96, 109-110, figs. 121-40. For a bedroom attributed to Balla, see Maurizio Calvesi, *Le Due avanguardie*, Milan, 1967, p. 88, and Enrico Crispolti, 'Dada a Roma' (*Palatino*, 1967, no. 3), p. 319.

(7). For Balla and the theater, see *Futur-Balla*, vol. III, pp. 80-93, figs. 56-110.

(8). The original manifesto on clothes has been published in Fagiolo dell'Arco, *Omaggio a Balla*, pp. 55-57; cf. also *Futur-Balla*, vol. III, pp. 96-97, 110, figs. 141-52.

(9). For Marinetti, see: Francesco Tommaso Marinetti, *Teatro*, edited by G. Calendoli, Rome, 1960; *Teoria e invenzione futurista*, edited by L. De Maria, Milan, 1968.

(10). For the Futurist film, see: M. Verdone, 'Ginna e Corra,' op. cit., and additional documentation in *Futur-Balla*, vol. III, pp. 86-88. The scene of the marriage between Balla and the object was probably invented by Marinetti (at the end of his novel *Mafarka il futurista*, the hero, unaided by a woman, gives birth to a winged mechanical son).

THE BEGINNING OF MODERN RESEARCH,
1930-1940
Leonardo Benevolo

The opening statement of the program of this exhibition: 'The emergence during the last decade of Italy as the dominant force in design has already influenced the work of every other European country and is now having its effects in the United States,' should be supplemented by two further considerations:

1. Italian design is limited to single items and small environments and does not extend to environments in the wider sense. These much-admired items and microenvironments may be found in some of the ugliest cities in the world, where they spoil rather than improve their surroundings.

2. Italian design is not applied to all types of objects and microenvironments, but only to products that yield a profit. It therefore depends on the private organization of production rather than the private organization of consumption; and it affects only that part of the domestic landscape that is dominated by the existing social system in Italy.

These limitations cannot be attributed either to the Italian character nor to the remote past; they can be explained only by the events of more recent history. Neither Italian designers nor consumers were particularly enamored of small environments and commercial products, but these were what they were compelled to design and use almost exclusively.

Although a combination of old and new causes brought about the present situation, here we shall concern ourselves only with trying to reconstruct what took place from the 1930s on. Obviously, prior to that, design felt the pressure of economic and social interests, but it was unable to oppose them in any way, because design developed within institutional forms that had been created precisely for the protection of those interests. Design was restricted either to technical research circumscribed by a priori limitations (and concerned with mechanical elements rather than with their relationship within a functional system), or to research within the field of art, enjoying unlimited freedom within that field, but quite distinct from everyday life. In the 1920s, however, the modern movement in architecture began, and with it the beginning of scientific research on the artifacts that make up the background of our daily existence — a research that went beyond, and counter to, the traditional tasks allotted to design. This kind of research could be independent of established interests — that is, it might either comply with them or oppose them. It may therefore be of interest to trace the changes that took place in these relationships, which became the normal events of individual and social life. The 'history of art' came to an end, and the history of human environment began — an environment that might be either agreeable or unpleasant for those who had to live in it.

Since receiving his degree in architecture at the University of Rome in 1946, Leonardo Benevolo has combined two aspects of his career; he is a practicing architect with an active studio in Rome, as well as a noted architectural historian. In the latter capacity, he has taught at the universities of Rome, Florence, and Venice; in 1969-70, he was a visiting lecturer at Yale. He is the author of several books, including a history of modern architecture that has been translated into several languages, and another on the origins of modern town planning, which in 1964 received the Libera Stampa award.

Modern architecture began in Italy in 1926, when a group of recently graduated architects came into contact with the international movement. Some of them — Luigi Figini, Guido Frette, Sebastiano Larco, Adalberto Libera, Gino Pollini, Carlo Enrico Rava, Giuseppe Terragni — founded Group 7 and came before the public collectively (figs. 1-4). They exhibited in Stuttgart and Monza in 1927, in Essen, Milan, and Rome in 1928, in Breslau in 1929, and in New York in 1931. As has been observed, their following of Walter Gropius and Le Corbusier was in part the result of their cultural training, which was general rather than professional: Pollini, for instance, was interested in music and philosophy, Figini and Terragni in painting; and it was their very detachment from institutional and codified professions that enabled them to understand a proposition that far transcended these. The first works presented by Group 7 were large-scale architectural projects, intentionally summary in character, like Terragni's gas

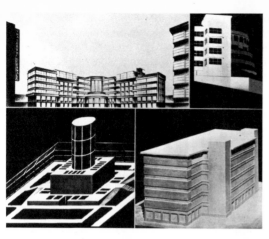

1. Gruppo 7: first exhibited designs. 1927
2. Luigi Figini and Gino Pollini. Writing desk for their own studio. 1929
3. Piero Bottoni. Armchair designed for Thonet. 1930
4. Piero Bottoni. Dining room. c. 1931

works and a garage for five hundred cars by Figini and Pollini. As their masters had taught, the same method was followed in designing both large and small items; but the young Italians had neither the means nor the time for the kind of patient apprenticeship available at the Bauhaus, which began with the design and manufacture of very simple objects. Since they lagged six or seven years behind the Germans, they could not afford to concentrate on any but productive opportunities. Now such opportunities presented themselves and had to be taken advantage of. The different phases of research — study, experimentation, discussion — overlapped at a feverish pace. Every chance had to be exploited, although none was to be taken at its face value.

It is hard to realize nowadays the difficult conditions under which the first modern architects were obliged to work. Italian architecture and industrial design were hopelessly behind the times. Giovanni Muzio in Milan and Marcello Piacentini in Rome were regarded as innovators in architecture, and their buildings, being less ornate than the usual ones (e.g., the 'ugly house' in Milan and the Corso Cinema in Rome), were considered scandalous. Gio Ponti was already the leader in industrial design, having transplanted to Rome and Brianza the taste and kind of organization of Josef Hoffmann's Wiener Werkstätte. Modern designers had to struggle not only against traditional eclecticism but also against a retarded form of avant-garde. Within the field of industrial design itself, they encountered either prohibitive competition or dubious protection. It suffices to mention that the outstanding Italian designer, Franco Albini, worked in Ponti's studio both before and after taking his degree, and not until about 1932, after meeting Edoardo Persico, did he come to realize the difference between the true modern tendency and that of the traditional avant-garde.

The first important example of modern Italian design was the Electric House that Figini, Pollini, and Libera, together with Piero Bottoni, erected for the Edison Company at the IV Monza Triennale of Decorative and Industrial Arts, in the summer of 1930 (pp. 312-14, figs. 33-37). It was a small building in a park, similar to a one-story villa, but with the living room on two levels, giving access to a roof terrace. Around the living room were arranged two bedrooms, a bathroom, a kitchen, a dining area (separated from the living room by two curtains), and a small servant's room. It was equipped with every electrical appliance known at the time, in order to show how such modern devices could make life easier even in a house with only one servant (which was then regarded as the indispensable minimum for a modest middle-class family). Both the architecture and the furniture harmonized with the very simple and efficient mechanical equipment, so that the visitor's interest would not focus on any single item but on the entire setting. The large living room could be enjoyed on two levels and was linked to the exterior space through glass doors on the ground floor and the terrace on the upper level. The other rooms flanking the living room were reduced to the smallest size compatible with their respective functions.

A comparison between this house and the foreign models available at the time (such as the studies by the German theoreticians Alexander Klein and Bruno Taut, and their practical application in public-building projects in Frankfurt and Berlin, or the experiments of the Bauhaus, which had been summarized a few months before in the German exhibition at Paris in May, 1930) clearly show both the backwardness and the advances of Italian as against international research:

1. The experiments made in Germany during the previous decade had

5. Giuseppe Pagano and G. Levi-Montalcini. Design for lady's bedroom, Villa Gualino, Turin. 1930
6. Giuseppe Pagano. Interior, Turin. c. 1931
7. Piero Bottoni. Writing desk. 1932

resulted in a range of building types that could be fabricated by mass production. Some of them had already been executed on a wide scale (row houses, apartment houses of three to five stories), others had been developed as theoretical studies, alternative to these (houses ten to twelve stories high, with stairs and elevators or galleries, and with common facilities). The building method, the cost, and the social purpose of each type had been clearly established. The displays made as demonstrations for exhibitions summarized many previous researches; the Paris exhibition of 1930 was based on the eleven-story high-rise devised by Gropius and Marcel Breuer for the Spandau district in 1929, and later perfected in several other projects; the concept of the high-rise had been justified in theory by Gropius in a paper that he presented at the CIAM conference in Brussels in 1930.

Nothing similar to this had been done in Italy, and the Electric House of 1930 may be considered the first approach to many previously unexplored problems. The choice of the building type was incidental (a detached one-family house, its equipment depending on the commission — in this case, given by an electrical company) and was meant to publicize both the house as a whole and the individual pieces of apparatus. It was intended for the general public rather than for any particular social class, and it represented the most comprehensive scheme possible, not a concrete and repeatable proposal. (The official statement declared: 'Keeping in mind the financial side, the architects have designed similar types of houses in a lower price bracket.')

On the other hand, the designers clearly expressed their theoretical conception of a kind of dwelling far from usual at the time, but of great importance for future developments. Taking as their point of departure a series of mechanical apparatus (familiar in type, even though here isolated and automated), they used them to abolish the ceremonies of traditional domestic life. What they showed was a different kind of domestic landscape, suggesting a different mode of life, freer and richer in possibilities. The size of each of the less important rooms was carefully calculated with respect to the specific function it served — cooking, laundering, ironing, sleeping, bathing, and so on; but they all converged on a central point, the large living room, which communicated with the space outdoors, where social relationships, encounters, and cooperative ventures took place. This was only a summary exploration within a world of new types that still had to be categorized and evaluated. It departed from the well-tested mainstream of typological research in Europe and might be compared with the models for collective housing being developed in Russia at the time by Ginzburg and his associates. The latter were theoretical proposals for a still-undetermined class of users (generically proletarian rather than bourgeois), and although the method of their construction and operation was still unknown, they were to be developed in the future.

2. Each item of furniture shown in the experimental houses in Stuttgart and at the German exhibition in Paris was the result of long studies and progressive refinements (for example, Breuer's tubular metal chairs and armchairs). None of the pieces in the Electric House reached such a degree of refinement, and some — such as those by Libera and Frette in the living room and master bedroom — were even somewhat crude. But all this furniture was conceived as elements within a well-integrated ensemble and manifested a clear desire to create an environment rather than a collection of individual objects. In all the rooms, and particularly in the children's room by Figini and Pollini, and in the kitchen by Bottoni, harmony was achieved by the use of similar shapes, repeated on different scales. Throughout

8. Figini, Pollini, and Baldessari. Craja bar, Milan. 1930
9. Felice Casorati. Entrance hall of architectural exposition, V Triennale, Milan. 1933
10. Albini, Palanti, Camus, and Masera. Library interior shown at V Triennale, 1933

the house as a whole, harmony resulted from grouping all the less important rooms around the central living room and was completed by the landscape outside, which was visible from the windows and surrounded the terrace.

This approach, also, was based on a theoretical principle. It was a way of putting into practice the 'architecture as total environment,' proclaimed by Mondrian: 'The house can no longer be a collection of rooms made up of four walls with openings for windows and doors, but should be constructed of an infinite number of colored and colorless planes, related to the furniture and other objects, none of which will be of any value in and for themselves, but will be constituent elements of the whole.' The house, in turn, will no longer be 'a place of refuge and shelter, but part of a whole, a constituent element of the city.' The absence of any specifically artistic sense, which so disconcerted critics of the time, was due to the renunciation of art as used in traditional environments. 'Art is a mere substitute, as long as the beauty of life is inadequate [but] it will disappear as life becomes more harmonious .... Then painting and sculpture will no longer be needed, because our lives will be art made actual.' Precisely herein resides the original political motive underlying the struggle between modern research and established society: such research endangers the distribution of power on which control of the environment has traditionally been based and refuses to be relegated to the marginal role allotted to 'art.' This program, which Mondrian rather naively imagined to be within reach (in 1927, he declared 'with a little goodwill, it would be possible to create a sort of paradise'), was soon shown instead to present many difficulties and was destined to remain for a long time only a theoretical statement, though today it is still the basis of most advanced research in environmental design.

If we consider the research carried out by Group 7 within the framework of the Italian situation, it appears even more risky and unattainable. The best experiments of the older architects — for instance, buildings and interior decorations by Giuseppe Pagano and Levi-Montalcini in Turin — repeated models already widely known throughout Europe in the previous decade (figs. 5 and 6), and did not greatly influence current production, which continued to follow the old eclectic patterns. The propositions advanced by the young members of Group 7 were either rejected out of hand or accepted with curiosity as contributions to the repertory of 'styles.' *Domus*, the authoritative periodical edited by Ponti, in its write-up of the Monza Triennale of 1930, praised the Electric House for its technical innovations, although it commented that 'criteria of this sort lead to results completely remote from traditional lines, so that it is quite premature to state whether this can be regarded as Italian art,' and it gave more unqualified praise to Muzio's great entrance hall ('The noble Italian marbles, the delicacy of the colors and design, make this room the outstanding environment of the whole exhibition') and to the vacation house by Gio Ponti and Emilio Lancia, which had pediments over the windows and hangings decorated with bird motifs. Meanwhile, exhibition installations served as theoretical demonstrations and afforded Group 7 a means of presenting their ideas. But they had far less opportunity to proceed from exhibitions to city planning, and they had to struggle hard in order to avail themselves of any concrete opportunity.

Between 1930 and 1933, the modern Italian movement played all its cards; it lost the more important stakes but won some minor ones. In the end, although its scope was slightly wider, from that time on it had to work within rigid limitations that were hard to overcome. The first completed works — buildings like Terragni's Novocomum

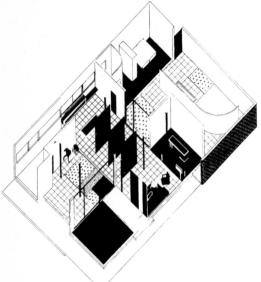

(1929), the De Angeli Frua offices at Milan by Figini, Pollini, and Luciano Baldessari (1930), Terragni's Vitrum shop at Como, the Craja bar by Figini, Pollini, and Baldessari (1930; fig. 8), the SISI school in Turin by Pagano (1932), and interior decorations by Bottoni (fig. 7), Pagano, and Terragni — were private commissions but nevertheless provoked general discussion. The new architecture offered an alternative to the traditional style and hoped to win the battle with large-scale public works; both critics and architects called it the style best suited to the Fascist regime. In 1931, Pietro Mario Bardi published a report on architecture that was clearly intended for Mussolini; modern architects organized themselves into a large association that embraced the whole of Italy, the MIAR (Italian Movement for a Rational Architecture). In 1931, Bardi also arranged an exhibition of this movement in Rome, openly polemic in its attack on traditional architecture, and succeeded in getting Mussolini himself to inaugurate it. The Fascist syndicate reacted swiftly, calling for the disbandment of the MIAR, but critics such as Bardi and architects such as Pagano and Terragni continued to carry on their campaign within the Fascist movement. In 1932, Terragni began plans for the House of Fascism at Como; in 1933, the Tuscan group led by Giovanni Michelucci won the competition for the railway station at Florence; Eugenio Montuori participated with Piacentini and Arnaldo Foschini in planning the University City at Rome; Libera, Giuseppe Samonà, and Marco Ridolfi won three of the four competitions for the new Roman post offices; at the V Triennale, organized at Milan in Muzio's new Palace of Art and directed by Piacentini and Ponti, modern designers participated and were successful to a great extent in influencing the general trend of the exhibition. By this time, modern architecture was being publicized and supported by several magazines, such as Bardi's *Belvedere*, *Casabella*, for which Persico wrote from 1930 on, and *Quadrante*, founded by Bardi and Bontempelli in 1933; and it was documented in the voluminous book by Alberto Sartoris, *The Elements of Functional Architecture*, published by Hoepli in 1931. These successes made the argument become even more heated; the slogan 'The Triennale against Rome' was invented; and on May 20, 1934, there was even a parliamentary debate on the pros and cons of modern architecture.

The entire program of modern architecture, including the field of industrial design as well, depended on its relationship with the public authorities. If modern architecture were to be in a position to create a total new environment, it would have to proceed from models to actual production, from small-scale production to mass production, and from the sheltered environments of private lives to the open environment of the public life that contained and coordinated them. This entailed the planning and completion not merely of private houses but of government-subsidized housing developments, public buildings, and city plans as well. It involved designing not only decor for private houses and deluxe articles of furniture, but also equipment for schools and offices, inexpensive mass-produced furniture, mechanical appliances (telephones, radio sets, etc.), which were now in use everywhere and manufactured by large industrial concerns. In short, it meant winning the competition with 'technicians' and 'artists,' who still held the monopoly in their respective fields.

In 1931, in his foreword to Sartoris's book, Le Corbusier wrote: 'As yet, we have hardly begun. Up to now, we have been allowed to build houses, but insofar as the city and social life are concerned — and these are the bases for houses — we are barely babbling the first words .... We must give our theoretical and practical enterprises a far more powerful impetus. The academicians who have reacted up

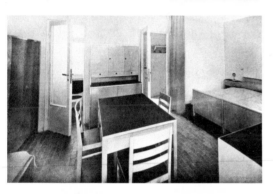

Opposite:

11-12. Figini and Pollini. Villa-studio for an artist shown at V Triennale, 1933
13. Villa-studio for an artist: Dining-room.
14. Pagano and associates. Steel high-rise, V Triennale. Living-room. 1933

Above:

15. Steel high-rise. Children's room.
16-17. Griffini and Bottoni. Low-income housing project, V Triennale. 1933
18. Annoni, Comilli, and Masera. Interior for school shown at V Triennale. 1933

to now have been frightened far too quickly; they have no idea what still lies in store for them.'

But the prevailing rules on how to succeed or fail are fixed in accordance with traditional cultural patterns, and if modern designers were to accept them, they would be putting themselves in a false and weak position. Each was expected to prove that he was a bonafide technician or artist in the accepted sense, and also to prove himself Italian, which was obligatory at the time. They therefore had to disguise what was actually the most important feature of the new method, in which distinctions of this kind were quite impossible. They represented modern architecture to Mussolini as 'Fascist architecture,' and thereby not only associated modern architecture with a misguided political program but had to concede that it was just an alternative style to any other kind of style. Thus they were forced to do battle on their opponents' terms and were necessarily doomed to defeat, for modern architecture was an anomalous style (its premises not yet fully established, and full of surprises), while traditional architecture could offer a whole gamut of normal, well-known, and time-tested styles.

The V Triennale took place in 1933, while these arguments were at their height, and it clearly manifested all the still-unresolved ambiguities. The exhibition had a new setting within the park of Milan, in the Palace of Arts built by Muzio in a style skilfully balanced between the classic and the modern ('modern in an Italian manner,' as Piacentini remarked in the magazine *Architettura*). While its appearance was slightly modern, to permit the use of modern architecture in some minor displays, its structure was completely classical, being symmetrical and dominated by large formal salons (entrance hall, grand stairway, ceremonial hall), which overwhelmed and isolated everything presented in the exhibition galleries. This overpowering classical setting was accentuated and endorsed by the artists whose paintings and sculptures adorned the main rooms: Marino Sironi, Giorgio de Chirico, Carlo Carrà, Massimo Campigli, Achille Funi, Arturo Martini, and Marino Marini.

Within this ambiguous locale, an extensive and heterogeneous documentation on the experiments of modern architecture and industrial design in Italy and abroad was on view. It included an architectural exhibition arranged by a 'neutral' commission (Sironi, Ponti, and Pietro Aschieri, among others, with Agnoldomenico Pica as secretary); a hall dedicated to the CIAM (installed by Pollini and Bottoni); one-man shows of the work of Gustave Perret, Konstantin Melnikoff, André Lurçat, Frank Lloyd Wright, Josef Hoffmann, W.-M. Dudok, Walter Gropius, Ludwig Mies van der Rohe, Erich Mendelsohn, Adolf Loos, Le Corbusier, Antonio Sant'Elia); an exhibition of interior design (arranged by Ernesto Griffini); an exhibition of typical works either already completed, under construction, or still in the planning stage; a display by the recently created Italian Faculty of Architecture; an architectural bookshop; and an exhibition of 'decorative and industrial arts,' including a selection of Italian objects (some by young modern designers) — vases, rugs, cutlery, and chairs of all types. In addition, there were sections devoted to foreign countries; exhibits of bronzes, glassware, ceramics, and textiles (these were organized by Baldessari); and a series of rooms decorated with the products of the most important furniture manufacturers, in the style promoted by Ponti and his followers (even those by the young designers Franco Albini, Renato Camus, Paolo Masera, Giancarlo Palanti, Asnago, and Vender were in this same taste; fig. 10).

The modern architects' ambition to create a new environment by blending architecture and decoration could not be realized in the

displays within the Palace of Arts; they therefore devoted their best efforts to the exhibitions in the park, a series of pavilions with unusual names: house on the gulf, recreation houses for an aviator and for a land agent, weekend house for a married couple, country house for a scholar, hut for twelve skiers, small mountainside inn. Several modern architects undertook individual schemes of this kind, which corresponded to potential private commissions: an artist's vacation house on a lake (Terragni, Pietro Lingeri, Dell'Acqua, Mantero, Giancarlo Ortelli, Ponci, Mario Cereghini, Giussani); villa-studio for an artist (Figini and Pollini); summer house (Pagano, Aloisio, Antonio Ghezza, Cuzzi, Levi-Montalcini, Enrico Paolucci, Ettore Sottsass Jr., Turina); house in the colonies (Luigi Piccinato). Other presented collective schemes for mass production: two high-rises in steel (Pagano, Albini, Camus, Giancarlo Palanti, Mazzoleni, Giulio Minoletti, and Luigi Carlo Daneri, Luigi Vietti); a group of dwellings for housing developments (Griffini and Bottoni); an interior and equipment for a school (Annoni, Comolli, and Masera; fig. 18).

Such essentially disparate themes precluded the buildings from constituting, or presenting as examples, any new urban or regional environment different from what already existed: each had to find a separate place within the extant setting. Even the most ambitious and experienced designers resigned themselves to this state of affairs and had to be content with making their presence felt within the allotted limits; they opted for enclosed schemes, which they made coherent on the interior, thus avoiding an impossible confrontation.

Within this wide range of projects, the two most extreme examples, and the most interesting on many counts, were the villa by Figini and Pollini and the housing developments by Griffini and Bottoni.
The villa-studio for an artist (figs. 11-12) was a detached one-story residence like the Electric House of three years before. The general scheme of the Electric House — a dwelling related to its outdoor surroundings — was here made more explicit: the villa was composed of enclosed and open spaces harmonizing with one another but having a minimum relationship to the outside environment, so that it would not be out of place either in the city or in the country. This work, too, was related to researches being carried out abroad, particularly to the courtyard houses (Hofhäuser) of Ludwig Mies van der Rohe, and it was undoubtedly the most refined architectural work exhibited at the V Triennale. It was also a decidedly expensive house, meant for a limited, clearly defined class. Within the precinct of this house, the bedrooms, the large living room, the studio, and the three courtyards formed a coherent sequence, and the furniture, including chairs by Breuer (fig. 13), was carefully coordinated with the architectural framework; Melotti's large sculpture of a horse and rider was discreetly placed, to preserve the balanced decor.

Harmony was also achieved by renouncing any relationship with the town and regarding the villa as a self-contained architectural island, to be enclosed and protected from the vulgarity of its surroundings, which it had no hope of ameliorating in any way; it was the epitome of the 'shelter' against which Mondrian had inveighed. (Figini and Pollini developed the same idea in the house which the latter built for his own use in the journalists' village in Milan, in 1934; it was a villa detached from the ground by 'pilotis,' and surrounded, together with its accompanying open areas, by a parallelepipedal enclosure. The artists stated that they wished to erect 'an anticity within the city.') The integrity of the architectural setting was further compromised by the desire to impart a preconceived character — Mediterranean, classical, sunny — that reintroduced traditional artistic qualities as consolation and compensation for an irremediable deprivation.

19. Persico and Nizzoli. Parker shop, Milan. 1934
20 Persico and Nizzoli. Hall of Gold Medals. 1934

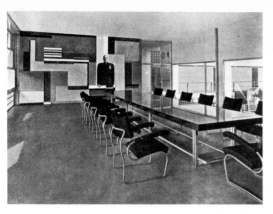

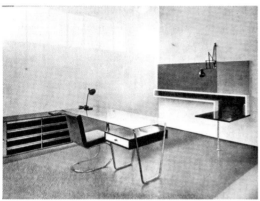

21. Terragni. House of Fascism, Como. 1936
22. Interior in housing exhibition, VI Triennale, 1936
23. Albini. Newsstand, Samples Fair, Milan, 1939

The steel high-rise by Pagano and his associates (figs. 14-15) was a theoretical alternative to the villa by Figini and Pollini: a seven-story building with large luxury apartments, having no open areas, but with large windows overloking the town. The apartments were given a degree of privacy by the simple expedient of providing a stairway and an elevator for every two units. This building was influenced by the steel high-rise that Gropius had worked on from 1929 to 1930 (and which was realized for the first time by Brinkmann, van der Vlugt, and van Tijen in Rotterdam, in 1934). It nevertheless did not purport to be a potential choice for public housing but was only a technical rationalization of what was being produced by private enterprise — which, however, had not the slightest desire to be rationalized and went right on using with great success the traditional types of building that composed the chaotic townscape of the kind Figini and Pollini sought to exclude from the precinct of their villa. The theme of housing as a social phenomenon thus remained vague, and the same indeterminate quality was apparent in the interior design, which was far more generalized and commonplace than that in its precedessor.

The housing projects by Griffini and Bottoni (figs. 16-17) had their counterparts in two well-known models already in extensive use throughout Europe: communal houses served by stairways, and those served by exterior galleries. The two architects sought to adapt these international models in order to overcome the backwardness of Italian research with respect to that in Europe. In 1931, Griffini published the first systematic book on inexpensive houses (*La costruzione razionale della casa*, published by Hoepli in Milan). The building at the V Triennale was underwritten by the Istituto Case Popolari (Institute for Public Housing) of Milan. Bottoni took charge of its interior decoration, which was simple but in excellent taste, keeping the size of the furniture carefully suited to the dimensions of the rooms. As a result, these rather bare interiors were just as harmonious as the luxurious ones in Figini and Pollini's villa-studio. This demonstration might have been successful, and could have modified the complex townscape, as it had in Frankfurt and Berlin, if it had led to a vast public-housing program. Instead, it came at a time of a crisis in construction, during which the Institutes for Public Housing reduced their programs; in fact, their sole reason for sponsoring modern architects had been that they were the only ones capable of showing them in a good light at the Triennale, but they had done so with no guarantee of further commissions, because in actuality they preferred traditionalists who raised no complicated problems. For example, neither the Institutes for Public Housing nor the furniture manufacturers had any intention of trying to relate the interior decorations to the houses themselves (and it is sufficient to recall that the only important modern housing development built in Italy between the two World Wars — and a very modest one in comparison with those in other countries — was that in the Via Filzi, Milan, completed by Albini, Camus, and Palanti in 1939).

The architects' intention still remained that of 'identifying exhibitions with real life,' as Francesco Monotti wrote in *Quadrante* (June 1933); and he went on to say:

Anything can be, or rather *must* be, the subject of an exhibition, and nowadays we see that even houses are shown in considerable numbers at the Triennale. Our salvation lies precisely in widening the concept of an exhibition to the utmost extent possible. Little by little, the whole of Italy must become an exhibition, that is, a model for all other countries, in every form of life, and thus in creation and action. The trend which has just begun promises to become even more far-reaching. Already, in many cases, the exhibition has gone beyond the walls of a building or pavilion, has come out into the garden, has

24. 'Useful Presents for Christmas': radio; Exacta camera; Olivetti typewriter, sewing machine, refrigerator (from *Domus*, December 1936)
25. Figini and Pollini. Prize-winning radio-phonograph. 1933
26. Albini. Installation of exhibition of antique goldsmith's work, VI Triennale, 1936

abolished doors, and made use of the airy transparency of glass. Now it must enter into the actualities of everyday life, and it is up to men of goodwill to make its path easier.'

A realistic evaluation of the V Triennale, however, shows that the plan of building an integrated new environment for human life was still not one that could be proposed in its entirety: it could only be confined to a privileged part of the environment, reaffirmed as a theoretical proposition, or postponed to a distant future.

The position of modern architects was weakened by their attempt to invoke the traditional sanctions of 'art,' for by doing so they placed themselves in the marginal orbit traditionally assigned artists. The approval and endorsement sought by traditional artists and critics seemed, however, either equivocal or even harmful. Professional artists were the custodians of the very qualities that the everyday environment rejected, and which it was their task to foster separately as a 'substitute' for a beauty excluded from life. Their works at the Triennale in 1933 — the paintings and sculptures, architectural embellishments such as the archways by Marino Sironi at the main entrance, or Felice Casorati's 'free-standing pilasters' at the entrance to the architectural exhibition (fig. 9) were still in complete conformity with the academic structural framework.

It nevertheless could be considered quite an achievement for the best architects to have succeeded in standing out against such a background, and to have presented portions of their program, which was enough to frighten the traditionalists even if it did not really amount to an organized demonstration. Modern research did not succeed in becoming a movement: the battle of movements was over, and the battle for position was about to start. In fact, 1933 marked the beginning of a slow retreat, the disappearance of opportunities and the stiffening of opposition.

This retreat could be divided into two phases. From 1933 to 1936, there was a period of confrontation, during which modern design struggled valiantly to oppose tradition, and to supersede it by showing what it could do on public or official occasions. From 1936 to 1940, there was a phase of accommodation between the two tendencies, during which modern designers, though definitely rejected by the establishment, still worked uninterruptedly in a marginal field and so were able to produce even more convincing results. The VI Triennale of 1936 and the VII Triennale of 1940 epitomized these two phases and may here serve as reference points.

Persico, who with Pagano became the joint editor of *Casabella* in 1934 but died suddenly two years later, tried to take the road of complete opposition. From 1933 to 1935, together with Marcello Nizzoli, he produced some memorable interiors in Milan: the two Parker stores (fig. 19), the Hall of Gold Medals at the 1934 Air Force Exposition (fig. 20), an advertising structure of metal tubing in the Galleria. They were the exact opposite of what was being done in Italy at the time; they opposed eclecticism with a rigorous choice of forms, confusion with order, facile solutions with restraint; and they seemed to put into practice a program outlined eight years before by Group 7: 'The number of elements used should be kept to the absolute minimum, and they should be refined and brought to maximum perfection and the abstract purity of rhythm.'

In the same spirit, Persico designed the covers and typography of *Casabella*, reacting to the superficial modernism pervading Italian publishing by stressing the severity and sparsest essentials of letters and figures. Figini and Pollini followed the same line in industrial design. In 1933, they won a competition for the design of a radio-

27. Albini. Metal chair shown at VI Triennale, 1936
28. Albini. Wooden armchair for Villa Pestarini, Milan. 1938
29. Albini. Armchair shown at VII Triennale, 1940

phonograph, sponsored by the National Gramophone Company (fig. 25), and in 1936, a competition for a writing desk, sponsored by the VI Triennale. The lines of the prototypes were very simple and cleancut, to differentiate them from either period furniture or the banal streamlining of current production. But they remained exceptions, viewed with skepticism by manufacturers.

Terragni and Pagano, on the other hand, took the line of compromising with the authorities. The former designed the House of Fascism at Como (fig. 21), proposing a modernization of the rituals with which the ceremonies of the Fascist regime were enacted — rallies, meetings, and the cult of party martyrs. Architecture and decoration became distorted and quite commonplace to adapt themselves to this content; for example, the tubular metal furniture became full of useless curves, and the embellishment of the walls of various rooms was mannered and artificial.

In 1934, Terragni, Ligeri, Figini, Pollini, and the firm of BBPR (Gianluigi Banfi, Ludovico Belgiojoso, Enrico Peressutti, and Ernesto Rogers) entered the competition for the Palazzo del Littorio in the Via dell'Impero in Rome, to show that modern architecture could be monumental when occasion demanded. Pagano offered to cooperate with Piacentini and the academicians in great public projects — the University of Rome (1932-1935), the 1936 Triennale, the Universal Exhibition at Rome, planned (from 1937 onwards) to take place in 1942. As the editor of Casabella, he simultaneously backed both the cultural argument and the political compromise, showing himself to the end a programmatic optimist. For example, in 1936 he wrote: 'The plans for the VI Triennale in Milan and the 1937 Exposition in Paris demonstrate that the doctrines of avant-garde architecture have also become part of the heritage of great official manifestations.' As far as Paris is concerned, however, it is well known how repeatedly Le Corbusier failed in his attempts to influence the plans for the Universal Exposition. Pagano had somewhat better success with the VI Triennale in Milan. He had an official position on the organizing committee, he erected a pavilion attached to Muzio's Palace, and he managed to arrange for Persico, Palanti, and Nizzoli to design the decorations of the ceremonial hall. But this recognition won by the leading modern architects only made it more difficult to differentiate modern from academic architecture, since they were now inter.   gl in practically every section of the exhibition. The ceremonial hall, which was quite elegant, but symmetrical and just as useless as all places made for show, gave credence to the notion that modern architecture was just a new variant of the traditional styles of decoration.

Once again on this occasion, any new contributions by modern architects must be sought in a few marginal projects: a housing exhibition organized by Albini, Camus, Clausetti, Ignazio Gardella, Mazzoleni, Giulio Minoletti, Gabriele Mucchi, Palanti, and Mario Romano, which presented a series of typical rooms for residential hotels, middle-class houses, and low-income housing projects, mainly designed by these same architects (fig. 22). Giolli wrote in Casabella that the unadorned and severe interiors of these projects provided the true alternative to the usual overabundant exhibition of furniture. In the park were exhibited the first seriously studied town facilities, for example a permanent playground for children by Mazzoleni and Minoletti, and a beach hut by Cosenza.

After 1936, when the futility of a confrontation with the academicians had become an accepted fact, modern architects found the time and concentration required to analyze the elements of the new landscape and to carry out the studies and experiments that they had

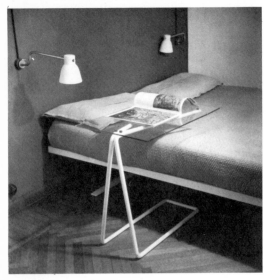

30-32. Albini. Interiors, architect's own house, Milan. 1940 (armchairs in upper figure are from model shown in fig. 29)

been obliged to neglect while engaged in the polemics of the preceding years. It was at this time that the best works of Italian architecture and industrial design were produced — the high-level exceptions that only served to demonstrate the low level of average production, without being able to remedy it.

Between 1935 and the beginning of the war, some exemplary individual buildings were completed (Gardella's dispensary at Alessandria; Bocconi University in Milan, by Pagano and Predaval, with its highly appropriate decoration; Nervi's hangars at Orbetello; some houses designed by Daneria in Genoa and by Ridolfi in Rome); as well as many praiseworthy interiors by Albini. To these can be added the INA pavilions at the Milan Fair and the Fiera del Levante in Bari, in 1935; the exhibition of antique goldsmiths' work at the 1936 Triennale (fig. 26), two exhibitions at the Brera Gallery in 1941, by Figini and Pollini, and the Olivetti booth at the Milan Fair of 1935, which marked the beginning of their cooperation with that firm; they were to be followed by Nizzoli's buildings for Olivetti at Ivrea, and the Montecatini pavilion for the Milan Fair in 1939.

From 1935 to 1940, the first Italian objects on an international level were designed by Albini: chairs (figs. 27, 28, 29), a deck chair, and a steel-and-crystal newsstand shown at the 1935 Fair, financed by the publishers of *Domus* and *Casabella* (fig. 23). Others designed types of school furniture; Clausetti and Romano designed some for the Paolini firm, which was shown at the Milan Fair in 1937, and Irenio Diotallevi and Francisco Marescotti did some for the VIS, shown in 1939. Throughout the 1930s, the furniture designed and produced by the Umanitaria School in Milan was of particular distinction for its restraint and practicality. *Casabella* published designs that were actually produced as well as those that remained as theoretical studies only, such as the minimal house by Minoletti (1937), the horizontal house by Pagano, Diotallevi, and Marescotti (1940), and low-income houses by the latter two.

It was during this period that the first objects of Italian industrial design, revitalized by the consistent application of modern design, made their appearance: Olivetti typewriters designed by Nizzoli after 1936; the Lancia Aprilia car, with its body work, of 1937; the Phonola plastic radio set of 1938 by Caccia-Dominioni and the Castiglioni brothers; the Cisitalia car of 1940, with bodywork by Pinin Farina. They were the first products of research that was to develop more fully in the postwar years.

The VII Triennale, in 1940, proved once and for all with unmistakable clarity the predominance of the academic, and the marginal character of modern architecture in Italy. The architectural exhibition, organized by Piacentini, showed large models of the E 42 zone in Rome and the new Via della Conciliazione. Its eight sections were installed by Piacentini himself, together with his pupil Lodovico Quaroni and other young designers. The firm of BBPR was in charge of the section devoted to landscape gardening in the city. Pagano directed the international exhibition of mass production, and Bottoni the modern housing show, once more in collaboration with the Milan Institute for Public Housing. Thus, two typical sections of modern design were shown separately, as was proper, but they were shown confusingly amid the multiplicity of other accessory exhibitions.

Italian design, which missed the opportunity to influence the everyday environment on a large scale, had already become an elitist movement, capable of producing separate pieces and small scale settings of high quality. Postwar events were only partially to alter this situation.

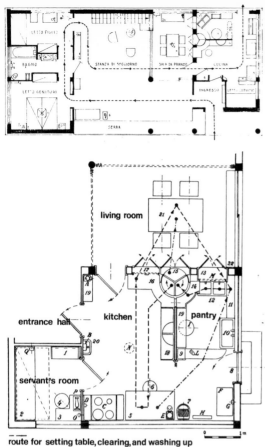

route for setting table, clearing, and washing up

Legend: 1. metal clotheshanger; 2. folding bed; 3. chest of drawers; 4. stool; 5. table; 6. stool; 7. bench with drawer; 8. service entrance; 9. broom closet with built-in ironing board; 10. washtub; 11. counter for soiled dishes; 12. double sink; 13. pass-through for soiled dishes; 14. draining board; 15. revolving shelves for dishes; 16. cupboard; 17. pass-through for food; 18. sideboard; 19. shelf; 20. water basin; 21 table; 22. cabinet

Electrical appliances: A. coffee grinder; B. sterilizer for water; C. range; D. wall lamp; E. motor for blender, mixer, juicer, etc.; F. refrigerator; G. wall lamp; H. heater; I. washing machine; L. electric iron; M. wall lamp; N. ceiling light; O. wall lamp; P. sewing machine; Q. wall lamp

33-34. Figini, Pollini, Frette, Libera, and Bottoni. Electric House, IV Triennale, Monza, 1930. Ground floor plan and diagram showing electrical apparatus

An 'All-Electric House'

(translated from an article published in *The Edison Review*, 1930)

Who doesn't know the annoyance caused by a badly organized house, due either to the lack of a good housekeeper or by the hard work a good housewife must put in to make the house comfortable at every time of day? Only machines can save everybody from this kind of slavery; and nearly all the work done nowadays by women at the cost of great sacrifice, patience, wear on their health, and a loss of freedom, will in the future be carried out by obedient, precise, and impersonal machines. Women would then be free to look fresh, serene, and smiling, and to enjoy their lives, instead of being tired out by some heavy job or irritated by a domestic crisis. They would be neat at any time of day and could devote more time to their children.

Is this a far-off dream? Those who have visited the 'Electric House' at Monza know that the dream could become reality. The setting for it, made of bricks and mortar, and finished off with plaster, metal, paints, and modern materials, is already there. In any type of Electric House — designed for every purse and need — one can find the comfort and modernity he desires for his home.

The villa, created through the generosity of the Edison Company by a group of Rationalist architects of Milan (Figini, Pollini, Frette, and Libera of Group 7) and Bottoni, contains all the most up-to-date electrical apparatus and household appliances (figs. 33-34).

The arrangement of the rooms, the classical lines, and the absence of any unnecessary decoration make the house itself the most modern that can be found, from an architectural point of view. Our dream comes true in the harmonious use of colors, brilliant paints, metal, and rubber and linoleum used as flooring or as covering of various surfaces.

The silhouettes of the trees seem to enter the room with the light and air that come in through the wide windows (figs. 35-36). Inside, a small entrance hall leads to a large living room on two levels, divided by curtains from the dining area. A wall made of double glass through which the surrounding landscape can be admired converts the room into a greenhouse. As most of the day is spent here, the adjacent bedrooms are small, with built-in wardrobes.

The parents' and children's bedrooms are lit by the most modern type of lamp. We can also find in them footwarmers and electrically heated carpets, fans that blow hot or cold, scented or unscented air, stoves, hair curlers and dryers, an electric iron, and a portable kettle. A small suitcase could supply the traveler with all the comfort and quick service that, at certain times and under certain circumstances, cannot be provided even in first-class hotels.

The bathroom is faced with a new type of rubber and is full of electrical appliances: the lights are built into the small cabinet above a new Rational model of washbasin; there is a stove, an immersion heater for instant hot water, and an electric hand dryer; the toilet has an automatic deodorizer. An electric hair curler and dryer have also been provided for the woman's use.

The maid's room is so small that a very efficient model of bed was designed to fold into the wall. Here, too, we can find a stove, and an electric sewing machine with a speed-control switch that permits all kinds of accurate stitching.

It is in the kitchen (fig. 37) and living room, however, that the maximum development has been achieved. The electrical appliances are so closely linked to the preparation and cooking of food, its passing through into the dining area, and to washing up, that a separate description is required to show how their position is as important

as the fact of their existence. Distances have been kept to a minimum to spare the woman working there any unnecessary motions.

The arrangement of the items and their general position in relation to the living room and dining area are such that the various operations can be carried out smoothly without going back and forth. It is sufficient to look at the movement of crockery: revolving shelves serve three rooms (kitchen, dining area, and pantry), thus solving the problem of taking out and putting away crockery and glassware in three different rooms simultaneously. On the small plan (fig. 34) are marked the possible routes of crockery and cutlery when laying the table, serving the food, and clearing away. The kitchen cabinet has a new type of celluloid rolling shutters and sliding doors. Inside it, in metal containers, is kept the uncooked food, which is prepared on the kitchen table. There is a small mixer, designed to power a large number of appliances: meat can be minced and then mixed with vegetables, spices, etc.; dough can be kneaded to the right consistency for cutting up into noodles; eggs, cream, and milk can be beaten and blended with other foods; ice can be crushed, vegetables liquidized, and in short, any food prepared with much greater precision than by hand.

There is an electric grinder for coffee, and in the living room a coffee pot and tea kettle, both electric, as well as kettles and immersion heaters, egg cookers, electric toasters, and food and plate warmers. All these marvelous appliances can be used to prepare and keep hot breakfast, snacks for the children, and tea for the grownups. Other meals are cooked on the electric stove in the kitchen, which has a range with four hot plates having three temperature settings each, and quite a large oven.

In the adjacent pantry, a small area is set aside for laundering and ironing. Soiled laundry is put into the washing machine with soap and after a time is removed to a small rinsing basin, before being put back into the machine for spinning and extraction of all water. Ironing is done on the door of a small broom closet; when pulled down, it acts as a padded ironing board. A most essential part of woman's work in the home is therefore accomplished in very small space. All the cleaning of floors, carpets, and upholstery is also done with electric appliances.

Our visit to the Electric House is now completed. Our visitors, however, should not regard this house as being out of the ordinary; keeping in mind the financial side, the architects have designed similar types of houses in a lower price bracket. Most of the appliances could be used in them, and the dream house could thus become a reality.

35. Electric House. Exterior
36. View from Terrace
37. Kitchen

ITALIAN DESIGN, 1945-1971
Vittorio Gregotti

An historical treatise on the subject of 'design' demands particular critical caution and a very special method of approach. Either one accepts the idea of attributing everything to a single cause, reducing the role of design to that of a fundamental bearing structure — whether aesthetic, economic, technological, by social class, etc.; or one deals with the historical data by maintaining a precarious balance between the complexity (in the cultural, not the constructional, sense) of the materials involved in designing single objects, and the effort required to organize creatively the new nature made up of these objects that people the 'real' nature they overlay.

One must therefore trace the long journey, from the realization of a need to its satisfaction, that constitutes the basis both of culture and of the planning process of design. This takes one along a hazardous course that is, moreover, subject to a twofold interpretation. On the one hand, there is the theory that would attribute the principal cause of modification to that capacity for renewal inherent within the discipline of design itself. This view is held by those who regard the methodological changes in the designing process, the capacity of the morphologies produced (and often self-reproducing) to become a support for social imagination, and the innovation of spatial concepts and relationships as fundamental instruments for a new cognitive relation to reality and a new interpretation of social relationships themselves.

On the other hand, there are those who maintain that the basic causes of modifications in design should really be sought outside the discipline, in the phenomena of that transformation due to the evolution of an industrial civilization, the conditions of production and technology, and the changes in social and political relations. Thus, they seek the causes not in design's capacity to act by making proposals, but rather to react in response to the underlying problems and urgent social needs with which it attempts to cope.

These two theories are like opposite sides of the same coin. But, although they finally converge on the same plane, in dealing with concrete historical facts they suggest different interpretations of the relative importance of various phenomena, and in the last analysis they present different versions of the history of contemporary design. This double interpretation, however, definitely saves anyone who concerns himself directly with design from the temptation of treating its history purely as a sociological symptom; for such an approach would obviously be just as incomplete as if he were to believe that he should restrict himself to the false neutralism of methods, techniques, and systems.

Nowadays, when the characteristic subjective independence of creative thinking has become so undermined, in our opinion the important thing for an intellectual technician is neither to allow himself to be expelled from the realm of his own concrete action, nor to make that concrete action into a 'false coinciding' with the whole of reality, but rather to assert that this activity is first of all a dialectical treatment of the present, and the only sound basis for the vision of a collective future.

In a more technical sense, we must recognize that for the past ten years we have been in a new phase of design. Now that the limits that defined it have been done away with (beginning with the idea of mass production, the industrial nature of production, the scale of the object, and its relation to its locale), design regards itself as an institution with an essentially unitary method at every level. This applies not only to those activities typically associated with the idea of constructing an object, but also directly to every technological activity that purports to be a planning, formative, and to that extent creative, operation. This poses a new problem of becoming more specific, both

Vittorio Gregotti is a professor on the faculty of the architectural school of the University of Milan. His work as a practicing designer and architect is closely linked with his activity as critic and author. For many years, he was editor-in-chief of *Casabella* and has subsequently been responsible for ten special issues of the periodical *Edilizia moderna*, which he directed until 1967. His activity in organizing the program of the XIII Triennale at Milan in 1964 provoked a lively international controversy on architecture and its relation to society. Among his principal theoretical works are *Il Territorio dell'Architettura* (1964) and *New Directions in Italian Architecture* (1968).

315

1. Ignazio Gardella. Home for Three People.
RIMA exhibition, Milan, 1946
2. Paolo Chessa and Vittoriano Viganò. Home for
married couple. RIMA exhibition, Milan, 1946
3. Vico Magistretti. Demountable bookshelf and
chair. RIMA exhibition, Milan, 1946

as regards history and as regards one's work, and of attempting to put
in order and prune away some of that dense, continuous matter that
seems to have become destined to rule our entire physical environment.
If we are going to discuss design, we must know exactly what it is that
we are discussing.

Our account must furthermore take into consideration a serious
philological deficiency: the lack of a really exhaustive investigation of
the development of industrial production in Italy, carried out with a
point of view as clear, from the cultural standpoint, however different it
might be, as that of Siegfried Giedion when he wrote *Mechanization
Takes Command* some thirty years ago. Even though the organization
of Italian industry has to a great extent been imitative of preexisting
foreign models, it has nevertheless produced a series of original
designs, whose development and motivation are to a great extent
unknown. The story of our design is therefore one in which one of the
principal actors is present only as a ghost.

In the course of this article, we shall generally be speaking of design in
very traditional terms. On the one hand, we shall discuss the various
meanings that have been associated with the concept of design since
1945, and on the other, we shall chiefly be concerned with objects and
the relations among objects produced with industrial methods and
purposes, and designed with the creative intention of fitting them into
the 'tradition of the new' in contemporary visual arts and
communication. Finally, we shall try to bring out the connections and
relations among these objects, with respect to their use and the manner
in which they are consumed both culturally and technically; and their
reciprocal physical relationships, which, with respect to the
environment, constitute the basis and physical confines of their
intersubjective behavior.

The number of episodes to which we can refer when speaking of
Italian design between 1945 and 1950-52 is very limited, but sufficiently
precise to enable us to reconstruct a clear picture of its aims, its
progress, and its failures. We must begin by saying that up to 1946, the
tradition of Italian design had been completely formed and developed
along the lines of a culture closely linked to architecture. This, of
course, was Rationalist architecture, which after 1945 became firmly
associated on an ideological plane with the politically anti-Fascist
movements of national liberation, and with a strong tradition, centered
especially in Milan. It was still internationalist and elitist, and was at the
time concerned with the problems of standard elements and
prefabrication. At the same time, remnants of '1900' Art Nouveau
still harked back to the tradition of craftsmanship, defense of the
ideologically deluxe object, design as something intrinsically Italian,
and the unique quality of artistic expression.

And in fact, between 1946 and 1948, Rationalism seemed to be the final
victor, associated with that impetus toward the democratic rebirth of
the country, and capable of providing models and programs for the
great concept of a national reconstruction that would also entail the
reconstruction of conscience and the establishment of social justice.
Writing in the first issue of *Domus* after he had assumed its editorship
and given it the subtitle 'The House of Man,' Ernesto N. Rogers
declared: 'It is a question of forming a taste, a technique, and a morale,
all directed toward the same purpose — the building of a society.'
There were at the time other publishing efforts, such as Pagani and
Bonfanti's periodical *A*, and even (if only for three issues) the
revised *Casabella*, but essentially, during the two years that it was
under Rogers's direction, *Domus* remained the chief point of reference
for Italian culture in design. It became a means for establishing
international contacts, and as such it played a major role in ridding

4. Vespa 50 scooter. Piaggio. 1946 (first production; this model 1963)
5. Lambretta Innocenti scooter. 1947
6. Isetta car. 1953
7. Pinin Farina. Cisitalia car. 1947

Italian culture of the provincialism which for more than twenty years had closed it off from direct international influences. *Domus* was simultaneously the touchstone for the Milanese Rationalist tendency, which since 1945 had organized itself into an association called MSA — 'Movement for the Study of Architecture.'

Significantly enough, the first issue of *Domus* was devoted to five 'first aid' projects for furnishing war-damaged houses (submitted by the architects Gandolfi, Latis, Tevarotto, Mongiardino, De Carli, Chessa, and Zanuso). There was also a note on prefabrication (also dealt with in one of the numbers of the new *Casabella*), always considered from the standpoint of modernization of methods; as well as a discussion of objects, and notes on art. During these two years, the magazine was chiefly an organ reflecting several of Rogers's maxims, such as 'From a spoon to a city' (implying unity in methodology, irrespective of scale) and 'Design as the result of utility plus beauty.' In this instance, beauty was judged in terms of *art concret* and the relation between mathematics and the plastic arts, in accordance with the theories being propounded at the time by Max Bill — one of the figures frequently referred to in the magazine, largely because he represented the Bauhaus idea of unity between the craftsman and the intellectual, and among the painter, the sculptor, the designer, and the architect. By 'utility' was meant above all a right relation between materials and technology, between means and the use of those means. On the other hand, there was a very discernible preoccupation (evident in articles by Giorgio Crespi), expressed in a highly literary manner, with maintaining a connection between objects and meanings, within the context of a 'synthesis of the arts' rather than of an anthropological construct of cultural patterns.

Many of those who preached the value of the module as the foundation for design were more concerned with trying to preserve the ethical and social values originally inherent in it than with its use as a practical tool for a real stepping up of quantity; the subject was never discussed from the standpoint of the economics of production, but only in its pedagogical and moral aspects, with reference to both consumer and producer.

One of the most telling events was the exhibition of the RIMA (Italian Association for Exhibitions of Furnishings), which in 1946 organized a show, in the newly redone Triennale building, on the subject of popular furnishings. Nearly all the architects associated with the modern movement participated, and the ideas put forward, especially by members of the younger generation, were lively and enthusiastic. Above all, the themes of flexibility, of reducing the spatial requirements, and of doing away with rhetoric were emphasized. Ingeniously balanced between an almost total lack of any perspective about industry and the radical simplification of building procedures, and the desire not to lower the standard of design, the final results offered a kind of option — at times purely symbolic — for interior decoration designed, and at times even constructed, with the user's participation. Within the framework of an attitude expressing a kind of pride in modesty, the models proposed for emulation were derived from articles as humble and as lacking in history as the deck chair, the folding table, or a simple utensil. Highly typical of this approach were the dwelling for three persons designed by Ignazio Gardella (fig. 1), the home for a married couple by Paolo Chessa and Vittoriano Viganò (fig. 2), and the small, almost 'self-made' pieces of furniture by Vico Magistretti (fig. 3).

An introductory section organized by Carlo Pagani was meant primarily as a demonstration of the practical advantage of modern interior decoration: its ability to free women from domestic servitude, its

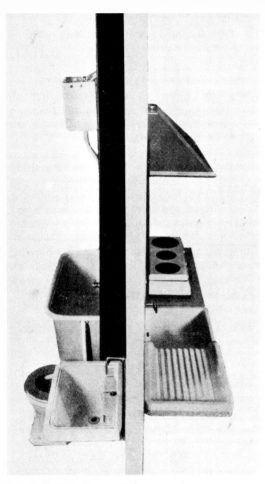

8. Togni bath-and-kitchen blocks. 1946
9. Renzo Zavanella. Pullman bus. Belvedere O.M. 1948

advantages in economy and flexibility, in short, in 'facing the problem of providing the masses with the furnishings they require, at prices they can afford, and shedding new light on plain, simple, and natural furniture in accordance with the needs of the man of today and his way of life, and not with the useless and dangerous tradition of decoration.' Thus, the effort was to provide a cultural, as well as disciplinary, alternative to the stylistic imitation that generally prevailed throughout the market as a result of the new civil situation. Needless to say, this naive hope, with its basically pedagogical outlook, supported, it must be remembered, by an authentic collective attempt at democratic collaboration among the anti-Fascist parties, was soon dashed, not only by the essentially conservative habits of the lower middle class in Italy and the proletariat itself, who attributed quite different symbolic values to the possession of furniture, but also by contact with the real forces, which very quickly expressed themselves in terms of power — in part because of the inability of the forces of design to respond with large-scale planning that would have enabled them to meet social needs quickly and effectively.

This is demonstrated by the story of the VIII Triennale (1947), the other important exhibition held during those years. The Commisariat for it set up by the National Liberation Front for Northern Italy made it possible to create an organization entirely different from previous ones; study committees were appointed, embracing a large number of members, who worked out in close collaboration a unified plan for the VIII Triennale. These committees had the virtue of trying to reestablish ongoing links between designers and society, deciding that the whole subject of the home was the most urgent and the most neglected one in Italy's postwar situation. The working parties transformed themselves into organizing committees, thus assuring continuity between the phases of analyzing the exhibition and actually bringing it into being. Within the prevailing climate in which the VIII Triennale originated, the unifying theme of its program made possible a convergence of forces that, despite their heterogeneity, nevertheless were able to operate within the framework of the whole project. Here again, the industrialization of building and furnishings for the masses were the central themes of the exhibition, except for a section devoted to individual items of furniture.

Highly important was the introductory section, organized by Diotallevi and Marescotti, which attempted for the first time to provide an unbiased estimate of the disastrous state of housing in Italy. But 1948, the year after the VIII Triennale, saw changed political conditions in Italy that dispelled a series of illusions. First and foremost among these was the hope of its being possible for culture to come to terms with the organizational structure of the country, so that it could operate at the very heart of national realities. The fact that the realistic aspirations of Italian architecture and design could be achieved only after 1950 (for which the economic and operational backwardness of practices within the discipline of architecture itself were partly to blame), or rather after the political collapse of leftist forces in Italy in 1948, and the progressively bureaucratic stance taken by the Italian Communist Party from that date on, completely altered the direction of designers' efforts. They wound up devoting their efforts not so much to the realities of national life as to an often sentimental vision evoked by the elements that composed it. Economic forces, the proletariat, methods of production, political relations, the designer's role in society, were all transformed into an imaginary projection of themselves. Realism, in fact, came to an end in Italy in 1950, together with the journal *Politecnico* and Elio Vittorini's efforts to give the Left a content compatible with modern culture; it ended with the compliance with Soviet Zdanovism, the rise to power of the

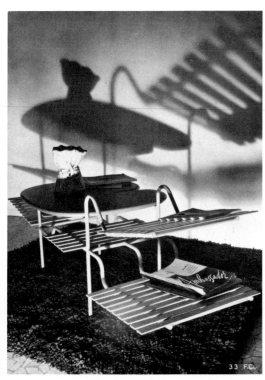

10. Marcello Nizzoli. Lettera 22 typewriter.
Olivetti. 1950
11. Marcello Nizzoli. Lexikon 80 typewriter.
Olivetti. 1948
12. Ettore Sottsass, Jr. Table in iron tubing and
wood. Fede Cheti furniture exhibition, Milan, 1949

Christian Democrats, and the formation by all parties of subgovernmental structures that were destined to last a long while.

Meanwhile, during the same period 1945-48, Italian industry itself had gone through the difficult process of reconstruction and modernization, on the one hand, and of reasserting its power, on the other. In the absence of any national program for converting wartime into peacetime production (such as that in England, for example), and any planning that would have established priorities among overall needs, the end result was that the old economic imbalance between North and South Italy was reestablished and even aggravated. But where industry was able to get underway again, it quickly achieved a high level of technological maturity, owing to internal reasons largely attributable to a natural bent and talent for improvisation — national characteristics that are at once virtues and vices. In part, it was also often owing to the intelligence of the old-style artisan type of worker, with his roots in a technical tradition that, while it may not have been at all up to date, was nevertheless many-sided, enabling him to work out systems that otherwise could have been achieved only on the basis of planned rationalization. In this way, some surprising successes were achieved in the field of design, which were decidedly advanced with respect to the state of the art among architect-designers.

I have in mind the invention (or at any rate the wide-scale application) of the Vespa 98-cc. scooter produced by Piaggio (fig. 4), and the Lambretta scooter (fig. 5). Based on two contrasting structural principles — the former having a load-bearing bodywork, the latter an open tubular framework — these scooters were a true typological innovation, which for many years characterized the Italian urban scene. Geared to covering short distances through narrow, winding streets, they were perfectly adapted to the country's topography and the features of the environment, as well as meeting the demand for a cheap form of motorized transport. Along the same lines was the Isetta car (fig. 6), which took up very little space, had an interesting front-opening system, and gave a notably economical performance. Also in the field of car design, Pinin Farina in the same period put into practice his principles of compactness and continuity in the plastic conception of automobile design, carrying forward ideas that had existed in embryo in some solutions of 1940 (fig. 7).

Likewise, before the 1950s, other ideas began to spread, such as studies for groups of bath-and-kitchen equipment (the Togni blocks; fig. 8); the first experiments in mass-produced kitchen units, designed by Magnani and produced by Saffa; the first experimental bath blocks, designed by Giulio Minoletti, and the systems of bookshelves with metal scaffolding (Feal and Lips Vago), even if these had already been tried out in other countries.

A quite extraordinary achievement was the new panoramic type of bus, for which the architect Renzo Zavanella served as consultant in 1948 (fig. 9); the result was a coach that in plastic terms was somewhat overabundant but surely displayed a concern (even if a purely stylistic one) with the idea of continuity among the various parts that was very rare in the production of public transport vehicles in Italy at the time.

The number of available objects of good design nevertheless remained very small, while any contact between industry and design was infrequent and wholly marginal. An exception to this was the Olivetti concern, which had an extraordinarily farsighted policy in the field of design that it had already been following in its program and products for over ten years. After 1945, it set an exemplary precedent in the collaboration of Adriano Olivetti with the graphic designer Pintori and the designer Marcello Nizzoli. To explain the extraordinary

13. Vittoriano Viganò. Armchair in bent plywood. 1947

14. M. Cristiano and L. Frattini. Chair in sheet metal with rubber joints. Crespi-Saga-Pirelli. 1964

320

sophistication in form and technical operation of such products as the Lettera 22 of 1950 (fig. 10) and particularly the Lexikon 80 of 1946-48 (fig. 11), we must go back to the models of the adding machines, which from 1940 on had already attained a high state of development in respect to the coherence in all plastic elements, the perfection of movable parts, and the arrangement of levers. In the Lexikon 80, the relation between the mechanical parts and their housing was solved, through the application of a die-casting technique, on a level of technology and design that was completely revolutionary in terms of current practices. This was also the result, on the one hand, of the inclusion of the designer in the process of production as a coworker and not as a mere embellishment, and his direct collaboration with those engaged in the process of fabrication; and on the other hand of Nizzoli's long contact during the 1930s with the most notable theoretician (and most sensitive artist) of all the Italian Rationalists — Edoardo Persico. Including the series of calculating machines that he produced during this period, Nizzoli certainly reached at this time the highest point in his career as designer. He was able to fulfil his own capabilities to the maximum, his point of departure being that direct, almost manual, training, which as we have said is one of the talents of the Italian worker, and which Nizzoli was able to use with great intelligence as a source of strength in his design.

Conversely, the architects, who were the ones principally responsible for most of the designing for homes (which, as we have noted, remained the key area in Italian production), developed along very different lines and approaches. Most frequently, the methodology of architecture was simply transferred to the making of objects. An opportunity for the younger generation to compare and become aware of different experiments and approaches was provided in 1949, when the firm Fede Cheti decided to sponsor the design of several models of high quality (fig. 12).

Thus, the architect began to strive to follow more technologically and productively advanced procedures. If he still felt tied to manual, craftsmanlike execution, it was with the embarrassment of one who has no other means at his disposal. The youngest members of the profession, stimulated by foreign examples (Charles Eames was published in Domus in 1947), tried to imitate foreign patterns based on very different kinds of technology: Castiglioni and Viganò (fig. 13) by using bent plywood in a manner intelligently applied to structure, Cristiano and Frattini (fig. 14) with seats made of sheet metal with rubber joints, Chessa and Zanuso with small armchairs in metal and other materials, which have a remarkable figurative aggressiveness; Magistretti with his stress on elegant typological inventions; Menghi (fig. 15), Canella, and Radici, principally concerned with the meaning of the contrast between the old and the new, and the use of perforated sheet metal, metal joints, and systems of balance. The experiments of the Castiglioni brothers were of a somewhat different order. On the basis of a long tradition of attention to problems of design (it was they who had developed the little plastic Phonola set in 1939), they proceeded to develop during this period designs for a number of pieces: a radio set (fig. 16), a table lamp in tube metal with fluorescent light, and various handles and lamps, all designed with surprisingly strict industrial methods.

By contrast, the architects of the Rationalist generation stubbornly persisted in taking the functionalist approach, which, despite many variations, is always at the root of their methodological procedures. Carlo Mollino and the group of his pupils in Turin employed a formal vocabulary based on complex curves, quite different from the purist Milanese tradition (fig. 17). Gardella and Caccia-Dominioni began to

15. Roberto Menghi. Libra Lux lamp. 1948
16. P. G. Castiglioni and L. Cacciadominioni. Radio set, 1945-46

express their true vocation of portraying middle-class dignity; Franco Albini developed the technological use of wood to the utmost limits of its expressive possibilities, working on the thin section, tension, and the reduction to black and white (see, for example, his splendid decor for the Zanini fur shop; fig. 18), with a use of space that exotically conveys the subtle poetry of his environments.

Gio Ponti, impelled by activism, but also by an intuitive understanding that led him to concern himself as early as 1948 with designing the Pavoni coffee machine (fig. 19) and the Visetta sewing machine, was the chief mediator in dealing with the innumerable artisans who worked on individual pieces and the use of manual skills in articles made for export. Ponti, whose style in design links him with the draftsmanship of Campigli and Fiumi, represents both figuratively and in a sociological sense the ideal bridge with the prewar middle-class culture that found its expression in the art of 1900; he stands out alone for his unexpected explorations of modern figurative art or, rather, modern 'style.' His most significant piece is certainly the very lightweight chair inspired by the traditional Chiaravari craftsmanship (fig. 20), which he designed in various versions for Cassina up until the definitive one of 1957.

The years from 1951 to 1954 were the critical ones for the growth and organization of Italian design. In 1951, at the IX Triennale (which opened in the midst of profound differences of opinion among members of its organizing committee), Peressutti and Belgiojoso of the BBPR firm arranged a section with the title 'The Form of the Useful' (fig. 21); with this presentation, design — in the methodological and international sense of the term — was officially inaugurated as a distinctive part of the most important manifestation of applied art in Italy. But the idea of design was beginning to insinuate itself everywhere in different parts of the IX Triennale. One section of experimental furniture included a group of objects designed by Albini, which consistently followed that principle of a reduction to essentials which is the earmark of his design. It included a number of demountable cupboards (the result of researches going back to 1948), in which Albini elaborated a technique of cutting and bending plywood that reduced the walls to very thin panels; the Adriana armchair, based on the contrast between its component parts and the elasticity of the plywood projecting over a light framework of heavy wood; and lastly, the rush armchairs — true modern reincarnations of the old artisan method of working this highly traditional material (fig. 22).

In all this, there seems to be a strained effort to graft the lacerated members of ancient craftsmanship and its skills onto new principles, figurative and constructivist rather than based on production and economics. The difference is largely this. For the younger generation, represented outstandingly and with great creative power by Marco Zanuso, the processes of production provided a source of formal inspiration rather than a difficulty to be overcome; they offered an incentive and an instrument for putting to the test the Rationalist world, which had always seemed confined to rigorous moralizing and closed off from the changes taking place in society. For Albini, what matters is precisely the strict morale of method and the exploitation of technological processes, to the extent that they give proof of rationality and the consistency between form and function as modalities for creating an object.

At this IX Triennale of 1951, Zanuso, too, showed a series of pieces of furniture destined to have a great future. Making use of foam rubber and nylon cord, and drawing inspiration from methods of automobile design (even to mimicking their performance systems), he gave a first impetus to increased contacts between the designer and industrial

17. Carlo Mollino. Rolltop desk and chair in bent plywood. 1946
18. Franco Albini. Zanini fur shop, Milan. 1945
19. Gio Ponti. Pavoni espresso machine. 1948-49

processes, facing all the risks of being subject thereby to capitalistic control, but nevertheless recognizing that without this updating, the designer would have lost the battle on every front (fig. 23). This implies a belief that there is a significant link between the ideas of progress and reason, and faith in a gradual transformation of society in a positive sense; it implies a renewal of the ancient trust in the capacity of processes of production to become rationalized, and confidence in their capacity to become socially useful as well — a confidence badly shaken by many disappointments in the recent past.

Beginning with 1951, after the IX Triennale, a chain reaction set in within Italian design. Albini remained associated with his former manufacturer Poggi; Gardella, Caccia, and Magistretti founded Azucena in 1949, a group that produced furniture and objects of high quality both in design and execution, and connected, both by their sociological outlook and their methods of production, with a clearly defined cultural milieu in a way that made their activity during the 1950s a kind of Italian Wiener Werkstatte (fig. 24).

Meanwhile, Zanuso's experiments gave rise to Arflex (associated with the large Pirelli company), with the specific purpose of producing on a small scale but employing industrial methods (fig. 25). Other developments of this time included the founding of Kartell, which for many years was to exercise leadership in the field of design in plastic materials (fig. 26); Solari, which collaborated with the BBPR design group in the field of electric clocks (fig. 27); and the activities of Pellizzari, an exceptional intellectual industrialist keenly aware of the importance of design.

These same years saw the founding of the magazine *Civiltà delle macchine.* Sinisgalli was to remain one of the important indirect links working within design and industry — first with Olivetti, later with Pirelli's department of public relations, and finally at IRI (Institute for the Reconstruction of Industry). For his cultural function (in his capacity of writer and poet, also), he chose the fields of science and technology as the privileged vectors of contemporary values, and by illustrating their communicative and expressive aspects, he made himself an intermediary between production and design, as part of a more general strategy for unifying scientific and humanistic culture. This was very important for Italy in the 1950s, at a time when the country was stalemated by a closed academic culture and a backward scientific development.

It should not be forgotten that during this decade there began an unexpected acceleration in Italy's economic and industrial development, and although imbalances in the revenues of different regions persisted, this growth opened the way for a series of long-awaited aspirations regarding services and consumer goods. Peace and national liberation, the end of destroyed cities and prostitutes, brought also, with the American invasion of Italy, not only chewing gum, powdered milk, and Coca-Cola, but also first and foremost the idea of 'comfort' and the mechanization of the home. The myth of the refrigerator was born, together with that of electrical household appliances; and above all, the competitive wish to attain prosperity and security began to seem not only possible but even attainable. Simultaneously, there was an ambition to gain social status, concretely visualized in terms of possessions that would in the future permit the maximum luxury for an underdeveloped country; and a rejection and opposition to the traditional principle of preserving and handing down from one generation to another useful objects, household effects, and utensils — a tradition that had been as much the heritage of the poor peasant as of the middle class with its tendency to accumulate and save.

20. Gio Ponti. Chiaravari chair in ashwood with rush seat. Cassina. 1949
21. Peressutti and Belgiojoso (BBPR). 'The Form of the Useful' exhibition, IX Triennale, Milan, 1951
22. Franco Albini. Margherita rush armchair, 1951

The progress of design and its diffusion served to contribute to, and carry further, this drive toward transformation in Italy, even if the achievements in terms of social mobility and mobility of the labor force were also the results of the revolution in communications caused by the spread of audiovisual media. It was therefore not only an outcome of the increase in incomes, but also of a détente in the political struggle, or rather a transformation of that struggle into one for wages and services, followed by a general crisis in values and customs. It was this transition from a popular culture to a mass culture that signaled the defeat of the hopes of design to participate in the realities of national life; for the dominance of the Italian Left in the early 1950s meant that the very instrumentalities that design could offer tended to be regarded with constant suspicion. That this suspicion was to a great extent justified was also owing to the inability of Italian left-wing culture to develop an adequate theoretical and practical basis that could have made design itself a serviceable tool in the progressive transformation of society. If the ideology of design finally became an integral part of the ideology of capitalist power, this was neither fatal nor inherent in the nature and contradictions of the discipline; if anything, it was the outcome of a collective incapacity to make design a weapon in the class struggle rather than rejecting it outright on a priori grounds.

Between 1952 and 1954, three events marked the level of maturity reached by Italian design: the X Triennale and the conference organized in conjunction with it, the founding of the magazine *Stile industria*, and the establishment of the Golden Compass award. This was followed, two years later, by the founding in Milan of the ADI (Association for Industrial Design), which still remains the only organization uniting all those who are connected with design in any way whatsoever.

Thanks in large part to the efforts of Zanuso, the X Triennale had a more coherent scheme than the preceding one and was intended to summarize its own approach by focusing the exhibition on the theme 'The Production of Art.' This was a considerable departure from the general tradition of the Triennale, which had been to emphasize architecture and trade, and the subject was so compelling as to influence the two subsequent expositions also. The organizers were clearly intent on setting underway a vast operation to encompass the developing strength of large- and medium-sized Italian industry and suggest to architects the possibility of their effective inclusion within the system of production, consequently — through the example of foreign countries, also — suggesting to industry that it might make use of design. Apart from giving a marginal role to the commercial and craftsman sections and setting aside a sizable space for individual articles of furniture, the heart of the exhibition — from a visual standpoint as well — was the industrial design pavilion, organized by the Castiglioni brothers, Menghi, Rosselli, and Nizzoli, as well as by Michele Provinciale and Augusto Morello, who were subsequently to play an important role in the establishment of the Golden Compass award.

'The subject of this key section of the X Triennale,' the catalogue declared, 'is industrial aesthetics, which in accordance with an American term that has now come into general use may be designated as "industrial design" and interpreted as the interpolation of the concept of "form" into the industrial process, or into the formal aspect of technology, the essential meeting-point between art and industry.' As is evident, the ideas were not very clear from a theoretical standpoint; but the conference held in connection with this X Triennale contributed decisively to a clarification of the issues. Participants in the discussions

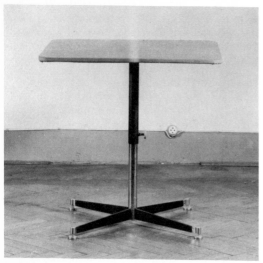

23. Marco Zanuso. Bridge chair. Arflex. 1951
24. Ignazio Gardella. Table with adjustable height. Azucena. 1949
25. Marco Zanuso. Lady armchair. Arflex. 1951

were Paul Reilly, Tomás Maldonado, Max Bill, the sociologist Pizzorno, Giulio Claudio Argan, Ernesto N. Rogers, the philosopher Enzo Paci, Konrad Wachsmann, Gino Martinoli, Walter D. Teague, and others. On this occasion, Argan put forth his celebrated theory on 'city planning as an extended form of design and the place for social redemption,' strongly upholding the aesthetic quality of design itself. This gave rise to a discussion, characterized by noteworthy and less important speeches, that was to have a significant effect on the updating and coming of age of Italian theory of design. The conflicts brought out between capitalism and design, and between art and design, in questions of method, history, and education, were somewhat restricted to the field of the planning and technology of production, without touching to any extent on interdisciplinary influences; it thus provides a clear example of the level of discussion before people became aware of the concept of a consumer society.

The second event was the founding of the magazine *Stile industria*, edited by Alberto Rosselli. Up until it ceased publication in February 1963, it remained the most constant and authoritative organ of Italian design, constituting together with the ADI a focal point for discussion, although such other periodicals as *Casabella*, *Civiltà delle macchine*, and *Domus* displayed a healthy interest in arguments on the subject of design. The shortcomings of *Stile industria* (apart from its unfortunate but significant title) were perhaps its neutrality in respect to ideological questions, its reluctance to take sides in the fundamental issues and encourage basic investigations on the conflict between needs and production — the dark spot in Italian design.

In 1958, the magazine *Il Mobile italiano* began publication; it focused its efforts in effecting a transformation, through furniture display rooms, large distribution outlets, and local cooperatives, in the great artisan hinterland north of Milan, which at that time was turning out almost a third of the national output. These efforts, however, finally wound up principally as an attempt to bring styles up to date, owing to the complete lack of a national plan that could clearly carry out a program of transforming the methods of furniture production and distribution in Italy. It must be remembered that just at this time, the national market was at the 'modern furniture' stage, completely swamped by Swedish products, which, in Italy as in the rest of Europe, had won a large share in the economy, thanks to the restraint and quality of their design, excellent systems of distribution, and a notable maintenance of standards of quality in their execution.

In 1959, what remains the best economic study on the problem of furniture production and consumption was published. Its author was Silvio Leonardi, an engineer who specialized in economic problems. In this work (*Produzione e consumo dei mobili per abitazione in Italia*, Milan: Feltrinelli, 1959), he clearly showed, by means of a painstaking nationwide survey, the general state and artisan character of furniture production, as well as its considerable productive potential; he systematically explored the sources of models used and showed the actual gap between the quantitative and qualitative nature of the demand, and the inadequacy of any organized response on the part of the designer.

The third event of this period, as we have said, was the establishment of the Golden Compass award by the large Rinascente department-store chain, primarily owing to the initiative of Cesare Brustio and Augusto Morello. The Golden Compass had been preceded by an exhibition held at the Rinascente under the title 'Aesthetics of the Product' (fig. 28). This exhibition met with such success and stimulated so much public interest that the Rinascente decided that the time was ripe for promoting the idea of design in Italy also. Thus there came into being

26. Gino Colombini. Plastic dustpans. Kartell. 1958
27. Peressutti. Electric clock. Solari. 1951
28. Rinascente exhibition, 'Aesthetics of the Product,' Milan, 1953

the Golden Compass award, which gave recognition to the producer and the designer of the prize-winning model, outstanding 'for its aesthetic qualities and the technical perfection of its production.' From 1959 to 1965, the Golden Compass competition was run in collaboration with the ADI, which in the latter year assumed complete responsibility both for the cultural aspects of the award and the details of its organization. For the past fifteen years, the Golden Compass, which has been awarded to more than a hundred products, has been considered as the most coveted recognition in the field of Italian industrial design, as well as having won undisputed international renown. Looking back at the awards today with the perspective of many years, one must frankly concede a remarkable consistency in the judgments of the various juries (leaving aside a few inexplicable exceptions, in which the objects chosen were not actually bad so much as merely marginal and insignificant). While on the one hand this might indicate a constancy in methodology and taste, at the same time, it also reveals a conspicuous indifference to the complex and even rather bloody history, from a cultural standpoint, of the ideas on architecture and design that succeeded one another in Italy from 1954 to the present.

The general reports issued by the juries also make significant reading, precisely because of the differences and similarities that they reveal between intention and actual practice in respect to the awards (often conditioned by the fact that material was researched and assembled in too hasty and inconsistent a manner). The reports allow us to interpret somewhat differently the significance of the award in relation to the real-life competition in terms of production, distribution, and quantity. Up until 1958, the stipulation that the prize objects had to be sold in department stores considerably limited the area of choice. Between 1958 and 1962, the juries' inclinations were rather to isolate the object from its context and dwell on its intrinsic qualities. This corresponded to the actual situation at the time with respect to the relation between production and consumption, which was still completely elitist at every level. From 1962 on, the juries' reports steadily denounce the constant bias toward the motivations of a consumer market, noting also the general transition from 'product-oriented' manufacturing to 'market-oriented' objects. They add to this a concern about the imbalance in the various branches of production affected by design, the lack of products worthy of receiving prizes in the area of utilities, and the link between the industrialized aspect of the country and the development of design, clearly indicative of the indifference of public authorities to problems of this nature. On the other hand, from 1962 on, the consumer objects that received the award became increasingly closer to the current level of production, so that finally they became virtually interchangeable. The two most recent versions of the competition have emphasized first and foremost the necessity for the Golden Compass to stimulate theoretical studies and research (while meanwhile the national and international awards have been abolished), thus underlining the present scarcity of theoretical and methodological investigations, the disorganized state of training, the lack of research institutes at the university level, the dearth of publications, and the disinclination even within the ADI itself to hold discussions on the profession of industrial design.

Nevertheless, the general atmosphere that gave rise to the Golden Compass awards was one of keen interest in Italian design on the part of both producers and a large segment of the consuming public as well, though in contrast to the somewhat more highly developed situation throughout Europe, this interest was still confined to a rather elite, progressive group of designers. The entire undertaking and the efforts involved in it were nonetheless successful in that they

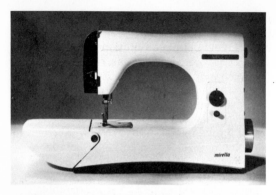

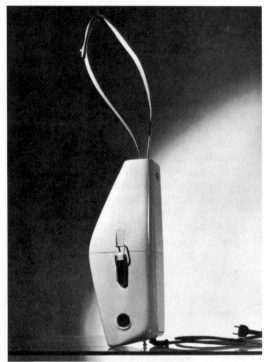

29. Marcello Nizzoli. Mirella sewing machine.
Necchi. 1957
30. A. and P. G. Castiglioni. Nylon vacuum cleaner.
REM. 1956
31. A. and P. G. Castiglioni. San Luca armchair.
Gavina. 1960

marked the beginning of the public's concept of the specific disciplinary and professional characteristics of the designer, who up to that time had been under the shadow of architecture; and above all, they established contacts with industry, even if in only a few limited branches. The Necchi sewing-machine company consulted Marcello Nizzoli, who between 1953 and 1961 produced for it a series of models, of which the 1957 Mirella remains the most faithful to Nizzoli's principle of plastic continuity (fig. 29); here, the entire design is cut off abruptly at the end, as if to provide a clear demonstration of the relation between the mechanism and its housing. The work that Nizzoli produced for Olivetti during those years is of less interest.

Besides working intensively and experimentally in the field of construction, the Castiglioni brothers produced a series of lamps and a portable nylon vacuum cleaner, very carefully thought out in respect to its practical function (fig. 30). From 1956 on, the number of products they designed notably increased: lamps by Flos (pp. 65-66) and Kartell, Gavina furniture (fig. 31), Ferrania projectors, and Phonola radio sets. In these, the approach became sharper and more precise, tending toward a reconquest of the fantastic image through an almost total reduction to the technical requirement and its ironic use. The Leuci light bulb of 1957, in which the pure filament within the glass is freed from its metal base, is almost a symbol of this tendency (fig. 32); while simultaneously an opposite trend began to appear — a feeling for assemblage that was to become a characteristic feature of subsequent production.

Gariboldi, after having won a Golden Compass award for Richard-Ginori with a stackable, compact set of dishes that showed a particularly intelligent grasp of typology (fig. 33), continued to work as a designer for that firm. Solari, meanwhile, was producing new models of electric clocks designed for it by Gino Valle and Michele Provinciale (fig. 34), and Valle also began working for Rex Zanussi, producing for that company a new line of domestic electrical appliances. This last was a branch of industry that expanded rapidly in Italy at the time and began to increase considerably its collaboration with designers.

Among the numerous producers of high-quality furniture that began to appear during these years, Arflex was outstanding, thanks to the constant work of Marco Zanuso, who offers the most dynamic and interesting example. He had begun collaborating with Borletti in the field of sewing machines, producing a model whose plastically severe and somewhat quirky form seems the antithesis of Nizzoli's models (fig. 35). 1954 saw the metal-sheeting chairs by Rinaldi (fig. 36), but it was above all Franco Albini's Luisa chair that represented the high point in the development of furniture at that time. Redesigned in numerous versions since the model in iron tubing was first produced by Knoll, the Luisa chair seems to symbolize the patient research and obstinate quest for perfection characteristic of Albini's working methods (fig. 37).

In 1957, Feal — later to develop a branch of industry devoted to lightweight prefabrication — produced a series of handles, ingeniously executed by the simple cutting of an appropriately designed outline; and in 1960, the same firm won a Golden Compass award for its design of a portable radiator (fig. 38).

In plastics, Montecatini's research institute produced, in 1957, an experimental bathroom unit designed by Alberto Rosselli (fig. 39); and Menghi designed a series of industrial containers in polyethylene, in which the design began to make skillful use of the difficult properties of that material. Above all, Kartell, in its lamps branch, and with the collaboration of its own designer Gino Colombini, came out with a series of small household objects of very high quality (fig. 40).

32. A. and P. G. Castiglioni. Light bulb. 1957
33. Giovanni Gariboldi. Stackable dishes.
Ceramic. Richard-Ginori. 1954
34. Gino Valle. Electric clock. Solari. 1956

In the field of lamps, also (which was to become one of the specialties of Italian design, even if at the level of tailoring rather than of lighting technology), some interesting examples were already being produced, in large part owing to Arteluce (fig. 41).

In transport (besides the Flying Dutchman boat, the beautiful Abarth car (fig. 42), and the little Falco touring airplane (fig. 43), the Golden Compass award went to the Fiat 500 designed by Dante Giacosa (fig. 44), after numerous aesthetic defects in its bodywork had been overcome. It demonstrates a desire to go beyond the equation design= beauty + utility to enter into the field of typological invention and the organization of production, with the intent of operating with regard to the quantitative effects of the product on the city and the environment, and of its 'widespread popular diffusion,' even if the attitude is still paternalistic and on the ambiguous plane of a social pedagogy poised between collective conscience and propaganda for a private product. On an international level, the Cisitalia car designed by Pinin Farina was selected in 1951 for inclusion in the 'Eight Automobiles' show at The Museum of Modern Art, which at this time began its active interest in the field of Italian design. Two shows of Italian design were held in London in 1956 and 1958, respectively, and in 1959 the Illinois Institute of Technology organized an exhibition on this subject in Chicago. Almost everything of any merit in the design field produced during those years, however, was clearly marked out for attention by the prizes and citations of the Golden Compass; and what is striking is the variety and range of what was produced by design at that time, giving rise to the hope — later to be considerably dashed — for a widespread public and private commitment to design throughout broad and diversified sectors of industry. It is sufficient to mention, besides the field of transportation cited above, the first Olivetti computers (Sottsass, fig. 45), the products of the Greco company in lighting, the school equipment designed by Castiglioni and Caccia-Dominioni, and such diversified sectors as sports equipment and optical apparatus.

But the years 1955-60 were also marked by a serious crisis within the sphere of architecture and design. In the preceding years, profound differences had been arising between the two leading centers, Rome and Milan. On the one hand, there was Bruno Zevi's advocacy of organic architecture, and on the other, the maintenance of a continuation of Rationalism; professional relationships were essentially with public agencies in Rome, and with private entities in Milan; the former concentrated on town planning, the latter on design; and what was becoming increasingly manifest was the close relationship between the territorial zones given over to industrial development and the development of design. Naturally, an important factor was the inequality in income levels and the unevenness of distribution; but there was — and still is — a virtual rejection of design on the part of cultural circles in Rome, with the exception of a few people such as Argan, the painter Achille Perilli, and, somewhat later, those who established the School of Industrial Design in that city. The old prejudices against the minor or applied arts still played a part, but even more to the point was the ideological resistance already mentioned.

Furthermore, in Rome there had been an aggravation of neorealist and national populist tendencies (in part related to what was going on simultaneously in literature and the film), and a revision of ideas about history and the neighborhood — in short, of tradition and previous historical and environmental conditions. One of the first manifestations of this crisis was the refusal by architects of the MSA to participate in the XI Triennale, which was once more organized around the topic of design. They cited as their double pretext the

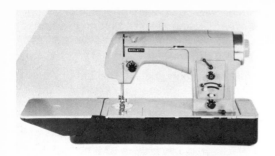

35. Marco Zanuso. Sewing machine. Borletti. 1956
36. G. Rinaldi. Duso chair in sheet metal. 1952
37. Franco Albini. Luisa armchair. 1954

inadequacy of the organizing body and their ideological opposition to the world of design as being the domain of capitalist power. This situation is significant for an understanding of concrete actions that have been taken by designers in their work; they have operated either as if motivated by a guilt complex or have tried to overcome this complex by obscurantism, cultural violence, or pleading professional necessity. The profession's lack of institutional recognition, as well as of bodies given over to teaching and research; its relegation to a marginal position by public agencies; and above all, the tentative nature of an activity then still carried on somewhat as an avocation (as a second line for graphic artists or architects), all too easily gave rise either to arrogant aggressiveness or to profound doubts about the actual social necessity for the profession and the possibilities of production.

In any event, the XI Triennale opened under the slogan of 'European Triennale.' This indicated a program intended to bring Italy up to European levels of production and consumption, and was, obviously, a program that affected only Northern Italy and a class of people who can hardly be regarded as belonging to a low-income bracket. The section devoted to industrial design was a highly effective gallery showing the most important designers of the world, paradigms of examples to be emulated by producers and designers. The foreign sections, in actuality, showed a greater concern with design itself than with the typical or high-priced object.

A second manifestation of the critical reversal of ideas on design took place in the North between 1955 and 1960, led primarily by the more youthful groups who worked in those years around the magazine *Casabella*. It is worth citing some of the reports presented on the occasion of the exhibition 'New Designs for Italian Furniture,' held in Milan in March, 1960, and sponsored by the *Osservatore delle arti industriale*, as a conclusive indication of the entire climate of opinion (fig. 46):

'The myth of the object as a secondary phenomenon of the "fetishism of goods" is a true characteristic of our times and its activism. It is to the merit of the Bauhaus that it made this point clear, and that it tried, by breaking up the elements of furnishing into their constituent parts, to bring furniture as "design" into a total program of production. The death of the "producer of ensembles" was a wholehearted attempt at surgical healing of the artist as decorator. Nevertheless, subject to the simple laws of survival, things assume forms that, in the corrupt state of distribution (when the ideology of method lapses into a simplistic statement of the problem), reduce contact to pure manipulation, without allowing for that added factor that can live on as experience and allow the user some freedom of behavior. Thus an interior put together as a collection of individual pieces manages nowadays to look like a cross between a furniture display room and a temporary film set; the ensemble always seems to give the impression that the objects within it are interchangeable, and hence lacking all character... But while there are some who believe that a synthesis can be attained only through unity of style, others, even if they are more revivalist in their approach, find that we are far from that taste for compatibility and tactful harmony of the Wiener Werkstatte's middle-class family. On the contrary, irony, treachery, and pessimism reveal hesitations and conflicts. When equilibrium is attained, it is by means of a state of tension between the divergent factors of a pluralist, stratified society that plunges off peremptorily in all directions. Thus persistently, against our will, renunciation of a sense of the whole, of a taste for the sweetness of life, often plunges us into the labyrinth of

38. Portable radiator. Feal. 1960
39. Alberto Rosselli. Bathroom unit. Montecatini. 1957
40. Gino Colombini. Plastic colander. Kartell. 1962

possible interpretations, where it seems increasingly difficult to find an impetus toward any one direction, expressive of a common effort.'

In reality, the effort was, as usual, freely expressed in verbal and stylistic terms and was only rarely truly brutal or malign. It nevertheless created an enormous international scandal and led to a public questioning of design as it was then constituted, basic questions addressed to its own conflicts. In terms of practical execution, the best pieces did not tend so much to reject modern technology as to fail to take it into account as an active and primary means of expression; they often demonstrated a procedure that was the reverse of simplification, inasmuch as it was governed by a rather naive interpretation of the industrial process.

Further, it should not be forgotten that these developments were directly influenced and inspired by the first generation of Rationalist architects and designers. On the one hand, there is a quest for the typical, as manifested in the work of Albini, for example: the use of natural wood as a material to be emphasized, of cloth, frames in relief, and the tendency to round off the edges of the pieces to enclose their shapes, rather than cutting them off abruptly. On the other hand, there is a throwback to the middle-class tradition of the nineteenth-century neoclassic style, as in the Azucena group (fig. 48). There is also the theory of reminiscence, which uses memory as a material in design as a defense against what is too new or disturbing: so we have a strong connection with the heritage of the fin-de-siècle — as a balance between the last continuous link with the fabric of society and the first formation of the idea of an avant-garde (fig. 49). Lastly, there is the desire to endow the object with a strong symbolic and communicative force, to oppose its reduction to the status of a mere tool, and to move in the direction of a concept of an inclusive whole, which to a great extent foreshadows, even though only in a stylistic way, a concern with the environment.

This phenomenon of revolt, like every polemical gesture, involved only a few people and was limited to a small group of young men who, for various motives, came together briefly in a transitory consensus that was almost immediately negated — and by others degraded to a style vulgarized by being badly misinterpreted.

The XII Triennale, therefore (which had 'Compulsory Schooling' as the somewhat disingenuous title for its central theme), had the difficult task of trying to piece together again the broken shards of a discipline full of conflicts and contrasts. I do not believe that it was sheer accident that this manifestation saw the vigorous emergence of two personalities who were, to some extent, new, and who were destined to have a significant impact on Italian design in succeeding years: Carlo Scarpa and Ettore Sottsass, Jr.

The influence of the former was wholly indirect. During the critical years, he had been designer for the Venini glass factories in Venice, and he had never designed a single piece of furniture for mass production. The lesson he had to convey was that of a lack of preconceived notions, and a highly refined mastery of his craft, filtered through a sense of history and a knowledge of Central European traditions (notwithstanding his clear inspiration by Frank Lloyd Wright). These have enabled him to survey and absorb the entire range of sensibility of the plastic and pictorial experiments of the avant-garde (fig. 50). Together with Franco Albini, Scarpa has been responsible for some of the best museum installations in the world. His extraordinary ability to establish clear spatial divisions by means of attention to, and invention of, detail, and his sense for giving continuity to the total environmental picture, have decisively affected the best Italian design.

41. G. Sarfatti. Lamp model 1055. Arteluce. 1954
42. Abarth 1000 automobile. Fiat. 1960
43. Falco F 8 L light airplane. Frati. 1960
44. Dante Giacosa. Fiat 500 automobile. 1959

For Sottsass, by contrast, the profession of design (even if interpreted in a wholly original way) has been the deepest and most continuous aspect of his activity. He entered this field through profound disillusionment with the state of architecture, with the presumptuous pretensions and paternalistic attitudes that are its distinguishing features. Since the outset of his career, he has been a painter and has taken an intense and active interest in contemporary literature, particularly American. He determined upon his present career after his first trip to the United States in 1955. The style of his interior design up until 1960 (culminating in his antirhetorical entrance hall for the XII Triennale; fig. 47) was based on the patient reconstitution of a series of fragmentary images, with pictures, fabrics, objects, and articles of furniture used like so much raw material in a vast spatial collage. 'Investigation does not center on the individual object of furniture,' he has written with reference to one of his decors of that period, 'but on the relationship and distances among the various pieces, their placement, and above all, on the materials. The pieces of furniture are only part of the whole, and coherence is attained through the colors and from the consistency between the materials used and the character of the design.'

Since 1959, he has worked in collaboration with the Olivetti company, his first work for it being the Elea 9003/1 computer, and during the past decade he has had the same impact on the form of Olivetti's products as Nizzoli had during the 1950s (figs. 51 and 52). The radical stereometric simplification which he imposed on his first experimental designs did not prevent him from also imparting to them the same marked vivacity that has characterized his activities in the world of modern art, particularly his interest in contemporary American painting. Thereafter, a series of experiments in electronics gradually led him to the idea of establishing a flexible and expandable system of coordination that would enable him to bring a general sense of order into the field of mechanized services. By means of interchangeable parts, it is possible to keep abreast of the rapidly developing dynamics of modern technology and allow for gradual acquisition of equipment on the part of the user. He therefore considered in its totality the problem of the microenvironment of this kind of work (data processing, etc.), which is becoming increasingly mechanized, and more and more like factory work. He had the opportunity to verify his own theories regarding the relation of this kind of work to its environment when, about 1962, he was commissioned to make a study of coordination among various firms producing electronic office equipment.
During the same period, he developed designs for many kinds of Olivetti machines: electric and other typewriters, electronic computers, etc., up to the latest Valentine model (p. 75). They are all bound together by a tendency for strong, controlled imagery that still does not lose its sense of pleasure in the play of colors and awareness of a plastic presence within an environment actually based on alienation; his effort is to establish the same normal relationship with a piece of apparatus that one feels toward the objects in one's home (figs. 53 and 54).

The same feeling of pleasure in colors also comes to the fore in Sottsass's activity in furniture design — but can we still use the word 'furniture' in this connection? They are violently colored presences, centers and elements with magical references, new altars in which everything depends not only on a symbolic relationship with the object but even more on a ritual approach (he has been profoundly affected by Indian and Far Eastern cultures as the cultures of peace and love). These are places meant to evoke within the dwelling a preideological way of life, in which love and attention take the place of manipulation and use (fig. 55 and pp. 50, 104-106).

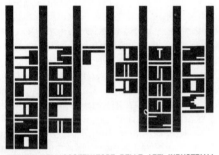

MOSTRE DELL'OSSERVATORE DELLE ARTI INDUSTRIALI

45. Ettore Sottsass, Jr. M/IV computer. Olivetti. 1958
46. Cover of exhibition catalogue, 'New Designs for Italian Furniture,' Milan, 1960
47. Ettore Sottsass, Jr. Entrance Hall, XII Triennale, Milan, 1960

I believe that Sottsass's influence has been particularly strong precisely because of the great attraction of his personality and his constant quest for a continuity between work and one's way of life, with no distinction between the world of activity and moments of play.

Similarly, though in very different ways, the XIII Triennale of 1964 represented an important turning point in Italian design and has had an extensive indirect influence. (Its theme, 'Leisure,' often provided only a pretext for what was actually shown). In the first place, there was a definite break with the theme of art production, which had dominated the preceding three of four exhibitions. Secondly, because a new generation now assumed the role of protagonist (comprised, for the most part, of the same group who had taken part in the 'New Designs for Italian Furniture.' Thirdly, because it brought up for discussion, and on an ideological and linguistic plane as well as on the plane of methodology and theory, many features that were to characterize Italian design in the ensuing period. The introduction to the first section declared: 'The focus of our attention has not been to depict a situation but rather to show the direct action that should arise from it, so that each individual, recognizing the falseness of the situation, can take personal responsibility for regarding this as the necessary moment in which to consider possible ways of giving a rational direction to the situation itself.

'To accomplish this, we have reduced to the minimum explanatory devices, interpretative labels, and guidance with any pretense at rationality that would indicate a division of the route into successive scenes, placing the visitor in the center of action, eliminating any distinction between him and the object, offering him a series of alternative choices, or rather forcing him to act, and attempting to transmit at every point the general meaning that we wish this section of the exhibition to impart, so that it may be understood at any level of interest and knowledge. Thus, all the elements selected as having a meaning in denoting a certain portion of leisure time are, to some extent, fragments of our objective world, whose meaning in other contexts is perfectly well known. We have tried to arrange these elements, not so much by isolating them or loading them with symbolic meanings, as by placing them in a different visual context that might imply making them appear in a new light. This is accomplished not so much by their repeated, unfamiliar, reciprocal juxtapositions, by gathering together a whole range of consumer products, or projecting and enlarging them "beyond their scale" "beyond the materials" that are appropriate to the object itself (typical devices used in advertising), but principally by setting them within a different spatial framework, which arranges and multiplies them, projecting them into some unknown urban future — in accordance, that is, with architectural procedures. No matter if, in our case, this space is constructed, so to speak, "in the negative," and made up of illusions and absences, in the effort to establish a kind of historical void, an absolute presence around the object and ourselves as spectators, applying a rigid linguistic method that tends to reduce to the minimum the component elements in play and increase to the maximum the labyrinthine ways of combining them.'

To begin with, what is clearly manifest here is the notion of the physical environment in its totality, its essential connection with the idea of behavior and action on the part of the user, and what this implies with respect to what is offered visually: enlargement of the designed materials, an indispensable lack of bias regarding the tradition of Rationalist composition, formal technology adapted to dimensions of huge scale, and composition by segments, making use of the principles of collage and assemblage as elements of design.

48. Gregotti, Meneghetti, and Stoppino.
Cavour armchair. Azucena. 1960
49. Gabetti, D'Isola, and Raineri. Armchair. 1956
50. Carlo Scarpa. Olivetti shop, Venice. 1958

Secondly, the XIII Triennale raised the crucial question whether control of the environment and opposition to its deterioration can be put in terms of the rational construction of individual objects. Nothing guarantees that the sum of a series of well-designed objects will yield a good environment. Crowding, poor placement, overlapping, or too sparse relations can wipe out the effect attained by a single object. Finally, this Triennale gave prime importance to the subject of consumption as the basic component in the process of giving the object its form. It was not a question, therefore, of merely taking up again the theory advanced by Reyner Banham many years before, which regarded the language of design as a 'popular art,' but rather of considering all the 'aesthetic' aspects of design in terms of communication.

These were the topics dealt with during the following year and given explicit theoretical form in *Edilizia moderna* (no. 85). A process that had already been underway in Italy for a long time, but which had its definitive formulation only after 1953, thus became defined and clarified, and hence became something to be used in the process of design. Such theorists as Gillo Dorfles, who has written extensively on the subject of consumption, and Umberto Eco, who has devoted himself to visual communication, have made important contributions that have clarified this new tack in Italian design. Dino Formaggio's book on the aesthetics of phenomenology (*L'Idea di artisticità*, Milan: Ceschina, 1962) advanced the theory of artistic quality as having a significant technical function and was of signal help in clarifying the distinction between the artistic and the aesthetic, replacing the notion of 'beautifying' with the concept of communicative design.

During these years, the number of theoreticians who have dealt with the subject of design has notably increased (Frateili, Menna, Morello, Van Onck, and others), as has the pressure for the creation of an institution for teaching and research on design at a university level. Venice was the first to found an experimental institute, which was soon disbanded; next Florence (where the institute is still functioning); and finally Rome. All are in the somewhat ambiguous position of having been transformed from former art academies into schools of design. In Milan, many programs were drawn up, and many meetings held, with no success. Meanwhile, in architectural schools, teaching of various kinds of industrial design held (as it still does) a wholly marginal position.

The principal theories and hypotheses in the field of design were recently discussed in a symposium sponsored by the ADI in 1970. Roberto Guiducci propounded the theory that it was necessary to transform industrial society into a society of services (i.e., activities that do not entail the production of industrial products through the use of natural resources, such as research, communications, data processing, etc.), and emphasized the role of design in this transition. Maldonado outlined the possibilities for including and integrating design within the general study of the environment. Spadolini discussed the actual situation in Italy today, and Ciribini the possibility of including design within research based on systems analysis. In his summary of the proceedings, Gregotti examined the concrete possibilities for including design within the ambiguous framework of university reform. In general, the meeting presented the usual depressing spectacle of ideas that become absorbed at a theoretical level before there is any possibility for concrete experimentation. Once again, we see a discipline pursuing a direction that might finally lead to carefully thought-out reflections and productive advances, while on the other hand, there are the needs of a country in which many internal

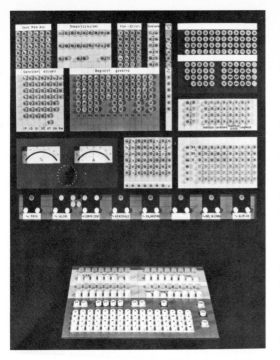

51. Ettore Sottsass, Jr. Elea 9003/1 computer. Olivetti. 1962
52 a) and b). Ettore Sottsass, Jr. Elea 9003/2 computer. Olivetti. 1962

imbalances are still unresolved; and an exchange between the two types of experience seems virtually impossible in practice.

What the XIII Triennale presented in the form of negative criticism has been put in positive terms by Tomás Maldonado, who for several years has been an integral part, and active protagonist, of the discipline of design in Italy. In his little book *La Speranza progettuale* ('The Hope for Design'), he attempted a vindication of his active, professional approach on a didactic and theoretical plane, and by using Marxist methodology provided a new political slant, based on his previous experience in directing the Ulm school of design. It is the record of a profound crisis arising from his own desperate attempt to avoid separating theory from practice and make his theories and his actual work appear consistent (or, at least, purposely contradictory). Maldonado's active practice in Italy, including his general responsibility for the appearance of the Rinascente group of department stores, lasted barely three years. This was sufficient to demonstrate how it might be possible to apply in a comprehensive way the methodology of design to an entire sector — that of large-scale distribution. He succeeded in imbuing with a dynamic unity the entire range of the Rinascente public image — from graphics to architecture, from equipment to the principles of the planning and location of stores. Included in the group was a branch for food distribution, which entailed the study and experimental completion of a new architectonic image (fig. 56); and a chain of low-price department stores for which a point-of-sales manual was written that analyzed and organized all the possible alternatives for the various elements (fig. 57). Finally, there were plans for the Rinascente centers, with standardized equipment, and for the firm's central offices. Gui Bonsiepe, Gino Valle, Tom Gonda, Vittorio Gregotti, Eduard Zemp, and many others were called upon to collaborate in these projects, which unfortunately came to a halt when the company changed hands. Nevertheless, they were important as representing systematic experimentation of a type completely new in Italy, the only precedent being that provided by Olivetti's thirty-year-old tradition.

In recent years, the work of Franco Albini has also moved in this same systematic and environmental direction; during this period, he has produced very little in the way of objects. Since this type of design has become an activity basically dominated by the connection between production and consumption, his interest in the object seems to have waned considerably. This has not stood in the way of his producing, on the other hand, the most complete and severe example of environmental design for a public service (together with those by Sottsass and Maldonado, already mentioned). Beginning in 1960, he, together with his collaborator Franca Helg, and Bob Noorda for graphic design, was commissioned to design the stations of Line 1 of the Milan Underground. The plan was conceived as a repeatable system based on the methodical use of prefabricated panels and the unification of certain syntactical elements — stairs, handrails, entrances, pavements, lighting, etc. The unifying thread throughout the whole project is provided by the signs, carefully studied with regard to their letters and numerals and their meticulous integration within the architectural scheme (figs. 58 and 59). The entire system has shown that it is able to absorb all the distinguishing characteristics of the different subsidiary elements (shops, kiosks, sales stands, etc.) that make up the complex and vital fabric of the stations, without losing its own identity. Seldom has the image of the subterranean life in these underground centers been so well conceived and strictly controlled, without the loss of any of that quality of complexity and contradiction that is its fundamental characteristic.

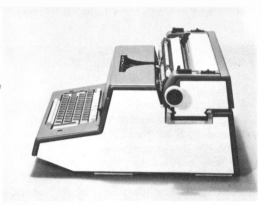

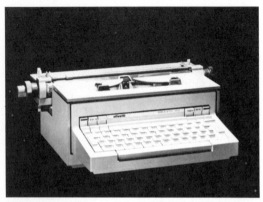

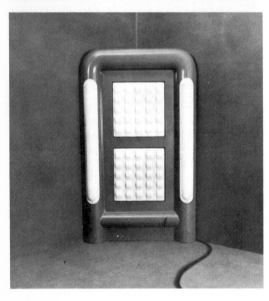

53. Ettore Sottsass, Jr., and Hans von Klier. Tekne 3 electric typewriter. Olivetti. 1964
54. Ettore Sottsass, Jr., and Hans von Klier. Praxis 48 typewriter. Olivetti. 1964
55. Ettore Sottsass, Jr. Cupboard in fiber glass. 1970

In this rather conventional historical account that we have been rendering, industrial buildings and systems of prefabrication should also be mentioned, even if only tangentially, in view of the attempt that has almost always been made through these methods to reduce the entire problem of architecture to the status of an object or a system. This is generally a somewhat outdated concept; prefabrication has not progressed noticeably, either in theory or application, since its first use thirty years ago. In Italy, three approaches to its use may be distinguished. The first, based on light prefabrication, has been used principally by Feal, especially for schools, and only occasionally for other types of construction. The design has traditionally been rather schematic, up to the most recent attempts, which are better articulated (the collaboration of Feal's architect Sacripanti is evident). A considerable number of theoretical studies have been made, such as the IPA system by Bellini and his associates, but have produced no important practical results. More interesting and original, as well as on a higher level of architectural quality, are the experiments in the use of wood prefabrication designed by Conte and Fiori (fig. 60).

The second approach is in the field of heavy prefabrication, represented in designs by Angelo Mangiarotti and Gino Valle. The former, who has also produced objects for the home that show excellent calligraphic qualities (p. 46), has based his research on structural systems executed with extremely precise design; several have had wide and successful application. The studies have usually been very well carried out with regard to engineering (thanks to Favini) and the methods used for erection (fig. 61). In this field, the work of Pierluigi Nervi, the studies on prefabricated canals by Silvano Zorzi, the reinforced-concrete bridges by Riccardo Morandi, and the iron bridges by Luciano De Miranda, have become world-renowned.

Valle, perhaps the most interesting Italian architect of the 1950s, has made a series of studies on prefabricated vertical panels for industrial buildings (fig. 62). He is also a very skillful designer; among his works are designs for domestic electrical appliances manufactured by Rex, the new Solari series of domestic and institutional electrical clocks (p. 72) and signboards (fig. 63), and very interesting projects for gasoline filling stations.

Returning to prefabrication, the third approach ranges from the studies of 'components' by Rosselli and Ciribini to the highly original experimental inventions of Renzo Piano, a young Genoese architect certainly destined for a brilliant future in this field. His approach, among all those we have mentioned, is the one most closely related to the problem of raising the methodology of design to the level of architecture.

Since 1960, Marco Zanuso, more than any other Italian designer, has shown the greatest capacity for continuous production and creativity; and in respect to quantity, also, he has produced a notable number of objects, while always maintaining a high standard of quality. Each of his designs seems to show his ability to make wise use of his previous experience. He is able to steer clear of the temptation to follow fashion, trusting to find a sure path for the development of his own work by adhering to the better aspect (and one more consistent with the middle-class ethic of professional seriousness) of the old Weberian function of serving as a mediary between capital and labor. His work may be accused of making a fetish of technological processes, of trying to make the process of the organization of production expressive in itself. This sturdy 'product-oriented' ideology, nevertheless, always keeps him from overstepping the bounds into

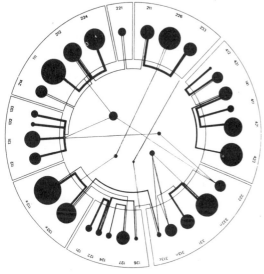

purely consumer-oriented stylization. Instead, the objects he designs often have a possible environmental use; for example, the delightful little child's chair of polyethylene, which is also a fantastic and ironic construction toy (p. 44); the collapsible schoolroom desk designed for the XIII Triennale (fig. 64); the series of television sets, pieces of equipment that are presences well adapted to the new family hearth, stripped of all rhetoric — pure apparatuses in miniature scale or plastically neutral, like the black set of 1970 (p. 69), or the little portable radio reduced to a single compact volume, as compactly jointed as a Chinese puzzle (p. 71); or the typological invention of the Grillo telephone of 1967 (p. 74). In the past ten years, even his best architectural works seem unified by a consistent methodology derived from his experience in design (in which the valuable assistance of his collaborator Richard Sapper should not be overlooked), even to the extent of involving an interchange of morphological types between the two branches of activity. He has combined his practice with such public activities as responsibility for two Triennale exhibitions, serving as editor of *Casabella*, teaching architecture, and being active in the ADI. In all these undertakings, he manifests a marked aggressiveness, always prompt to seize on new theoretical developments and relate them to his own work.

On an analogous professional plane is the work of Alberto Rosselli (fig. 65 and pp. 33, 38) and Rodolfo Bonetto. The latter had no formal architectural training and in this respect may be compared to a highly 'cultivated' artisan such as Marcello Nizzoli. His work is based on a deep and direct awareness of the techniques of connections and the technological processes of materials. His experience as a teacher at Ulm has certainly had an important influence on his work, leading above all to a severe and purist reduction to essentials in his use of forms (fig. 66 and pp. 49, 71-72).

The late young designer Pio Manzù also came out of the Ulm school. His collaboration with Fiat (which up to that time had been absolutely impervious to any influences from abroad) began with a little city taxicab (fig. 67), in part an outcome of his studies at Ulm, which might have opened up important new possibilities.

Mention has already been made of some characteristic aspects of the work of the Castiglioni brothers: their feeling for assemblage, their reduction to the extreme of technical elements, their sense of irony. In their recent work, these characteristics have been accentuated to the point of caricature (e. g., the screwdriver table, fig. 68) and characterization (the Poretti Splügen Bräu beer dispenser of 1964, fig. 69), and has even verged on the Dadaist 'ready-made.' In addition to their undoubted technological and formal skill as designers (evident, for example, throughout their very fine series of radio sets, phonographs, photographic apparatus, moving-picture cameras, and also apparent in their attention to the most minute details, such as light switches, bulbs, and plugs), there is a strong inclination to remove the object from its normal context, in a manner typical of the technique of bricolage. Thus, they combine a transformer, a fishing rod, and an automobile headlight, or a crossbow, a wooden pole, and a tractor seat (p. 102). Or, sometimes, in environments, they mingle the most delicate objects with the most humble and commonplace. The avant-garde technique of bricolage has only recently been taken over by design from planning procedures, but it offers some transitory effects that allow it to be used from a somewhat different point of view. Every means of lighting is utilized on the basis of its qualities of reflection, transparency, and mobility (pp. 65-66). There is also a liking for quotation; no other object has succeeded so well as the San Luca armchair (fig. 31) in picking up again the thread of the Futurists

56. Gregotti, Paulis, and Parmiani. SMA Supermarket. Rinascente. 1969
57. Point-of-sales manual. Rinascente-UPIM. 1969
58. Bob Noorda. Sign for Line 1, Milan. 1964
59. Franco Albini and Franca Helg. Underground station of Line 1, Milan. 1964

60. Claudio Conte and Leonardo Fiori. Structure of prefabricated wood panels, Brescia. 1963
61. Angelo Mangiarotti. Warehouse with heavy prefabricated elements. Mestre. 1963-64.
62. Gino Valle. Prefabricated panels. 1963

(Boccioni in particular) with such a knowing use of unusual methods; nowhere else have the specific qualities of our own cultural roots been so clearly related to an absolutely modern approach.

Despite the limited distribution of their products, the group working with Danese occupies a far from secondary position in Italian design. By means of small objects of high quality, still inevitably deluxe products, even if made only of paper, by ornaments that lie strictly within the range of visual and communications research according to the principles of programmed-art imagery, and by constructing them with exemplary procedures that imitate those of industrial production, they have revealed with microscopic clarity the contradictory tendencies now prevailing in design. But while for some members of this group, such as Bruno Munari (p. 131), a sense of play and creative irony often makes the reestablishment of contact seem a possible aim, for others, for example Van Onck and a large number of youths, this path of severe criticism more frequently leads to a total paralysis or refusal to produce work.

The quality and consistency of the work of Enzo Mari, perhaps the most representative of these designers (fig. 70 and pp. 76-77, 83, 89, 90-91), have been well documented in a monograph that he himself edited. It is interesting to comment on his position as exemplifying the attitude of the younger generation. For someone like Mari, the goal is to reduce the possibility of error in the way objects are used, rather than to increase their morphological combinations. Moreover, the very length of time that the work takes becomes in itself a polemical gesture against improvisation, a commitment to produce with dogged determination an antiproduct demonstrating the discrepancy between the time taken and the resultant object. It is a task of self-interrogation on the part of the designer and a questioning of the reduction of every visual activity to design. Perhaps in this there is a compensation for the bad conscience of every designer who is aware of his own position in society, obliged, if he is to be active, to renounce to some extent his own being (and with it abandon his general position of opposition); he must either choose the domain of collaboration or that of total silence.

But the psychoses of designers are probably not of major importance. It is more objective to ask whether it is possible, and if so, to what extent, for the process of design to establish not only an efficient method of control over the imbalances and increasing amount of things, people, and information (which no one at the present day can deny is essential for our survival); but also to achieve an incentive to progress, an exchange of views, and a critical distance with respect to reality, so that the profession of design neither winds up tending toward utopia (constructive or destructive) nor succumbs to being coopted.

The outcome may seem somewhat ambiguous. To the extent that creative thinking is inevitably incompatible with the criterion of maximum profit, the amount of 'negative thinking' (insofar as it is critical thinking) that is implicit in every authentic act of design can see into reality on only one plane. Working to determine the concrete forms, structures, and typology of things in an effort to achieve a balance between the criteria of their use and of their exchange value, and making them available on the market, are necessary aspects of the designer's situation, but they are not sufficient. In order to avoid identifying himself with value and possession, Mari finds it necessary to minimize the feeling of importance attached to buying and possessing something; this is accomplished, not so much by reducing the object to the status of a utensil, as by a terrorizing economy in the means of expression, and by imbuing it with the provocative presence of a radical image. But if the form of this radical image is schizophrenic, and seems to lapse into

63. Gino Valle. Electrical announcement board for airport. Solari. 1962
64. Marco Zanuso. Collapsible schoolroom desk designed for the XII Triennale, Milan, 1960
65. Alberto Rosselli and Isao Hosoe. Melior Pullman bus. Orlandi-Fiat. 1970
66. Rodolfo Bonetto and Naoki Matsunaga. HO 1 machining center. Olivetti. 1969-70

337

absolute simplification through the organization of the object's functional and structural qualities on different levels, this does not alter the reality of the very real condition of dissociation in which design operates: dissociation between design and the work, between the work and the social context in which it is situated, between the social context and the real pressure of demand.

Mari has a vision of revolution as the establishment of worldwide human communication, also to be brought about by means of design, and even at the cost of completely changing the language of communication, beginning with the alphabet.

The alternative theory of the younger generation is that of the environment, suggested symbolically or projected on a utopian plane. Here we are dealing with a true polemic against the reduction to formalism, the imperatives of consumption that, in a desperate quest for diversification (which also entails social diversification), constantly produces objects with such eccentric meanings that they lose all credibility (fig. 71). But we are also seeing, beyond this self-involvement with the same process of commercialization, an effort to discover openings through which space, relationships, and behavior can be transformed, and to achieve this by means of the nonconstructivist tradition of the avant-garde, which has always followed a path quite separate from that of architecture. But frequently the use of this tradition detracts inevitably from the true condition of alienation that provided the first avant-garde with its moral strength and replaces it with a cynical attitude toward even the most intimate aspects of subjectivity. The means used include various techniques taken over from the plastic arts — the 'soft' object, enlargement and inflation of scale, the happening; or involve a recourse to stylistic consistency governed by the rules of single-point perspective, to evocation of magical gesture and regression to childish fairytales, in the case of the UFO group (fig. 72), to the spatial rigidity of Archizoom (pp. 232-239), or to the religious terrorism of Superstudio. The multiplication of such experiments is also connected with an imitation of Pop discoveries (pp. 95-97), anthropomorphism derived from the feminine furnishings of Allen Jones (p. 98), and the noteworthy idea of the beanbag armchair (p. 113).

On the other hand, taking sculpture as the point of departure, there are strong influences of a return to *arte povera* (Mario Ceroli's experiments, for example; fig. 73) or to conceptual art. Finally, there is that tendency which, in reacting to neooriental overabundance in decor, ends up with total absence of the object. There are also commercial side benefits, of which no astute designer today loses sight.

In the last analysis, what does all this burgeoning of linguistics, theory, and methodology amount to, in relation to the realities of national life? What is the effective impact of the profession of design?

'Let's look around us,' Paolo Chessa has written. 'Let's enter any public office — a post office, for example. Then let's go to a military area, a barracks, or a seat of public administration. Looking beyond the pompous entrance halls, behind the ticket windows at stations or the counters at banks, let's go into the public lavatories; let's, in fact, poke our noses into what we may call all the national cesspools, and finally, without descending to the utmost limits of squalor, let's visit the thirty-five percent of dwellings of the Milanese (or those called Milanese) that statistics show are without any individual sanitary installations; or let's visit the homes of all the Italian porters, of the peasants, and of the overwhelming majority of workmen and office employees. What do we find?...

'In attempting to evaluate the respective conditions, and draw

67. Pio Manzù. City taxi. Fiat. 1968
68. A. and P. G. Castiglioni. Screwdriver table. Bernini. 1966
69. A. and P. G. Castiglioni. Splügen Bräu draught-beer dispenser. Poretti. 1964
70. Enzo Mari. 'Place for games' — collapsible playroom in cardboard. Danese. 1961

comparisons, it seems to me completely illusory to point to the statistics on the most recent dissemination of automobiles, for example, or the figures on the so-called "Italian miracle."

'Everyone knows, moreover, that the average Italian would rather boast of a Fiat 600 that he can park under the nose of his neighbor, or go to his tailor to have a complete single-breasted tropical suit made to his measure, or have a television set with twenty-one controls, or subscribe to the best reserved seats at the football stadium, than live in a civilized fashion. But — if this were not an astonishing hypothesis — what suggestions could the unfortunate average Italian get if he were to turn to his architects? What proposals, what plans, and at what prices? And we aren't speaking here of a small minority; we are dealing with thirty-six to thirty-seven million ordinary people as against three or four million of the privileged' (IN, Argomenti di architettura, no. 3, September 1961).

Nevertheless, since 1961, when Chessa wrote this, something has changed in the aspirations of the Italians. It is certainly no new revolutionary vision, but it has its basis in changing customs, even if only on a purely private level, even if only in the most highly industrialized urban centers, and even if it has generally come about through the introduction of a new style based on imitation of the ruling class, rather than on any actual change in the kind of dwelling. It is the latter that should really provide the sound basis for any significant development in Italian design for the home, which instead has flourished principally because of the success it has had on the foreign market, rather than within the country itself.

Something that must be taken into consideration, explaining both negatively and positively many things about the Italian designer, is his specific professional relationship with production. Almost all Italian designers practice their profession outside the production system, even when they have a prolonged connection with some company; it is even rarer for a designer to be engaged as a general consultant or part-time employee of a business enterprise; and it is rarest of all for a firm to have its own qualified designing office. The companies, for their part, generally operate in a largely imitative way with regard to design, guided by what has already been demonstrated to be profitable; or sometimes, as we have seen, they arrive at brilliant solutions by means of their own. The designer, therefore, is either given an excessive burden of responsibility (for marketing, typological invention, or plans for distribution), or else he is shoved into the field of promotion, coming under the advertising budget. In only a few instances is he organically linked to the operations of the firm. And only quite recently, in the production of household articles, and to the extent that because of specific times and customs, furniture and ornaments have become consumer products rather than durable goods, has the designer become of any importance in guiding a firm's policies.

It has been forecast that, on the base year of 1965, consumption of goods will have more than doubled by 1975 and quadrupled by 1985. This explains, at least in many respects, why the branch of Italian design that has undoubtedly enjoyed the greatest international success in recent years has been that of design for the home. Magistretti (pp. 33, 39, 41, 59, 65, 109), Aulenti (pp. 43, 109), Joe Colombo (fig. 74 and pp. 30, 45, 53, 62, 116-17, 123), Tobia and Afra Scarpa (pp. 28-30), Giotto Stoppino (p. 48), Cino Boeri (pp. 29, 61, 121), Frattini (p. 67), Asti (pp. 64, 88, 109), Vignelli (p. 77), and others, with the brilliant quality of their inventions, and the elegance of their plastic solutions, have in various ways contributed a true new culture of design for the home that will remain in the future the most concrete

71. Mario Galvagni. Experimental interior spaces. 1967
72. UFO Group (Bachi, Binazzi, Cammeo, Foresi, and Maschietto). Oversized temple-table on columns, plexiglass lamp, and rat eating table. 1969
73. Mario Ceroli. Interior. 1964

achievement of Italian design during this period. Each of them has had a special approach, for their poetic means of expression are vastly different, ranging from Joe Colombo's ingeniously contrived mechanisms for the future to Gae Aulenti's ability to shape an image for some one aspiring to the status of group leader of a modern, progressive society, always deriving the object from her solution to a specific context, to the refined and nostalgic technique of Tobia Scarpa, whose mastery of his art, like that of Magistretti, always manages to discover in an elegant way some authentic necessity of the moment, to the cold precision of Mario Bellini (figs. 75-76 and pp. 31, 48, 68, 70, 120. (Bellini's case is somewhat exceptional, however, because of his notable activity in other branches of design, including his extended collaboration with Olivetti, which has led him to have confidence in the idea that it is possible to develop a process of design that can really be integrated within the process of executing industrial objects.)

Impelled by these examples, therefore, in the past few years there has been a frenzied activity in the presentation of new items, the formation of new firms, the appearance of new lines of furniture, lamps, and other objects. There has even been the founding of a high-quality periodical, *Ottagono*, directly sponsored by producers; and there have also been exhibitions such as 'La Casa abitata' at Florence, a furniture fair, and Eurodomus. Every one of the women's magazines has its own column devoted to modern furnishing, and *Vogue* issues a special edition about the home. *Domus*, from the vantage of its long-established position in this field, tries desperately by distillation and sanction to differentiate its own role.

But in spite of all these contradictions, one can live in rather comfortable elegance; one is only obliged to denounce the duplicity of the situation. This duplicity is so ingenuous and open, and so receptive to social criticism, that there is not a designer busy making little lamps or paperweights who does not nowadays denounce (even on the children's hour on television), the 'repressive consumer society.' We are well aware that, for this very reason, Italian design in the past few years has simultaneously both enjoyed great popularity and been severely criticized. No one can deny that it offers a picture of felicitous forms that, thanks to its lively imagination, can with fertile inventiveness jump heedlessly over the actual difficulties of production methods that are often merely improvised. On the other hand, Italian design often clearly shows the worst defects of our national character: improvisation, superficiality, a weakness in respect to fashion, a passion for innovation at all costs.

With a few exceptions, the total picture is characterized by the most primitive aspects of consumption: minimum attention to the areas of production connected with public services, maximum development in the areas of private consumption most subject to rapid turnovers in style. As regards design for the home — the subject with which we are concerned here — Italian design today occupies the same position in the world market that Scandinavian 'good design' held in the 1950s. By means of stylistic modernization that made use of neomodern materials and types, Italy succeeded in replacing (on an imitative level, if not on the level of authentic production) the image of craft-produced, private-household, Nordic design with new standards of judgment of style, if not of culture. It has been an activity that sociologically could invite and reassure that precarious feeling of progress derived from the luxuriant flowering of consumption. It has become a far from secondary activity, supported by the daring of entrepreneurs and the aptness and brilliance of ideas that, despite their great diversity, nevertheless miraculously combine to form a consistent impression.

*'In my hungry fatigue, and shopping for images,
I went into the neon fruit supermarket, dreaming
of your enumerations!'*

wrote Allen Ginsberg in *Howl.* These verses might have been dedicated to Italian design of 1970, and perhaps in large part to the role of design in modern culture, to its present 'hungry fatigue shopping for images,' its mixed condition of development and decline, its success, and its loss of a sense of reality.

[Editor's note: Unfortunately space did not permit inclusion of the extensive bibliography prepared by Mr. Gregotti, with the assistance of the architect Marica Redini; typewritten copies may be consulted in the libraries of The Museum of Modern Art, New York, and of Centro Di, Florence. The illustrations for the article were assembled with the cooperation of Dott. Redini and Graziella Clerici.]

74. Joe Colombo. Candy air-conditioner. 1970
75. Mario Bellini. TCV 250 video console.
Olivetti. 1965
76. Mario Bellini. Logos 270 data-processing
machine. Olivetti. 1970

INTRODUCTION

The phenomenon of Italian design is remarkable not only for its prolific production but also for the body of criticism on the subject that it has generated in Italy. This analytical operation has not confined itself to an evaluation of the formal properties of the objects produced but has dealt especially with the sociocultural context of design and the historical processes that it undergoes.

In order to provide a frame of reference for the objects and environments in the exhibition, a number of distinguished Italian critics were asked to analyze, respectively: the role played by design in the country's economic development (Ruggero Cominotti); the relation of urban planning and housing — or of its absence — to the production and consumption of household furnishings (Italo Insolera); the external and internal influences on modes of thought that have affected the transformations of modern Italian design (Carlo Giulio Argan); the manipulation of design in the service of consumption, and the resulting dilemma of designers (Alessandro Mendini); the significance of the counterculture positions adopted by groups of radical designers (Germano Celant); the metamorphoses in the ideology of Italian design that have resulted from aesthetic and socioeconomic pressures (Manfredo Tafuri); and the aesthetic and political premises of the emerging counterdesign groups (Filiberto Menna).

To these critical approaches, many others might have been added. It would have been interesting, for example, to examine more closely the role that Italian magazines devoted to design played during the last decade in erecting the façade of Italian design; or to give a critical description of the active participation of the craftsman modelmaker and the Northern Italian industrialist in the actual process of giving physical form to a designer's proposals. There might also have been a detailed chronicle of the long struggles to reform design education in Italian universities and academies.

The critical methods encountered in these essays are almost as varied as the topics with which they deal. (Perhaps the only thing they have in common is their lack of complacency about the situation of design in Italy today and their skepticism about the prospects for substantial amelioration in the near future.) This diversity is in part due to the rapid changes that occur in many aspects of contemporary art and thought, not least in the field of design. Thus, the critics have a growing suspicion of the long-range strategies of classical polemic warfare and prefer to deal with a constantly shifting situation by dynamic guerilla tactics.

Such avoidance of stratification and rigidity in their critical stance has had several advantages. On the one hand, it has given the critics some partial successes; on the other, it has allowed the designers, in some cases, to recognize weaknesses and correct their own positions. By the same token, this absence of fixed terms of reference has also allowed criticism against consumption, for example, to be itself transformed, in turn, into a subject for consumption.

But things are not over yet. The skirmishes continue; the war is still on.

ITALIAN DESIGN IN RELATION TO SOCIAL AND ECONOMIC PLANNING
Ruggero Cominotti

The title of this essay might be considered ambiguous, or even misleading; one could object that it attaches too much importance to something that so far has had only a limited effect on Italy's overall economic and social development. Actually, planning began in this country some time ago; first with the five-year programs of the big state-owned industrial concerns, and then with the formal drawing up of plans for the entire national economy, dating back roughly to the end of the 1950s and beginning of the '60s. But, as is often the case in politically and socially fragmented countries, the coalition that presides over the birth of a new policy does not follow it throughout its operative life, and the most lofty and essential undertakings end up by having to fight vigorously to establish their position and assert their authority. This is not to imply that the Italian economy has not developed according to a fairly recognizable pattern, which economists and experts have analyzed with a clarity all the more impressive in view of its valiant efforts to conceal their inability to alter appreciably the course of events.

Italy's economy has changed remarkably in the past twenty years, and the visual aspect of the land, its towns and its countryside, have perhaps changed even more dramatically. A further change, more subtle and less obvious, has taken place in that rigid network of relations and roles linking together the members of the Italian family. Three factors chiefly contributed to economic development during the 1950s: cheap manpower, a strong foreign market, and a large backlog of unsatisfied demand for durable goods. Of these, the two latter have been directly influenced by a design capable of imposing a new style not only on behavior (through clothes and shoes, for example) but also on family life and structure, both of which were deeply affected by the greater mobility due to automobiles and by the beginning of a different role for women, whom machines had freed from some of their daily chores.

The Italian's love of novelty and of formal beauty in the things surrounding him has given an enormous impetus to industries (some of which sprang up out of nothing after the war) that now export household appliances, automobiles, and clothes to Europe and overseas. The elegant form of design given to television sets, refrigerators, and many other household commodities produced by Italian industry has contributed notably to the birth of large establishments in parts of the country in which there had been none before, and to the transformation of some particularly enterprising factory workers or carpenters into big industrialists. Of course, this was not only a question of design, and Italian refrigerators did not become the terror of all European manufacturers merely on the basis of their good design. There were other important contributory causes, such as factories built from scratch and equipped with the most modern methods, techniques, and materials, enabling wide-scale production to begin immediately; and there was cheap labor. Design has nevertheless had an enormous importance, by making producers habitually adopt the most rational solutions and turn out goods that fully satisfy the consumers' slightest requirements; it has also given them that slight edge over their competitors that was necessary if they were to maintain themselves in mass production.

Thus, Italy has managed to realize her dream of achieving a favorable balance of international payments, of becoming a sort of workshop importing raw materials and exporting manufactured goods that manifest not only know-how and technical and organizational ability, but also the capacity to invent forms derived directly from age-old practice in craftsmanship, long-standing contact with simple forms and natural materials, and the accumulated skill of artisans. Design also affected the primary and probably most important factor,

Ruggero Cominotti is an economist now living in Turin, where he is director of SORIS, a company established by the Italian Institute for Furnishing (IMI) and Olivetti to engage in economic research and marketing surveys. He has undertaken many surveys for public agencies and private industry, particularly in relation to industrial economics, distribution, and marketing. He has carried out two studies for the European Economic Community: one on the prospects for research and industrial development in the field of space aeronautics in Europe and the United States, and another on the potential demand for large computers in Europe up to 1980.

cheap manpower, inasmuch as it provided the Italian worker with an ideal to be attained, an acceptable objective on which to pin his hopes. The new design of certain goods and public projects was perhaps the most powerful force for change, setting forth a visual symbol of the new existence, the final goal on the painful road from the peasants' huts in the Appennines to the industrial towns; from a condition that was both lonely and subject to excessive social control to one that offered a higher income, greater access to services, and the possibility of participating, in however indirect and subordinate a way, in the great theater of modern life, with its public enactment of dramas and comedies. In this connection, we may recall what Alberoni says in *Consumption and Society*: 'In the case of the automobile, the significance of the revolt is still clearer: for the family, the car is the concrete symbol of its autonomy, its privacy in the face of collective rules and regulations, of its territorial and social mobility, and of the promise of new experiences not imposed from without; for the man who drives it, it is the symbol of his status as a free and independent worker answerable to no one; for the woman driver, it is the sign of her emancipation from domestic bonds; for the youth, it is the symbol of his freedom from authority; for lovers, it represents a foretaste of their homes and freedom from control in their sexual lives' (1).

Good design has made Italians accept thruways, motels, and service stations; it has given recently urbanized peasants the feeling of being in a big, beautiful theater in which everyone can play a role. The restaurants overlooking the highways, the service stations, at once simple and overdecorated, the mirage of brightly colored machines and objects in complete contrast to the traditional dark brown furnishings in the homes of Italian families (particularly those of peasants), have all played their part as sovereign instruments of an economic policy aimed at the radical, often brutal, modernization of the country, and the creation of an industrial basis wider and more solid than the weak one of the period between the two wars.

Possibly the designer may not have realized how essential his function was nor have understood the importance of his work. With very few exceptions, he accepted without protest his role as propagandist for a new world and as exorcist of the devils that might have frightened overtimid newcomers. Actually, during the country's economic boom, Italian designers came up with no alternative proposals nor any suggestions for a different approach to economic development. Design is still seen today mainly in its relation to business, as a modality for differentiating various commodities and lines of products, of increasing market penetration to the maximum, of enhancing the competitiveness of goods and encouraging the demand for replacing them before they are outworn. The tendency is still to ignore the obvious fact that the sum of these effects at the management level has in the past had remarkable importance at the national level, and will become even more important in the near future.

Just as the present exhibition at The Museum of Modern Art seeks to spot a new trend and look to the future, rather than to evaluate the past critically, so with respect to the economy we should also look to the future and assess the challenges and possibilities that confront Italian planning and design, now that the great 'spontaneous boom' of the 1950s and early '60s has collapsed.

Economic planning is a decision-making method developed by industrial concerns to choose their objectives, and the procedures and means for obtaining them, as well as to prepare programs for various sectors and establish techniques and instruments for their control. It has caught on because it provides a response to the characteristic features of modern industry: the long lead time required to think up something and produce it, the high investment in relation to unit cost, and the impossibility of switching quickly from one sector to another without loss. Planning is not, therefore, the sum of the programs of a firm's different divisions; on the contrary, its essential function is anticipating what others will do, or as technical jargon has it, forecasting the market trend; and in the light of that, defining what one wants to do and coordinating all actions to that end.

In a market with several producers, a steady influx of new products, and an abundance of many similar wares successively offered to the bewildered consumer, a company carries out its program through innovation and the condition basic to management — diversification. In planning, the firm must foresee the new trends in goods and services that will prevail in its market; and it must obviously try to prevent its competitors from controlling those trends. Innovation, the cornerstone of the whole planning structure, thus becomes fundamental; and design becomes important to the extent that it is the principal instrument for renewing not only the outward forms but also the functions of the products, and for establishing relations with the users. Anyone who can read the signs can see in the programs of the respective firms how each of them intends to contribute to changing the physical and human environment, and the pace at which it is prepared to improve this environment by means of products that offer a harmonious balance between function and appearance and succeed in winning consumer acceptance.

We have already mentioned the effect that the putting into practice of these companies' programs has had on the Italian economy by encouraging the urbanization of the masses expelled from agriculture. But, still with an eye on future possibilities, we must come now to the main theme, which concerns the transfer of planning techniques from industrial firms to the state, and from the sphere of production to that of civil and political affairs.

The modern state has such power to control and coerce the individual citizen, and wields so much influence over the economy, that it plays a dominant role in society. This is particularly true in Italy, a small country with few resources, where the state's ability to amass great sums through taxation has always been of prime importance in determining the rate of national economic development. The government's adoption of a planning system means above all the strict coordination of all public activities, and the directing of all its operations to a single goal in order to achieve strategic objectives. This elementary definition of public planning is one acceptable even to those who believe that a reduction of the state's activities is a necessary guarantee of liberty for the private citizen and works to his advantage.

In Italy, with the adoption of a planning system at the government level, the state began only quite recently to define and coordinate its objectives and to make a long-range evaluation, well in advance, of what it wanted to achieve, what means it would adopt, and what effect its actions would have on the daily life of its citizens.
It is not surprising that, once the boom of the 1950s and early '60s was over, national planning should have concentrated mainly on the economy and culture, on the way of life of Italians and their ability to reach a higher standard of living without completely losing the few positive qualities still left in their existence. The 1971-75 plan emphasizes what can be generically described as urban problems: leisure, municipal transport, schools, professional training, social welfare, public health services, etc.; in other words, it seeks to point the way to transforming the gains won during the boom by directing the new potentials of Italy's economy toward a higher quality of life. Factors and phenomena once taken for granted now come to the fore,

requiring an attention to design rarely paid to it previously by the state, either in Italy or other countries. Certainly, the characteristic design of London telephone booths was, if not precisely accidental, not too greatly influenced by a specific concern with its aesthetic and functional effects within the framework of a coordinated effort by the state in this regard. Nevertheless, the telephone booths and the shape of the Underground coaches, like the service stations of bar appointments in Italy, are important elements in the urban setting and significant components of the country's 'visual civilization.'

Public authorities in Italy have always paid great attention to the visual appearance of towns. 'Among those matters to which the men who undertake the city's government should turn their attention, its beauty is the most important,' wrote Salimbene da Parma in his thirteenth-century chronicle of Siena (2). All medieval chroniclers stress the interest the governments of the Italian communes took in the outward appearance not only of public buildings, but also of the towns' open spaces, squares, and streets. The miraculous building and city planning that took place in medieval and Renaissance towns was the consequence of a steady concern on the part of the authorities, and their ability to coordinate successfully the activities of their citizens.

Coordinated and planned action by public bodies, either national or local, assumes enormous importance when the problem with which the plan proposes to deal are as great as those in Italy. First of all, there is the opportunity to adopt the very solutions that made Italian towns what they are, and that are indissolubly linked to the Italians' concept of a city: the square as a town parlor, the street as an extension rather than a denial of family life, and that complex line of large and small buildings that make up what Ivor De Wolfe calls 'the Italian townscape,' one of the best examples of 'spontaneous planning' (3) which retains a common but not monotonous urban pattern throughout, from the line of the roofs to the shape of the gateposts. Faced with the necessity of building at least two million new rooms a year, the problem of not spoiling the townscape to the point of banality and bad taste requires an effort in design not limited to objects, but embracing as a complex whole 'the greatest of all human activities, the building of a private, specifically humane environment (town), against all preconceptions of Nature' (4), and also not restricted to the planning of the individual elements — the buildings — that compose it. The towns inherited from our ancestors are perhaps the best imaginable to ensure a maximum of what De Wolfe calls 'urban felicity.' Modern technology and planning must strive to keep the new up to the standard of the old, or at least not to destroy what already exists. On the other hand, two million new rooms per annum call for an enormous effort, not only in town planning but in the entire field of design. We must invent a new style of street furniture; to quote De Wolfe again: 'These are the bread-and-butter furnishings of the street, the objects of everyday service. In common with lamp-posts, bus-stops, hoardings, shop displays, parking meters, railings, pavings, fascias, street-names, neon, they determine eventually its character — so inevitably that you would expect street furnishings to be an expertise practiced by highly qualified specialists instead of what it is, an anarchy' (5). We must also invent a style of home furnishing that can bring high-quality design within the reach of all consumers and compensate urbanized peasants for what they have lost in open space and light.

So we are faced with the necessity of finding a positive solution to the problem set forth by Giorgio Giargia and Adriano de Maio in their 'Note on Design': 'Italian design in furnishing must now choose between two alternatives. The first is to remain limited to the élite and be involved only with a few firms having a small output characterized by semicraft fabrication, distributed mainly in metropolises to a fairly narrow clientele, and exporting through a few big agencies. The second is to widen the market considerably, both at home and abroad, which would require centralization and specialization in production and a greater impact at the distribution level' (6). The alternatives so well described by Giargia and De Maio are applicable not only to furnishings but to the whole of Italian design. If design intends to follow the guidelines and opportunities presented by national planning, it cannot avoid choosing the second alternative, although it is by far the more difficult.

An approach of this kind should be consistent with the definition of housing as a 'social service,' and with the inclusion in the national plan of 'pilot projects' involving problems such as those in Venice or a new town on the mainland side of the Straits of Messina. As we have said, the prime target of the national economic plan for 1971-75 is actually 'modernizing and strengthening the civil infrastructure ... needed to make a modern industrial apparatus work.'

The overall approach to economic planning in Italy fosters a significant relationship with design, viewed comprehensively as the visual expression of ideas about society and its development. Actually, economic planning in Italy began with the drawing up of a broad, all-embracing scheme for the economy that brought together in one coherent whole all the basic data concerning the country's future development. But the abstract nature of a scheme encompassing such a broad field necessarily reduced its effectiveness, making it somewhat remote from the real problems of social development. A comprehensive, coherent plan assumes importance only to the extent that there exist agencies and methods for carrying out its premises effectively; that is, only if the structures of production and society are capable of moving in the desired direction. It is not surprising, therefore, that the new plan has followed a different line, characterized by various ways of defining problems and methods of intervention. The purpose of the 1971-75 plan 'is not to set forth the proposal for a program completely defined in every respect, but rather to urge everyone concerned with planning to express concrete opinions on diagnosis, problems, objectives, proposals...' (7).

Accordingly, the 1971-75 plan is divided into numerous sections; here, we must refer to the one on 'pilot projects' that attempts to deal concretely with such individual problems as the organization of health services in a poor region of South Italy, or the creation of a system of harbors for tourists. The pilot projects represent, on the one hand, a change in planning technique, in that they mark the final abandonment of abstract, all-encompassing economic surveys and the pinpointing of specific problems, proposing concrete study for their possible solutions On the other hand, they open up a very wide field to Italian designers, if they are willing to give up luxury projects to work on less rarefied themes — for instance, drafting the physical and functional requirements for the multipurpose cultural centers to be erected in Emilia, designing a new kind of collaboration between agriculture and industry in the earthquake-stricken valleys of Sicily, or redesigning metropolitan areas in which uncontrolled building has caused far more damage than any earthquake. The designer's direct and open commitment to carrying out the planned program may bring him into contact with a much ruder and less manageable reality than designing unique objects for houses of the wealthy; but it may also bring about a complete revolution in his work and enable him to put his abilities and talents — hitherto only partially utilized — at the service of the whole country.

Italian planners and designers will not only have to give careful consideration to the urban landscape but will have to pay equal attention to the countryside, if, as Emilio Sereni says, 'the rural landscape ... is that form which man consciously and systematically gives to the natural scene, in the course of, and for the purpose of, his productive agricultural activity' (8). The 1971-75 plan foresees a further reduction of the farm population; the extent of this will be determined primarily by the role that agriculture manages to find for itself within a changing society.

Changing cultural patterns, an emphasis on the intensive and rational exploitation of arable land, the stress laid on cattle raising and the primary processing of produce, all require designing a new type of peasant, that is, a new wealth of objects and designed solutions to replace the traditional tools that were suitable only for underdeveloped, labor-intensive technologies.

If we succeed in accepting this challenge, salvaging the values implicit in craftsmanship; if we manage to endow modern objects with life and warmth, so that they can survive outside an urban environment; then at last we shall be able to say that Italian design has successfully accomplished one of its most difficult tasks.

Finally, the 1971-75 plan of course considers various sectors of industry, for which it sets very ambitious goals. The three factors that created and maintained the boom of the 1950s are now played out, but industry will have to continue developing at an average rate of increase of over 7 percent, roughly doubling its output every ten years. The plan lays emphasis on outstandingly innovative sectors, such as chemical products and mechanical engineering.

In general, it sets industry the task of considerably increasing employment and exports. The latter will have to to go up, particularly as regards machinery, chemical products, and vehicles, and to a lesser degree, textiles. Faced with the difficulties of the international market, which always particularly affect sophisticated manufactured goods, Italian industry will obviously have to make an effort specifically based on planned innovation and design. Formerly, some countries enjoyed permanent advantages in international trade; but today, as Austin Robertson has pointed out, 'differences of comparative advantage are becoming narrower and shorter lived' (9). The advantage goes to those who can maintain for their products a margin of superiority, however small, through innovation and the continual updating of their products. As labor costs level out internationally, and as know-how and manufacturing techniques in major industries become diffused among all countries — in short, with the disappearance of every fixed advantage that favors one producer over another, it will increasing be design that makes the difference and gives that small margin of advantage enabling the producer to keep and widen his share in the international market, thus expanding his scale of production and reducing his costs correspondingly. Increasingly, it will be design that determines who will win out in international competition.

But Italian design is faced with a challenge even more serious and important for economic development — and not that of Italy alone: on the one hand, the creation of an industrial environment in which workers can clearly see the function and utility of socially coordinated work; and on the other hand, the endeavor to remove from factory work and heavy industry that inhuman and alien quality inherent in it today. Probably, in the remote ages of human history, the transition from a hunting to an agrarian culture took place with the aid of a complex set of rites, symbols, and exorcisms to accustom people to a completely different environment (open fields instead of forests or prairies), and to a type of labor equally different from their traditional one (work that was continuous, systematic, and settled instead of intermittent, haphazard, and itinerant). In *The Myth of the Machine*, Lewis Mumford speaks explicitly of the 'ritualization of work' (10). The transition from agriculture to the production of goods and services may not yet have been completely accepted. The abandonment of an immediate, direct relationship with nature in favor of a relationship with other people — a working group — may still provoke shock, the same shock that South Italian laborers experience when they have to become factory workers almost overnight.

The modern worker may still reject the industrial landscape, just as the ex-hunter, neopeasant rejected the open fields and took shelter in the woods or brush. It is difficult to discover what kind of ritual might enable us to overcome this unconscious obstacle, which all too readily finds a way of rising to the surface. Given the prevailing culture, we could try to give to design — the creation of recognizable forms — the task of inventing for labor an environment capable of exorcising the alienating touch that industry carries with it, and of restoring to collective labor its necessary sense of purpose. Then, and only then, shall we be able to consider the brutalization of the worker in Chaplin's *Modern Times* a thing of the past; and only then will planning and design, hand in hand, be able to write finis to the paleoindustrial age.

(1). *Consumi e Società* (Bologna: Il Mulino, 1964).
(2). Quoted by Daniel Waley, *The Italian City Republics* (London: Weidenfeld and Nicolson, 1969).
(3). *The Italian Townscape* (London: The Architectural Press, 1963).
(4). Ibid., p. 22.
(5). Ibid., p. 119.
(6). Presented at the Club Turati — ENI Congress on 'Innovations and Strategy of Industrial Development' (Proceedings published in Milan: Etas Kompass, 1971).
(7). *Documento programmatico preliminare — Elementi per l'impostazione del programma economico nazionale 1971-1975.* I, p. 6.
(8). *Storia del paesaggio agrario italiano* (Bari: Laterza, 1962).
(9). Quoted in Andrew Shonfield, *Modern Capitalism* (Oxford University Press, 1965), p. 40.
(10). Lewis Mumford, *The Myth of the Machine: The Pentagon of Power* (New York: Harcourt Brace Jovanovich, 1970).

HOUSING POLICY AND THE GOALS OF DESIGN IN ITALY
Italo Insolera

For almost a century, housing in Italy has increased at a dizzying rate (a phenomenon that began almost two hundred years ago in Western and Central Europe, England, and the United States, and in Africa within the past twenty years). In Italy, two factors accounted for this increase: the general growth in population, and internal migration from countryside to city and from south to north. In 1871, the population was 26,000,000; by 1971, it had risen to 54,600,000 — almost double. In 1871, 21.5 percent of the population lived in cities with over 20,000 inhabitants; by 1971, this figure approached 50 percent. In 1871, only eleven cities had more than 100,000 inhabitants: six in North Italy, two in Central Italy, and three in South Italy. In 1971, forty-three cities had over 100,000 inhabitants: twenty-four in North Italy, seven in Central Italy, and twelve in South Italy. Because of large migrations, in Italy today there are sections in which the population is greatly increasing (cities, the North), while simultaneously elsewhere it is decreasing (rural areas, especially in the South); there is a dearth of urban housing, but farms in the countryside are abandoned.

Under these circumstances, any statistical average for the entire nation lacks all significance; for example, the figure that there are 56,000,000 *vani* for 54,600,000 inhabitants, or more than one *vano* per capita. (*Vano* rather than 'room' is the basic unit used in Italian statistical data on dwellings; it includes not only rooms properly speaking, but such other quarters as kitchens, bathrooms, and other service areas, according to their size. Inasmuch as *vano* denotes something that is socially nonexistent, it is a completely meaningless term.) In cities, however, there are far fewer than one *vano* per person, and for the decade 1970-80, the total need has been estimated at several million.

Thus the problem of housing, together with that of employment, has come to assume top priority during the past century. But the Italian government's policy in confronting these problems has, surprisingly enough, remained almost unchanged, irrespective of changes in regime. The ruling class in Italy has always been completely conservative in its attitude toward housing, and consequently its housing policies have been based on conservative standards. It is logical, therefore, that official policy should have fallen far short of attaining any of the needed objectives, even judged from a purely quantitative point of view. Given the doubling of population, increased urbanization, and internal migrations, a dynamic policy would have been required, certainly not one intent on maintaining the status quo both of the economy and the power structure.

Housing construction today still remains predominantly in the hands of private enterprise, just as it did a century ago. In 1970, the public share in the total amount of housing built was 6 percent, as against 94 percent for private enterprise; in terms of the amount invested, the proportion was even more depressing — only 2.8 percent of public funds as against 97.2 percent of private capital. Furthermore, the private sector has shown a proportionately steady increase since 1951, when 31.5 percent of building represented public housing, and the percentage of public investment was 25.9 percent of the total. The paucity of public initiative may be accounted for by the fact that characteristically, in this field of endeavor, it has always been complementary to private enterprise and non-competitive in nature. Public housing has been confined to those unprofitable segments of the market which private enterprise had no interest in cultivating. Moreover, during those years in which the government was most involved, housing policy was regarded as an aspect of the employment program rather than of town planning (1). In the earlier postwar years, the primary objective was to provide jobs for the greatest

Italo Insolera, who received his degree in architecture at Rome in 1953, has taught in the architectural schools of Rome, Florence, Venice, and, most recently, at the University of Geneva. He is a practicing architect with a studio in Rome; in 1971, in collaboration with Pier Luigi Cervellati, he was awarded second prize in the competition for the new University of Florence. In cooperation with city planners, economists, and naturalists, he has made a number of studies of plans for the coasts of Sardinia and Tuscany and is coauthor of a five-volume work, *Coste d'Italia* (1967-71), sponsored by ENI.

possible number of masons and other unemployed persons, and to give work to the traditional construction companies. Any policy looking toward the industrialization or standardization of building methods was overlooked; consequently, the number of dwellings could be increased and costs of construction reduced only at the expense of a decline in quality. Twenty-five years after the war's end, the same considerations still determine the public role in housing; while poor quality and small quantity have resulted from giving the greatest possible scope to private enterprise.

Furthermore, housing construction is only one type of investment and must compete with other investment possibilities; private capital will invest in houses only if this is more profitable than industrial, commercial, or financial enterprises. Funds therefore become available only for that portion of the construction industry that can guarantee the greatest return, irrespective of its degree of priority within any national housing policy. For private enterprise, building is subject to the same principles of investment and profits that govern other businesses; for the government, it still remains part of the employment policy.

Small wonder, then, that Italy's housing policy remains quite barbaric, and that the problem still seems to be in a state of barbarism. Although technological means for attacking the problem at its roots were available, it is symptomatic that a large private concern purchased from the Soviet Union an important patent for heavy fabrication but yet has never used it to construct so much as a single house (2). The innumerable housing types that have been studied, planned, and built in the last quarter century are all alike, and except for a very few deluxe private houses and a handful of experimental public projects, they all look as if they had been produced by the same real-estate agency.

In a typical Italian home, the essential elements are obviously the living room, the kitchen, the bath, and a number of bedrooms proportionate to the size of the family. The living room is distinguished from the bedrooms only by its slightly larger size, and perhaps by a balcony. The kitchen must adjoin the living room but be completely separate from it, this being based on the assumption that food is prepared in the kitchen, and that there is a dining area in one corner of the living room. Such a plan was appropriate to the life style of a middle-class family some time ago, presupposing the presence of a servant who was rigidly separated from the 'family'; but although this is no longer the case in middle-class households, the plan has remained unchanged, regardless of whether the family always eats in the kitchen, or whether the person who prepares the meal in the kitchen will thirty seconds later be eating it in the living-room dining area.

From the point of view of the Italian real-estate market, the ideal dwelling consists of a central corridor with rooms opening off it to left and right. Studies on housing made during the Rationalist period taught us that the corridor can be narrow, enabling the builder to save several square feet. Of course, we failed to draw from these studies the lesson that these square feet could be put to better use by enlarging the bedrooms and living room, or that the corridor might be done away with altogether, and the floor plan differently organized.

The culture of housing and that of neighborhoods have developed as quite separate disciplines, each almost invariably ignoring the other (3). In any event, the question of neighborhoods arises only in the case of public housing and a few large private developments. Otherwise, houses are built side by side, and the neighborhood is finished when the last one has been erected. The houses show no concern with the relationship between exterior and interior, with whether sunlight enters through the windows or does not, or with the possibility that the view from the windows might constitute a first participation in a wider social ambience. On the contrary, one's home — my home — is regarded as the means whereby I can shut myself off from my neighborhood and from the city; it is a means of defending and isolating myself from an oppressive landscape and a society that grows out of social barriers rather than social contacts.

Within the house, the family drama and domestic folklore still qualify and determine one's life style. The objects that fill the dwelling and give it character are conditioned by this pattern. As regards country houses, they have changed from those that existed fifty years ago only to the extent that the way of life itself has changed. The most common and widespread transformation has taken place in the kitchen. It is worth examining, because here two factors confront one another — tradition on one hand, and industrial requirements on the other.

Cooking methods have changed. Gone are the traditional Italian utensils: the big copper pot, the home-made apparatus for making pasta, the wood- or coal-burning stove, the numerous jars for preserving foodstuffs according to prized, atavistic recipes. Today, two objects dominate the kitchen: the refrigerator and the gas or electric range with oven, dishwarmers, etc. These two objects condition all the others, as they also condition the overall image of a kitchen made up of industrial products, whose 'design' was born with them, since it never existed before nor had any traditional model as a point of reference. The same is true of the less important objects, from the saucepans to the electric blender, the dishwasher to the automatic juicer, etc. A whole new technology has become necessary (urbanization would never have been possible with a fireplace in every apartment), and industry has found a tremendous market that has resulted in exploitation and manipulation.

The kitchen is modern of necessity, in part because it cannot be anything else (there is no such thing as an antique refrigerator, and the ancestral hearth cannot be moved from the country to the city), and in part because that connection between convenience and form which is the basis of contemporary design is unconsciously accepted even by the most conservative housewife, who on the other hand takes an absurd degree of satisfaction in proving to herself that she is 'correct' even when she rejects everything 'modern' outside the kitchen.

If the kitchen can — indeed, must — be modern, the living room (or rather, the 'parlor' or 'drawing room') has to be nonmodern, which means that its form must be nonmodern and is frequently only a shoddy imitation of some haphazardly invented old thing. The tableware used in the kitchen may be of stainless steel, functionally styled; but for dining, the service must be of silver, with 'baroque' or 'Sanmarco' embellishments. The false nonmodern style required in the living room and bedroom (absolutely mandatory in the large master bedroom, perhaps a little less so in that of the children, where a certain 'sporting' character is permissible) is the outcome of an outmoded way of life, or simply of that terrible void of leisure hours spent in the family circle, where for lack of initiative habitual forms persist unchanged.

Not even the industry of electrical household appliances that has overthrown the customary image of the kitchen has succeeded in effacing the image of the 'drawing room.' The radio, for example, disguised itself for decades by aping traditional forms, and only became an authentic and widely distributed object of design when transistors liberated it from electric outlets. The little transistor radio leads us to another point. For the first thirty years of its life, the radio had to have a fixed place and therefore took on an appearance matching its surroundings in the main corner of the living room. Subsequently, it has become a little box that can be put anywhere, but, unlike an ashtray, it plays a dominant role. The fixed position formerly occupied by the radio

From the film *La Classe operaia va in paradiso*
('The Working Class Goes to Paradise')
Produced by Euro International Films
Script by Ugo Pirro and Elio Petri
Directed by Elio Petri
Photography by Sergio Strizzi
Massa ... Gian Maria Volonté

Massa, a Milanese metal worker, after an accident at work considers selling everything, giving up his job, and emigrating to Switzerland. Partly in earnest, partly without really believing what he is doing, he makes out an inventory of his kitchen and living room. Around him are the objects typical of a 'good' home.

Interior of Massa's house. It's evening. Massa is completely absorbed in making an inventory of the domestic appliances of his kitchen, cutlery, furniture, pots and pans. He holds in his hand a sheet of paper and a pencil, with which he marks everything down.

MASSA (talking to himself):

Refrigerator — cost 105 ... present value, 10,000 lire. ... Dishes — they'd fetch — they cost about 1,000 lire apiece, more or less ... useful only to the junk man ... you'd get back a package of washing powder! ... current value, 300 lire.

Massa goes forward and opens the cutlery drawer.

Cutlery ... so old ... no value.

He looks around him. The kitchen is very untidy.

Gas-stove — spent 55,000 lire for it — a month's salary ... you can throw it away ... pots and pans — a week's salary, at least ... so old — worth nothing! Furniture: 120,000 lire from a Sicilian!

Massa is surprised and even a bit frightened, as if he had been caught stealing.

has now been taken over by the television set. Portable TV models are not comparable to transistor radios, for actually, irrespective of its position, the video image always finds its own place and establishes a relationship that cannot be anything other than visual for those who live in the house and use the set. In any case, the most common TV model is the large, fixed one, for portable sets are too expensive, and their but in the living room, mahogany or walnut is obligatory.

Television was quick to free itself from the temptation to assume a fake-antique guise. Its principal formal element is the transmitted image, always changing, always isolated from the domestic environment into which it is interpolated. The casing that surrounds and encloses it was soon reduced to the essential function of serving as a link between the televised image and the furniture. What really counts is the material: plastic is allowable in the children's bedrooms, but in the living room, mahogany or walnut is obligatory — that is, the same material as the rest of the furniture.

Those who lived in a traditional old house had a relationship with every one of its components, from the walls to the furnishings. Their ancestors or they themselves had built that house according to their own needs; they had completed it by making one piece of furniture or another; and they had transformed it through a slow process of settling and selection. The relation between the house and its inhabitants grew out of usage, through which it was continuously validated.

By contrast, a city apartment is bought (or rented) once and for all. Others have built it, and in doing so have imposed their plan on the lives of the future inhabitants. There is a break between the house itself and its furnishings, for through force of circumstance part of what should be an integral process has been delegated to others.

The tenant reacts by devoting his attention to the interior decoration, furniture, and ornaments, for only through these can he differentiate the cubic feet he occupies from those occupied by others and express in his home his own personality and individuality. These are the things that he will be able to take with him when he moves, allowing him to preserve some sense of continuity and reassert his identity in spite of the disruption caused by the removal.

After having received his house, either from the state or a landlord, the tenant takes a kind of psychological revenge through interior decoration; this allows him to enter into competition with the creator of the house, whom he defies by visiting a furniture store. Here he encounters no policies, no taxes, no laws; everyone is free to sign as many bills as he wishes in order to buy whatever furniture he chooses. The absence among the populace at large of any policy regarding social environment, and of any culture of dwellings, deprives the tenant-decorator of any standard except that of form. His problem is to select the form of objects that are already predetermined by the position, use, and size of the *vani* that constitute his home. What these objects will be can be deduced both from the pattern of his family's way of life within the household, and from what the market has to offer. He is confronted with finished products to be bought and placed within a finished product, his home — which in turn is part of still another finished product, the neighborhood. Certainly, it is possible to act as one's own interior decorator and create objects that can be used in a nonstandardized way; but this is either very expensive, obliging one to renounce industrial products and resort to a cabinet maker, or else it requires powers of invention that in turn could be attained only by adopting a different mode of life — and ultimately, this would also entail rejecting one's house and its neighborhood.

What does Italian industry offer in the field of interior decoration? Three different categories can be distinguished.

The first consists of completely machinemade products that can be manufactured by no other kind of process. It includes electrical appliances, either utilitarian (e. g., refrigerators, washing machines, dishwashers, stoves, blenders, grinders, juicers, etc.) or for other purposes (television set, radios, record- and tape-cassette players). We have already mentioned how such utensils and apparatuses have transformed habits and images within their respective domains. To this category also belong sanitary equipment (bathtubs, washbasins, etc.), and in general all household utilities.

This is the only category that comes into being through a process of 'industrial design' and has shown itself to be in a position actually to influence people's taste by resorting to mass advertising — an indispensable factor for industrial production. It is, in fact, the task of advertising to decree that a product is outmoded and obsolete whenever the cycle of production demands that the market be infused with new life. This requires frequent changes in the form of the object; at a certain moment, the problem of design is no longer that of expressing a technological innovation but instead becomes that of expressing something 'different,' which in spite of its diversity nevertheless retains enough recognizable features to distinguish it from the offerings of competitors.

The second category includes items that do not have to be machinemade, such as handcrafts, as well as objects produced with the collaboration of artisans, and those made wholly or in part by industrial processes. This is the largest field, and the one that prevails almost without challenge in the many Italian periodicals devoted to decoration — all of which are associated with some firm or advertising agency. Chairs, lamps, tables, bookshelves, wardrobes, beds, armchairs, sofas; science-fiction inventions side by side with classical types; traditional and new materials; good taste and bad taste; unique pieces and those that are mass produced. Usually, each piece is signed, and these signatures help to raise the price of the commodity, especially if they are by foreign designers for whose output a given company has sole distribution rights.

Almost all the articles (except of course electrical appliances) that are responsible for having established the reputation and success of 'Italian design' fall into this category, which must therefore be regarded as the pilot group in Italian home decoration. Its role is nevertheless somewhat limited, inasmuch as objects in this class are almost always expensive, sometimes extremely so. It is true that the nonmodern, non-industrially-produced objects in the third category are also often expensive, but on the other hand they are 'safe,' and not only from the standpoint of the purchaser, who lacks any concept of 'design.' Here, one can sit at one's ease in the same tasteless armchair in which one's father sat, in front of a tasteless table equally lacking in any sense of period. Bad taste never grows old and always has an assured market — a boon for the consumer as well as for the producer By contrast, the latest article of furniture by the best 'designer' has a quite different destiny. Its success depends upon the public's good taste — rarer by far than bad taste. Moreover, even at the peak of its success, it is subject to the same principles of capitalistic enterprise that require its eventual substitution by something else and that promote its aging, obsolescence, and depreciation.

Cut off in this way from the culture of dwellings, 'design' ends up as a style, a characteristic of a quality market that is inevitably élitist. Design establishes different forms within the middle-class pattern that it serves precisely to characterize; relying on taste, it is consequently conditioned by it. But isn't what we call 'taste' a prerogative born of certain social and cultural relationships that are restricted exclusively

Daylight. Massa stands in front of the long mirror of the dressing table.
MASSA: *How much can this mirror be worth — 10,000 lire ... the bed? around 10,000 lire ... the cupboard? 20 ... 30 ... with all the furniture, maybe I could get about 100,000 lire. ...*
He makes a face at himself in the mirror and walks around the room, looking at his furniture and bric-a-brac.

Finally, he sits down on a sofa covered by a sheet of plastic.

MASSA: *I'll sell the lot! Switzerland, emigrant ... a pair of lovers — 8,500 lire; 1,500, if I get that .. two days' salary ... small gilded table with a foreign print in majolica — 30 hours of hard work; 5,000 lire ... painting of a clown — 30 hours; 24,000 lire ... radio-book,* I Promessi sposi *with Indian music — 5,000 lire ... maybe 10,000 in Switzerland, 15,000 in Milan. ... crystal vase ... plastic almond tree — 2,000 lire; it even brings bad luck ... various animals ... if I ever find the guy who made this stuff! ... 'magic moment. Magic candlelight — with Ronson candles' ... Screw me! it I could only find the guy who made this rubbish!*

The alarm clock goes off.

The alarm, the alarm!? I wonder why? ... a dog-radio ... Let's see if it still works ... dog-radio — a dog! To think someone actually made it! ... Hours, days, weeks, months, piece work — all at a given rate ... till death, bah!

Massa remains thoughtful, sitting on the sofa in front of the table decked with two 'Ronson candles.'

to the cultivated middle class? And isn't it an imitation of this, propounded as a social objective, that we find mirrored on a wider scale in the recent suburbs of our cities? Equally detached from industrial production and from mass consumption, design has become merely a matter of style; and this is what differentiates it from the final category of objects, those in the third group.

If it is the second category that monopolizes the pages of interior-decorating magazines, it is the third that predominates in mass production. Furniture stores are most frequently located in the outskirts and the developments fringing the larger cities, where new housing has created the greatest demand for furniture; and the merchandise they display belongs almost entirely to the third category of objects (while, by contrast, the second category reigns supreme in the smart midtown shops). In the third group, just as in the second, one can find everything needed for the house, to suit every taste and meet every requirement — but not to accord with every style, simply because there is a complete absence of any style whatsoever, even though the dealer may advertise his line as 'English' (with few curves), 'French' (with many curves), or 'Swedish' (with no curves at all), employing terms that would scandalize any specialist in antiques. Here there is no longer a question of 'design' nor of industrial production, even though this kind of furniture is manufactured by processes that have nothing to do with true artisanry. What is involved is generally a centrally controlled commercial operation, and the exploitation of the threat of unemployment among cabinet workers.

Decoration reduced solely to the function of choosing forms thus corresponds to 'design' charged with no responsibility other than that of providing forms. The only exceptions are products resulting from some exigencies of use or of manufacture; and here the word 'design' still preserves some of its true meaning and significance.

It is not surprising that 'Italian design' has probably had greater success outside Italy than in its land of origin, given the fact that a middle class capable of appreciating the value of style in a piece of furniture, a slipcover, or a lamp is more numerous (and more affluent) in Central Europe and the United States.

But assuredly we must not believe that an inadequate, mistaken, or even nonexistent housing policy in Italy can be rescued by designers operating as interior decorators. Nor can we believe that Italian design is really an effective force or a commonplace for the Italian household; it is the exception rather than the rule.

And so it will remain, as long as one of the basic misconceptions of architectural thinking during the 1960s persists, which is that the quality of form is capable of solving problems that precede and follow formal facts. 'Design' is not simply a means of giving a different form to products that are always fabricated industrially. It is part of the quest for a different kind of habitat, within a society no longer conditioned by consumption and profits. But at this point, we diverge and range far beyond the capitalistic solution to the housing problem, and beyond the question of interior decoration in middle-class Italian homes.

(1). The most important public housing law is No. 43 of February 28, 1949: 'Plan for increasing employment throught the construction of workers' houses.'
(2). This is the Montedison firm; the patent is 'silicalcite,' which an establishment in the vicinity of Naples has been using for the past six years to produce very modest articles for the most traditional purposes.
(3). Even in the most serious studies and projects undertaken in Italy during the past twenty-five years, the question of orientation has been willfully overlooked, in spite of the fact that climatic conditions in the summer are extremely arduous throughout most of the country.

IDEOLOGICAL DEVELOPMENT IN THE THOUGHT AND IMAGERY OF ITALIAN DESIGN
Giulio Carlo Argan

Since the Second World War, the so-called 'Italian style' in certain branches of industry (furniture, mechanical and household equipment, automobiles, and clothing) has enjoyed undisputed prestige on the international market. This has been due to its high quality of design, the appropriateness of its types and functions, its precision of manufacture, and its moderate prices. The style was not distinguished by any single, unmistakable characteristics, nor did it offer any sensational novelties. The quality of production was nevertheless well above the average, owing above all to the insistence upon a collaborative relationship among designers, technicians, agents, and entrepreneurs. In other words, Italian products seemed to be well designed at a time when, in almost all the rest of the world, good design was passing through a crisis.

Systematic research in the field of industrial design began in Italy about 1930, thanks to some young artists associated with the Rationalist movement in Europe; it received support from a small minority of forward-looking industrialists. This research, however, encountered great obstacles. Italian society was highly stratified; there was a great gap between the already largely industrialized North and the South in respect to their mode of life and economic status. The program for highly skilled production did not arise from any widespread demand, but from proposals for education and reform put forward by an intellectual élite. Modern schools for technical and professional training simply did not exist, still less any that offered aesthetic education. The channels of distribution among department stores — indispensable for the planning of production — were weak and, in any case, reached only the big cities. Finally, the political situation was wholly adverse. The Fascist regime looked upon the backwardness of the working class as a guarantee of its own stability; it was opposed to any advanced research, obstructed the transmission of information and the free exchange of ideas, and favored a low type of academic art. Cultural isolation and provincialism were aggravated by strict economic controls: foreign products were boycotted; Italy's privileged but technologically backward industry was protected; small local concerns, handcrafts, and spurious folk art were all encouraged. Because the study of a system of design equally applicable on any scale — to town planning, building, and industrial production — was an outcome of democratic ideology and formed part of a comprehensive program of education and social progress, it was considered suspect, at the least. Walter Gropius, Le Corbusier, and other outstanding figures in the field of modern art and culture were, according to the official line of Italian culture, all dangerously subversive. As for the majority of industrialists, since the government protected them against the demands of the working class, they had not the slightest desire to switch from a regime of comfortable protectionism to one of free competition; and they realized that the new method of design sought not only to improve the quality of products but also to transform the entire system of production and consumption. Under these circumstances, Italian research in the fields of architecture and industrial design developed principally along analytical and formalistic lines. Although this was a limitation, it also provided some immunity against the compromises that researchers in other countries were obliged to make, if they wished to ensure the immediate success of their products on the market.

After the disaster of the war, the basic idea of design seemed to be that of a force directed toward the economic and cultural reconstruction of Italian society. It was a means of modernizing the obsolete industrial structure and its technology, of reestablishing relations with the civilized world, and of bringing the Italian standard of living up to that of the more industrially advanced democratic

Since graduating in 1932 from the University of Turin, his native city, Giulio Carlo Argan has been professor of the history of art at Turin, Palermo, and, since 1959, the University of Rome, where he holds the chair in the history of modern art. He has also held administrative posts in the Ministry of Fine Arts, with responsibility for the direction of national museums and galleries. His critical writings have received several awards, and he was formerly president of the International Association of Art Critics. Currently, he is codirector of the periodical *L'Arte* and director of *Storia dell'Arte*, which he founded. His many publications in various fields of the history of art and architecture include the recent works *The Renaissance City* (1969) and *L'Arte moderna 1770-1970* (1970).

countries. It was also a means of transforming the traditional Italian middle class into an industrial middle class. It must be emphasized that the so-called 'Italian style' was not a mere status symbol but represented a genuine desire to raise Italian society to the level of international life and culture; it expressed the Italians' wish to appear less provincial and in fact as international as possible. It is nonsensical to attribute the formal qualities of this design to any inherent artistic inclination or historical tradition of the Italians; on the contrary, it is a manifestation of a tendency to break with tradition and participate on an equal footing with contemporary culture, without any national limitations. To the extent that industrial design tends to establish an international style of mass production, one might say, paraphrasing a famous witticism of George Orwell, that Italian style aspired to be the most international of all. And precisely because, intentionally or not, Italian style set itself this goal, it has retained its ideological and educational aims, which in other countries have been subordinated to profit making.

This ideological and educational purpose has given rise to a kind of critique that has now become fully a part of industry and is used for the sake of competition in the open market. Italian formalism is simply an attempt to establish aesthetics as a factor conditioning and determining production. This quality must not be merely an added and complementary factor — one among other comprehensive considerations — but must constitute the final objective of the whole process, the synthesis of all guidelines for the entire scheme. One may define the problem of Italian design in the later postwar period as follows. Inasmuch as industrial products have an informative content, the search for an aesthetic quality that has a meaning beyond that of being merely an element in the whole is predicated upon the concept of industrial design as a system of information and aesthetic communication. If design always tends to have an aesthetic purpose, it is clear that aesthetic designing, whether specifically artistic or not, is in fact an aesthetic act. Accordingly, what is looked for in the aesthetic quality of the product is not its mechanical operation nor its utilitarian purpose, but the aesthetics of its design process, the correctness of the method followed in the course of designing it. The mental and physical behavior of the designer therefore becomes projected into that of the user or buyer of the product; the object then becomes only a means for imparting a standard of behavior. But if the object is nothing but a reflection of the design process, the rule of behavior it communicates is a designed behavior, or a design *for* behavior. The object teaches one how to act according to a plan for action: its human and social significance consists in the fact that, since behavior is a way of life, in designing objects one designs life itself. A clearly planned life is one that is genuine and self-contained. On the theoretical plane, Italian design has made the point that the methodology of design is also the ethic and deontology of industrial production.

Seen in these terms, the question of Italian design takes its place within the broad development of the theory of design as formulated after the First World War by the Bauhaus. Design is not merely one among many artistic trends but is the characteristic aesthetic activity of a democratic society; or, rather, it is the process of changing the meaning and value of aesthetic experience to coincide with the transition from a hierarchical, authoritarian structure of society to a functional one. Because in Italy this transition took place during the years following the end of Fascism, in that period of crisis Italian design proposed once again a return to the original, strict principles of design. This approach was seen as a revival; but what the world needed after the war was precisely this revival of an intrinsic constructiveness, and of the educational and informative aims of industrial production.

A revival always follows a crisis, as a critical revaluation of an historical development and a seeking for the causes that brought about the crisis. Design was going through a crisis, and in view of its link with democratic ideology, its crisis was probably symptomatic of one within democracy itself. A crisis of this kind can be caused by an abnormally rapid rate of industrial growth, that is, by the impossibility or inadequacy of social control over industry, which in a democratic society should constitute the typical means of production. Capitalistic management of industry had twice led to war; design, a planned system, seemed to be a nonrevolutionary means of managing industry for social and no longer purely capitalistic ends. Design, in short, tended to be looked upon as the typical peacetime industry, and peacetime industry to be regarded as the model for all industrial production. It is no mere accident that interest in design became accentuated immediately after both World Wars, at a time when the vast war industry was under the accusation of having been primarily responsible for these disasters.

In Germany, the industrial design movement came into existence with the Weimar Republic and was then abolished by Nazism, which considered it socialistic. In Russia, the movement was born of the Revolution, or, more precisely, of Tatlin's Constructivism; but it was suppressed by Stalinism as social-democratic, revisionist, and bourgeois. Quite independently of any political strictures, there were internal causes for the crisis of the movement in the postwar period. It was so conditioned and determined by political trends that it became too strictly theoretical. In contrast to the all-embracing concept of design as the leading factor in the transformation of society, in Northern Europe, especially Scandinavia, design was only vaguely populist, proposing the development of empirical artisan skills through modern processes of design and production. But its range of expansion was limited and did not go beyond a certain type of communal organization.

The seeds of the Bauhaus when transplanted to the United States failed to bring forth the expected fruit. Within this different social and economic context, the program lost its mordant polemic character. Design was reduced, on one hand, to visual education, and on the other, to a technical service that large-scale industry was prepared to accept only to the extent that an aesthetic factor might render the product more pleasing to the public and thus speed up sales. Because of the conditions under which it developed, Italian design — at least at the outset — did not have to face difficult problems, such as the relationship between an aesthetic program and industrial technology, or between the quality and the quantity of goods produced. At the heart of such investigations lies an almost theoretical problem: that of defining the new relationship between the object and space, the thing and the environment. Mass production deprives the object of its own predetermined position in space, eliminating every relation of proportion or perspective between the thing and space, the part and the whole. Among the extreme solutions to the problem are either the assimilation of the object into the spatial structure, or else the elimination of any a priori spatial structure, so that the object can be considered in itself, with respect to its form and function. These hypotheses had been developed by Futurism and *l'arte metafisica*, respectively. In Futurist paintings, objects decompose and disintegrate within the general dynamics of space; in the 'metaphysical' paintings of de Chirico, objects are isolated and do not communicate with space. Italian designers, who are always fully aware of developments in the field of art, observed these two hypotheses and formulated a third: that an object does not depend on any given, a priori spatial structure, nor does it isolate itself from space, but rather it defines space by its very presence. In an industrial society, the environment is no longer created by nature but by industrial production; it is itself a consumer product. It therefore becomes necessary, first of all, to determine what kind of relations can be established between the object and its immediate surroundings. The first kind, obviously, is the relation between the object and the consumer. It has been proved that when industry lacks the methodological and ethical control provided by design, it gives rise to an untidy, overcrowded, chaotic environment, measurable only in terms of quantity or volume. Such an 'alienating' environment inevitably ends up by making life psychologically and even biologically impossible. In the program of the Bauhaus, the designer was meant to play a direct role in production, designing aesthetic objects that, apart from differences in technique, were the equivalents of objects of the past. According to the Italian approach, the designer influences production through the discipline or 'ceremony'. that governs the correct use of the object as a component of the environment. Since the process of organizing or structuring the environment is one of the definition — or, more frequently, redefinition — of its component parts, Italian design from the outset has taken the form of a typically linguistic operation; and this explains and justifies its formalistic character.

The alienating confusion of the environment manifests itself in two typical phenomena that indicate a deterioration in sensibility and taste: 'kitsch' and 'styling.' They appear successively, the second seeming to be an inversion of the first. Kitsch disguises, whereas styling emphasizes, the functional aspect of the product. Abraham A. Moles's masterly psycho-sociological analysis of kitsch (*Le Kitsch: L'Art de bonheur*, Paris: Maison Marne, 1971) can also be applied to styling: both are attempts to render industrial products 'artistic.' The negative aesthetics of both are caused by an excess of the artistic, that is, by an accumulation of conventionally artistic associations. Kitsch and styling are not symptoms of a degeneration of industrial design; under its present management, industrial production tends spontaneously toward these decadent forms, that is, to the falsification of the environment as a consumer product. Design imposes a strict system that tends to eliminate such distortions, and, precisely for this reason, while it seeks an aesthetic goal, it refuses artistic means to attain that end. This is not only a question of taste but of linguistic propriety. As regards their informative function, hence that of constituents in the environment, both kitsch and styling are sham, because the messages they transmit are redundant, confused, and often intentionally misleading. Design, as a process that mitigates and corrects these spontaneous distortions of the system, seeks to use the industrial apparatus as a means of imparting correct information. The positive aspect of Italian formalism is that it has shown that the proper function of design is neither artistic nor technological, but specifically linguistic.

In Italy, also, the history of kitsch and of 'antikitsch' design is inseparable from that of industrial development. It begins, that is, when industry ceases to be an instrument and establishes itself as the principal, or even the sole, index of social progress. In the nineteenth century, industrial progress in Italy was held back by the struggle for independence and political unity. Thirty years after the country's unification, industrialism seemed to be the surest means of placing Italy on an equal footing with the more advanced nations of Europe. The Turin Exposition of 1902, organized in a masterly fashion by the architect Giuseppe Sommaruga, was the impressive and typically kitsch ceremony whereby Italy repudiated her traditional provincialism and associated herself with the modern Art Nouveau movement. Apart from the naïveté with which Italy hailed industrial progress as the creation of 'artistic genius,' in that exposition Italian architecture first came to grips with the problem of an exhibition complex homogeneous in style, psychologically striking, and immediately communicating certain set ideological ideas. The use of advanced techniques of construction — for showmanship rather than for functional purposes — revealed a desire to postulate 'modernism' as the typical psychological attitude of society at the time. The aim was to create an environment that communicated meanings and was, therefore, intrinsically semantic; the whole layout of the exhibition and the architecture of the pavilions were meant to constitute a system for transmitting information. In contrast to the eclectic, weighty monumentality of the new government buildings in Rome, true and appropriate temples of bureaucracy, the industrial middle class of the North, in what was then the chief industrial center of Italy, sought to create a lively, dynamic, ephemeral, and promotional type of architecture.

At this time, also, the painter Leonetto Cappiello was making an original contribution to international modernist taste through the graphic art of his advertisements. Pictorial advertising was then a recently invented form of art, which radically changed the structure of the image by transforming it from a static representation into news and endowing it with a striking visual impact that had a very definite psychological function. As a means of expression, what it actually expressed was not so much the quality of the product it advertised as its modernism, its effectiveness as an instrument of the rites and ceremonies of modern social life.

The cosmic-mechanistic kitsch of Futurism was a polemic reaction to the flowery Art Nouveau kitsch that flooded Italy between 1900 and 1910. Despite its avowed enthusiasm for a mechanized civilization, Futurism never sought to establish a working rapport between art and industrial technology. Its interest was confined to the spectacle of modern civilization: the landscape seen from a speeding automobile, the bustle and din of factories, the dynamism that machines in operation impart to space, uniting the life of modern man and the motion of the universe in a single rhythm of sight and sound. The Futurists' view of the new dynamic relationship between objects and their surroundings is especially manifest in their stage design, choreography, and films, in which the human protagonist is always present and active. The environmental designs of Giacomo Balla (1916-1920) altered the interior decoration, but not the structure, of the middle-class home, which still remained just that — the sole difference being that it was conceived for a bourgeoisie that had, or pretended to have, modern ideas and a modern way of life. The furniture was designed in conformity with the customary types, although its shapes showed forms or outlines intended, in accordance with Futurist rhetoric, to be symbolic of the dynamism of the cosmos. The workmanship of the furniture was likewise conventional, the objects preserving their traditional structure, although somewhat simplified in the manner of the stylizations of the Central European Werkbund. Only at a later stage did this furniture become integrated into a unified spatial whole, through its outward shapes and pictorial decoration. Balla also designed large flowers of colored wood. His utopia was the 'Futurist reconstruction of the universe,' the transformation of the natural into the artificial, the designing of a new kind of nature created by man.

The following decade was dominated by a widespread movement for the revival of decorative or applied arts. Its center was the School of

Decorative Art at Monza, its guiding spirit the architect Gio Ponti, whose organ was the magazine *Domus*, which he founded and edited. The movement sought above all to influence the taste of the middle and upper strata of the industrial middle class, and to hasten what appeared to be a natural evolution from a craft type of production to that by small industries. Its influence was especially felt in certain traditional fields, such as ceramics, glass, textiles, tapestries, etc. The goal was to give back to the artist his inventive and guiding role in utilitarian production; and many artists (e.g., Arturo Martini, Massimo Campigli, Gino Severini, and others) willingly accepted the opportunity of collaborating in what promised to be a renaissance of the 'social' traditions of Italian art. Interest was focused on the object: the quality of materials, an almost neoclassical simplicity of form, and stylized decoration. Two apparently contradictory tendencies combined to form this deluxe kind of art, which then deteriorated into high-class kitsch: a return to archaic forms and the archetypal object, and a modern type of stylization whose simplfications could satisfy the prevailing 'European taste.'

The proposal for aesthetic participation, no longer viewed as a renewed collaboration of artists with the working world but as the establishment of a systematic program of industrial production, was made after 1930, thanks to familiarity with the work and writings of Gropius and Le Corbusier, the formal instruction of the Bauhaus, and foreign tendencies in nonfigurative art. The first group of Rationalist architects and abstract artists was founded in Milan. With the publication of the magazine *Casabella*, edited jointly by the architect Giuseppe Pagano and the critic Edoardo Persico, discussion no longer focused on the form of individual objects but on that of vast structures, such as cities, factories, houses, and on mass production.

Pagano was the leader of Rationalist architecture in Italy. Persico's criticism followed closely all the ideological currents of European culture; as an uncompromising anti-Fascist, he saw in the broad social and political premises of Rationalist architecture and industrial design an indirect line of dissent that appeared to be the only possible means of combating Fascist totalitarianism. The first Italian investigations in the field of design had a critical intent rather than an ideological program; they opposed official rhetoric with a strict method, and conformity with the consideration of problems.

By one of those paradoxes that appear quite frequently in Italy, Rationalist architects, though consistently excluded from the ambitious and disastrous schemes for town planning and official buildings, managed to assert their technical superiority in secondary endeavors — the staging of large exhibitions, and the decor of houses and shops. Italian design developed in the display booths at fairs and expositions before finding a place in industrial ateliers. Its function was to present and advertise simultaneously production systems and lines of merchandise; thus, it was necessary to invent temporary but striking structures, very clearly defined spaces, explanatory and informational devices that would be immediately effective, and installations that related the objects to their surroundings. In other words, a dialogue had to be developed by means of forms. It was logical, therefore, to revert to the earliest examples of architecture conceived as the construction of a meaningful, informative environment: to the projects of the Russian Constructivists, the Futurists, de Stijl, and especially, of course, the Bauhaus.

In spite of its official character, the Milan Triennale gave a vigorous impetus to this kind of research. Originally, it was an international exhibition dedicated to the decorative arts, but subsequently it became increasingly concerned with the broader themes of town planning, social architecture, and high-quality industrial production. It played an informative role, encouraging an exchange of ideas and experiences and providing an incentive to production, as well as an occasion for critical evaluation. Many industrial products were designed and manufactured with this competition specifically in view. No less important than the selection of products was the layout of their surroundings, the construction of an environment that could develop a discourse showing the purpose and necessity of every object presented. Closely related to the exhibition booths was the architecture of shops, agencies, and head offices of management. The need for a coherent environment led to a certain homogeneity in the method of design, and thus an increasingly strict correlation between the design of the object and that of its environment came to be imposed. The object could no longer be thought of as something located in space, but as a form possessing its own special properties. It was through working in this apparently subsidiary field of exhibition architecture that the first Italian architect-designers emerged: Franco Albini, Enrico Castiglioni, Ignazio Gardella, Ernesto Rogers, Carlo Scarpa, and the man who must be regarded as the very first Italian designer, in the technical sense of the word — Marcello Nizzoli. It was he who designed the Parker store in Milan, in 1934 — perhaps the first completely integrated environment of an informative kind, planned and built according to a strict method of design, which pinpointed the several aspects of the problem and resolved them with absolute coherence within a unified plastic whole.

The Olivetti factory at Ivrea, manufacturing typewriters and calculators, was the first firm to adopt systematically planned design as a strategy for its own business and marketing activities. 'Olivetti design,' a style connoting order and smooth operation, displays these qualities throughout its offices, laboratories, workshops, and showrooms, as well as in its products and the highly informative graphic art it uses to advertise them. Adriano Olivetti, the most enlightened of Italian industrialists, held Fabian socialist views regarding labor. He was the first factory owner in Italy to have a nonpaternalistic concept of business, to conceive of a company as an integrated, fundamentally democratic community, and to regard industrial production as a collaborative activity with aims at once economic, social, and cultural.

Calculators and typewriters, and later electronic computers, are part of the mechanical apparatus in management offices. Their particular function is not to produce anything but to speed up and organize the channels of communication required for management. These mechanical devices and how they work are of no direct interest to the person using them; they should neither condition his behavior nor make themselves conspicuous. The operator does, however, come into direct contact with the outer covering or shell of the apparatus, which has a dual function: it protects the mechanism and separates it from the user, who might be distracted by awareness of the inner workings and their motion. The typist keeps his eye on the sheet of paper in the carriage, just as the driver keeps an eye on the road; his movements must be accurate and automatic, not causing any problems. Typewriters, automobiles, sewing machines, telephones, and audio-visual devices all present more or less the same problem — that of their framework or housing. Following Olivetti's example, Italian design has made a particular specialty of studying the housing of machines; the purpose is not to condition the operator's behavior to that of the machine, but to achieve the most favorable conditions for coexistence and use.

Olivetti's designer Nizzoli was not an architect but a painter. He had been associated with the later stages of Futurism, had been a decorator, and created some excellent graphic design for advertising; he also took part in the vigorous revival of graphic art that logically followed the development of Italian design (the Boggeri Studio of Milan was of particular importance here). He then worked in the field of exhibition architecture, thus developing on an environmental scale the theme he considered fundamental — that of visual information. His first nonfigurative paintings, of about 1930, are clearly influenced by Malevich's Suprematism. For several years, Nizzoli worked with the critic Persico, and the two formed a veritable team; this was the first time that an artist and a critic collaborated in both the planning and the execution of design. The critic's role was that of the methodical man who checks, step by step, the cultural as well as the functional validity of the chosen forms. Owing to this experience, Nizzoli had no difficulty in becoming part of the management team when he first began to work for Olivetti. Instead of behaving like a 'creative' artist, who first invents the form of a product, or later makes it more pleasing and attractive, he acted like a technician specializing in form, allying himself with the technicians who specialized in the mechanics of the object, and following all the stages of its production. Of course, mechanics are not the direct concern of the designer, except as regards those parts that remain visible. But there is an internal and spatial aspect of the machine, just as there is of a living organism; and although we are not, and should not be, clearly aware of it, our intuitive sense of it is the sense of life itself. The grouping and distribution of the mechanical parts is the content of which the casing is the outward form. This, then, is the boundary between interior space, of which we have only an intuition, and exterior space, of which we are fully aware; and between the mechanical dynamism of an apparatus and the physical and psychic dynamism of a person. Nizzoli realized that the first and fundamental environmental relationship is that established between the object and the person through usage. For the one who uses it, a machine is first and foremost a transmitter of information about how the apparatus is functioning and the work it is performing. No less important than the measurable space taken up by the housing is the series of indicators needed for the proper operation of the machine — its keys, levers, and handles. Thus the outer casing has the further function of separating out, from among the many messages transmitted by a machine in operation, those that should be received from those that should not. Both in the Olivetti typewriters and in the Necchi sewing machines, Nizzoli's design is therefore basically one of signals, and, in the shapes and colors used, it shows a clear recognition of what is visually apparent. The essential elements of the design do not originate in the mechanical construction but in an analysis of all the possible relationships — visual, tactile, aural, and psychological — that enter into play when the object is in use. Everything that could cause annoyance or irritation must be avoided, but there is no necessity for attempting to establish any emotional relationship between the object and the person. There is no question of trying to make the work more agreeable through pleasurable tactile or visual sensations; the machine must be absolutely neutral, evoking no sensations or emotions whatsoever. The apparatus designed by Nizzoli generally have rounded edges, regular lines, carefully joined parts, nonreflective surfaces, and dull, neutral colors. The process of design is not the outcome of any a priori hypotheses about form but simply develops out of current types that need not be altogether rejected. Designing is regarded as a process of gradual habituation, so that the new object is presented as if it were already familiar. The first freehand drawings primarily sketched the movements of the operator, and the quick, obedient, and precise response of keys and levers to his hand. In the second phase — that of making a model of the machine and progressively simplifying it —

the habits of the operator and the wear and tear on the object were foreseen. Unintentionally, Nizzoli's style of design was completely opposed to the process of 'styling,' which tends to exert a stimulating, almost passionate, effect on the user.

The Italian design of automobile bodies, in which Pinin Farina was the pioneer, likewise aimed at achieving a psychologically tranquilizing effect; the rhetoric of power and speed was avoided, and a normal relationship of coexistence between people and cars established. The 'custom-made' car is obviously associated with the tendency to regard the car as a status symbol. It appears to be, and in fact is, a contradiction of two fundamental laws of industrial production: the principle of the standard model, and that of mass production. Since even in this case the design process requires the making of a handmade plastic model, craftsmanship of a sort is superimposed upon the partly finished industrial product, to differentiate and embellish it. Here, again, the 'requalifying' of the product seems in contradiction to the cyclic development of industrial technology. Actually, however, the intervention of this secondary, artisanlike procedure tends to counteract the trend toward 'styling' typical of every industrial production. The process, in fact, departs from the typology and morphology of mass production and tends toward a requalifying that does not alter it, but only makes it more explicit. As Gillo Dorfles has observed, the custom-made article actually makes the formal characteristics of mass production stand out more clearly: the custom-made car, instead of being called 'non-mass-produced' should more rightly be termed 'super-mass-produced.' For the most part, beginning with Pinin Farina, the making of automobile bodies has become an industry in its own right, which usually adjusts the production cycle of its own factories to that of the parent industry. The lines of this byproduct are determined by analyzing the process of usage and wear and tear undergone by the primary product; the aerodynamic outlines are not calculated to achieve maximum speed but derive from a study of actual usage and the average performance of the vehicle — they are the shapes that make for the best possible relationship among the vehicle, the space through which it travels, and the comfort of the passengers. Born of a period of experimentation and of what one might describe as an artificial aging process, the custom-made article is rarely an invention in itself, a new proposal, or a novelty; more often, it tends to be regarded as a potential prototype for a subsequent model of the original standard product. It thus serves as a check that prevents a model from becoming too quickly obsolete or too frequently traded in, and so protects the product from overrapid changes in fashion and 'styling.' This explains two divergent tendencies we find in recent researches on automobile design. On the one hand, there is the deluxe 'custom-made' model (such as the Lamburghini Miura or the Fiat Dino), which make evident in plastic form the shock of a vehicle shooting through space like a comet — a visualization of pure speed, regarded as the condition and psychic ambition of modern man. On the other hand, there is the complete abandonment of the mystique of automobile aesthetics, and the reproposal in concrete terms of its normal function as an urban vehicle, adapted in size and structure to traffic conditions (e.g., Pio Manzù's design for a taxi).

The linguistic character of design, with a consequent redefinition of types, has undergone constant refinement, thanks to the work of architect-designers like Enrico Castiglioni, Marco Zanuso, Alberto Rosselli, and Ettore Sottsass, Jr., who are more interested by far in finding and redefining meanings than in obtaining optimum performance from the object. Specifying the exact significance, the semantic content, of an object means removing from it the generalized, vague definition previously assigned to it through misleading analogies. Radio and television sets provide a typical instance; at first generally classed as furniture or ornaments, they have subsequently been reclassified as receiving sets for widespread information networks. Since, in the long run, mass production must be considered an information system, research no longer concentrates on the object itself, but rather on its role as a kind of sign. This naturally implies a new interpretation of the present condition and urbanistic function of the house, which is no longer thought of as the center of family life nor simply as a dormitory, but as the focal point of information circuits, by means of which one maintains contact with the world and spends one's leisure time in social pursuits.

Once it had become clear that systematic production must depend upon the control of consumption, it followed as a logical consequence that the problem of design should be seen as the problem of defending the public from overinsistent persuasion to buy irrationally, indiscriminately, and by quantity alone. It is noteworthy that, in Italy, the initiative was taken by those organizations most interested in the public's buying — department stores. The Rinascente-Upim not only called in theoreticians and critics of design (Morello, Tomàs Maldonado) to control and guide them in choosing merchandise, but they also organized the annual Golden Compass competition, which awarded prizes to high-quality products. This periodic, selective recognition, a direct outcome of the buying and consuming system, improved the quality of production just at a time when, owing to the rapid expansion of industrialization, the consumer market was expanding beyond all bounds.

Although the state's educational system is utterly conventional and continues to disregard the urgent needs arising from the country's industrial expansion (there are no public schools of industrial design in Italy), the demand for specialists in the field of industrial design has encouraged private individuals and companies to sponsor pedagogic enterprises. In Milan, an Association of Industrial Design (ADI) has been founded, its primary activity being that of defining the cultural significance of research in this field and formulating ethical and conceptual principles to guide the designer in his profession. That the widespread diffusion of enormous quantities of industrial products can either better or worsen our social customs and material environment is something that even the man in the street was bound to realize sooner or later. After the war, Italian cities were disfigured by irresponsible town planning and wild speculative building ventures, by an invasion of noisy and evil-smelling cars and motorcycles, and by vulgar and ugly advertising billboards. Faced with an environment that was both visually and aurally polluted, and psychologically alienating, the widespread dissemination of inexpensive industrial products for everyday use, carefully designed, and no longer seeking to reform or improve society but simply to serve specific purposes in daily life, seemed to provide a healthy restorative. The period in which the utilitarian automobile and household gadgets were regarded as outward symbols of what might be called the rituals of affluence had come to an end. Now it was necessary to descend to another level, that of the widest possible diffusion and expansion, with an atomic explosion of high-quality products that would forever dissociate them from any connotation of a privileged class and social advancement.

Kartell's production of household furniture made of colored plastic advanced a radically new proposition. This line of furniture is virtually unlimited, the prices are very low, the pieces are constantly changing, and their performance is exemplary. The shapes are not governed by any fixed or precise aesthetic notions but serve as typical signs. The 'thing,' besides not being of any great intrinsic worth, and therefore not requiring preservation indefinitely, is also light, resilient, and almost transparent. It has no existence of its own but is merely a transitory presence in the course of one's life; rather than being a 'thing,' it is a link that connects someone with the surroundings in which he exists. No longer do designers strive to create an object that may serve as someone's alter ego; they are trying to establish an infinite network of relationships, so that the screen dividing man from material reality becomes so extremely thin and elastic as to be almost nonexistent. The linguistic problem of design is no longer on the level of literature but has become a matter of compiling a dictionary. Carefully analyzing in detail the web of all possible relationships, the designer reclassifies them into new groups and assigns them new definitions, by means of new objects that are as flexible and freely interchangeable as words. Seen in this light, any class consciousness is no longer possible; in the last analysis, design becomes a fundamental element of mass culture. It is no accident that this kind of 'impoverished' design, which replaces the object's material value by its appropriate word-sign, should coincide with a parallel search for an *arte povera* that strips the art object of its mystery, does away with all strictures and all shortcomings, and in this way tends to decondition human behavior in order to reestablish freedom within the 'compulsory' framework of a consumer society. So-called *arte povera* (it suffices to mention Pino Pascoli's antiobject object, the suggestion of an 'ironic' consumption, and the creation of fantastic environments that use materials and things taken from everyday life) is in fact an antidesign design, aimed at validating and encouraging an anticeremonial type of consumption, free to the point of being quite arbitrary, and for this very reason no longer alienating man from his environment.

The present period in Italian design may seem to be — and to a certain extent, is — a time of crisis, if not of regression. Industry supports research only insofar as an aesthetic factor (which is often not even immanent in the product itself, but only in the advertising campaign that promotes it) may increase sales and encourage a rapid turnover. As the result of mistaking a higher standard of living for an actual change in the structure of society, the effort to educate and rebuild society by the 'object lesson' of industrial production has been abandoned. The most serious research no longer centers on the 'product,' but on 'visual' design, and runs parallel to the contemporary movement of 'programmed art,' which is based on a study of the physiology and psychology of vision. The attempt to organize culture on a visual basis also has an ideological premise: a product's aesthetic quality should not be a paternalistic concession on the part of industry, but a right demanded by the entire consuming public. Even before the war, theoretical and practical research in the field of 'visual design' had been undertaken by Bruno Munari, who might be considered the Italian Moholy-Nagy. Subsequently, other designers and researchers, especially Enzo Mari, have for the past twenty years been working in the same direction. The purpose of their research is to establish a gestalt, an intrinsic structure and organic entity of form — understood as the process of forming, and thus of organizing and constructing, the image. While form reveals itself through objects, it is not necessarily related to their practical function: the form of an object does not depend on its function, but rather its function adapts itself to the visual structure of its form. It is useless, therefore, to try to come to any preliminary accommodation with the economy and industrial techniques; in fact, such an accommodation is objectively impossible, as long as industrial technology still remains under capitalistic management. The aim of research can thus be only a formative and didactic one; only a higher level of culture among consumers could force the industrial system to produce an environment that is no longer negative and even biologically and psychologically harmful. Before taking shape on factory drawing boards, the models must first take shape in the minds of consumers.

We must therefore put into circulation objects that can act as formal models, which can contribute to visual education and — still more important — to the training of the imagination as the awareness of form. All the outlets of mass communication will be utilized for the dissemination of these objects: they will be 'published' and distributed through a system analogous to book publishing (cf. the productions of Danese). Since the purpose is principally experimental and didactic, the practical function of the object is a minimal consideration; the ideal object is a toy (Mari), the typical instrument for stimulating the imagination. The ultimate destinations of these model objects are schools and, beyond these, museums — museums no longer conceived of as storehouses for the heritage and artistic treasures of the community, but as centers and schools of aesthetic training. The final goal is to train people in the aesthetic interpretation of their material environment, so that they can recover their faculties of independent judgment and evaluation, which a consumer society tends to stifle by encouraging indiscriminate, prodigal buying. Research in the aesthetics of design is thus equated with scientific research, and freed once and for all from the chains that bind it to the technology and economics of capitalist industry, which has found expression in that inevitable compromising of a true 'industrial aesthetic' represented by 'kitsch' and 'styling.'

Italian design thus also comes closer to the research being carried out by artists, or, rather, by nonprofessional designers. The experimental analyses undertaken by men like Joe Colombo and Vincenzo Agnetti, regarding the possibility of organizing the material environment according to structural principles and restoring its order and spatial clarity (even though it be space thought of as a dimension of information and communications), have shown how the basic ambiguity of 'industrial' design is about to be superseded by research in pure, 'scientific' design — that is, a design no longer looked upon as a structural principle of objective reality, but rather as a principle of awareness.

THE LAND OF GOOD DESIGN
Alessandro Mendini

Design is the process whereby mankind should accomplish a conscious formation of the world's surfaces, in order to create symbolic and functional surroundings appropriate to his life. This formal definition, however, may be followed by another political one. Design is a conflict whereby certain groups enact the drama of an irresponsible formation of the world's surfaces, in order to establish centers of command that enable them to solve — by means of mankind — their individual problems of domination (not excluding the use of bombs). Western logic, in which we are atavistically steeped, makes us move alternatively from form to content and vice versa, always considering induction and deduction as seemingly opposite poles of the same axis; every approach to a problem implies its own converse.

For the sake of clarity and simplicity, we must decide which of these two definitions to adopt, and in this case we have chosen the first — taking those proposed by others for granted. These others, together with ourself, operate like an enormous team; and the larger this group is, the less power will be wielded by any one of its members.

Attention thus focuses on a consciously formed environment, which is the crux of several problems, some preceding and others following. This environment is constantly evolving, and is subject to heterogeneous, uncontrolled, noncommunicating forces. Its periods of transformation, which are usually those of degeneration, are very brief and impose very narrow limits on the scope of our action. Hence the need to avoid taking refuge in myths about the future, rushing along headlong propelled by technological optimism. Such reliance on the future only means a postponement of a possible problem, which, if it were ultimately to be solved, would no longer encounter the same environment that first gave rise to it.

Louis Kahn relates the following anecdote:
'It reminds me of a story... I was asked by the General
Electric Company to help them design spacecraft,
and I was cleared by the FBI for this.
I had all the work I could do on my hands,
but I was able to talk about spacecraft anyway.
I met a group of scientists at a very long table.
They were a very colorful looking lot,
pipe-smoking and begrizzled with moustaches.
They looked odd, like people who were not ordinary
in any way.

One person put an illustration on the table, and said,
"Mr. Kahn, we want to show you what a spacecraft will look like
fifty years from now." It was an excellent drawing,
a beautiful drawing, of people floating in space,
and of very handsome, complicated-looking instruments
floating in space. You feel the humiliation of this.
You feel the other guy knows something
of which you know nothing, with this bright guy showing a drawing
and saying,
"This is what a spacecraft will look like fifty years from now."
I said immediately, "It will not look like that."

And they moved their chairs closer to the table
and they said, "How do you know?"
I said it was simple...

if you know what a thing will look like fifty years from now,
you can do it now.
But you don't know, because the way that a thing will be
fifty years from now is what it will be' (1).

If a discussion of the situation of design in Italy today, and its possible

Alessandro Mendini, who received his degree in architecture from the Politecnico of Milan in 1959, is a practicing architect especially concerned with theoretical and critical studies. He is a member of the Dolmen group, an organization of Nizzoli Associati in Milan engaged in design, planning, and research in the fields of architecture, town planning, and visual engineering. Mr. Mendini is primarily interested in experimental planning that reflects the interplay of existential, formal, and political forces. From 1965 to 1970, he was editor-in-chief of *Casabella* and is currently a member of its editorial staff.

370

development, is to be fruitful, it is indispensable to analyze certain phenomena manifest in the present state of affairs, evaluate them critically, and recommend alternative lines of development rooted in our history. Since the Italian situation reveals a deep rift between management of the market and cultural demands, it must be analyzed with respect to these two factors, whose obvious mutual incompatibility leads to syntheses and reflections.

The Market

In the building trade, as well as in the production of objects, the most conspicuous feature of the Italian market is that it has failed to resolve consciously, through a gradual evolution, the problem of the transition from a craft process to an industrial one, nor does it favor organic coexistence of these two procedures, even in cases where such coexistence might be feasible. The compromise is evidenced by the practice of hiring unskilled labor, not bound by any contract or fixed work, because under these circumstances there may be surprising fluctuations of the market which require the making of adjustments. Aside from the socioeconomic aspect, we observe here, in any case, a situation that seems to be dragging itself out without any prospect of even a partial solution. The 'formal' consequences of this phenomenon — the subjects dealt with, i.e., the objects produced — are restricted, simplistic, and linked solely to a consumer economy. There is a total lack of interest in so-called 'social design,' in which the patron is the community, not a private individual, and in which the object fulfils a specific service and not merely a need stimulated by the object itself and the bombardment of advertising.

The key figure in this jungle, who exploits it but at the same time is dominated by it, because he expends his chief efforts on keeping up to date (the most beautiful, the newest, the most attractive) is the 'name' designer. His survival (and at the moment he seems in no danger of dying out) continues to foster a competitive type of professionalism. It is not possible to plan the proper kind of training for technicians required by the community; and therefore, for every one of these professionals who flourishes, there is another trying to keep the wolf from the door. Further consequences of the present system are styling, because every good idea seems only to provoke imitations and other versions made at a slightly lower cost; compromises to meet the lowest type of demand; superficiality; an inappropriate or simulated use of materials, which seems to be a major preoccupation of all the smaller industries; technocracy, because too often the object is considered only with respect to its technological yield with reference to false values, or with an evasion of its extra-economic components (although a few, Sottsass for example, do take these into consideration); a false type of 'progress,' because loading the object with images or processes associated with avant-garde technology is believed to bring industrial design into line with technological and scientific research (an attitude that only Joe Colombo has managed to adopt with sufficient detachment and irony to avoid falling into coarseness).

From the situation of compromise mentioned at the outset, we have therefore now moved on to the ascertainment of a productivity race, which, since it is unplanned, must basically be completely purposeless. Periodic saturation of the market and resulting periods of stagnation, corresponding with those prevailing in the general national economy, naturally lead to focusing attention on exports (particularly important in the field of automobiles, furniture, and domestic electrical appliances); and in this area the presumed superiority of 'Italian style' is exploited with subtle cunning. But the situation is at best ambiguous

Bel Paese cheese

Opening performance at La Scala, Milan

and in certain respects paradoxical: with the exception of large industrial concerns, such as Fiat, which have an autarchic organization — that is, internally self-sufficient — medium-sized and small producers export finished products, after having bought the raw materials from sources whose prices they cannot control, and after having commissioned the actual manufacture from still a third party. The producers, therefore, tend to play a kind of game with their merchandise, in which the sole choice left to the user is the superficial one of taste. The relation between industry and the purchaser is one-sided and always to the disadvantage of the latter.

The very serious deficiency of the Italian market must be strongly denounced, inasmuch as it also implicates industrial design in its acts of conditioning and violence. Enervated by its incorporation within a closed economic cycle, and concentrating primarily on technological-formal activity, the present system of design in Italy is certainly not in a position to consider playing an active role, still less the idea of self-management by the community.

Cultural Demands

Fortunately, the stupidity of this system and of the market has given rise to a lively and open cultural debate; its existence is an indication of vitality and not solely of dazzling loquacity, and in spite of numerous and obvious contradictions, it is something on which the country can rely. Briefly, it seems convenient to distinguish four groups that represent and epitomize an equal number of fields of research.

The first group confines itself to an historical kind of research on planning, which tends to define the problem of design as a specific discipline. It is summed up in a statement, made seven years ago by Vittorio Gregotti in an issue of *Edilizia moderna* devoted entirely to his writing, and just as valid and emblematic today: 'We may schematically distinguish at least four different phases in the history of design. The first is marked by the rupture of an essentially unitary system, a break between creative design and execution, with the advent of new technological and industrial production systems, thus linked to the problem of quantity and mass production. A second phase began when design became aware of the general problem of the applied arts, with the resulting controversy between craftsmanship and industry, and with the investigation of the terms on which to base the giving of aesthetic quality to industrial products. A third phase saw the problem of applied arts absorbed once again by architecture, and the attempt to reunite the technique of the object with the idea of functionalism. Finally, in a fourth phase, we can perceive the dissolution of this relationship, and the expansion of the idea of design as a controlling factor of the environment, applied to every dimension of the useful object up to the scale of the town, even to the idea of planning itself, to the extent that this coincides with those aspects of design that involve choice, foresight, and participation' (2).

A second group of researchers is made up of those chiefly engaged in the field of technology (for example, Mangiarotti, Zanuso, Valle); we cannot call them technocrats, however, since they are always concerned with formal themes and all their connections, even those involving certain choices that come to nothing. Here I would like to quote once more, this time from Spadolini's *Design and Society*, to bring out the point that even attention to technological processes may offer viable solutions, provided technology is clearly recognized as a means and not an end: 'Yet, amid this muddled agglomeration of information, hopes, and fears, justified and unjustified, one begins to perceive that only by following the trail of the innermost and truest meaning of industrial production can we perhaps be able to understand from the ground up the complex structure of present-day society; and this is obviously the indispensable premise, if one is to be able to operate within that structure, alter its characteristics by design, and above all — and this is surely of importance for designers — be able to act and move not only with the purpose of recovering man's ancient creative capacities but even of discovering new ones, still completely unexplored...' (3).

To cite only the best known among the instruments of this research, we may mention the theory of circuits proposed by Argan, or that of the general use of components (to which Spadolini devotes special attention). The latter theory calls for different types of intervention, the most important being that relating to open systems and the use of modular components in the industrialization of building. Although it can readily be accused of compliance with the consumer market, this theory nevertheless remains one of the chief indications of a willingness to permit the user some degree of participation in the processes of giving form to the environment.

The third group includes a large assortment of theoreticians with diverse and often contradictory cultural backgrounds, often in contact with researchers from other countries; proceeding from structuralism, and applying theories of signs and meanings and information theory, they have focused on a wide range of problems that all converge on design. Among them are Argan, Dorfles, De Fusco, Menna, König, and others, who have investigated the relations between design and the planning of buildings; the sociology of consumption, and the laws of consumption as the process of obsolescence; the principles of aesthetic communication and mass media. They have brought to light — not with moralistic intent, but as problems of awareness, so that they can be dealt with more responsibly — the phenomena of kitsch, of fashion, of the degeneration of taste, and of styling. They have put the subject of popular art into its proper perspective as the intermediary between political systems and freedom of expression, suggesting that it also be considered on the basis of its formal-figurative attributes; they have stressed the importance of the processes of design responsible for giving form to the environment, contributing to the definition and application of meta-design; with the aid of new social sciences, they have investigated the various mythological, monstrous, and caricature aspects of the object. To some extent, they have systematized the discipline of design in its various subdivisions, from packaging to graphics.

The fourth and last group would consist of those members of the profession who, while choosing to immerse themselves in the demands conditioned by the market, rather than trying to evade them, have nevertheless made positive contributions to some of the most problematic aspects of design. It is not out of the question, however, that by banking too heavily on their own expressive and formal capabilities as designers (behaving like character actors who rely on their own completely personal creativity, unaffected in any qualifying and integral way by the processes of industrial technology), they may have been indirectly responsible for the productivity race that accounts for all the previously mentioned hazards of mass production.

At this point, it may be of interest to compare the difference in attitude between the cultural community and those who have been called the real community. To the extent that the former seems to have been cast out by, and alienated from, the latter, it has had no effective ability to try alternative courses of action.

But the processes and fruits of the cultural debate, which we have schematically divided into four groups, come up against the problem of a pedagogic lack, which on the one hand involves the specific training of technicians at different levels of operation, and on

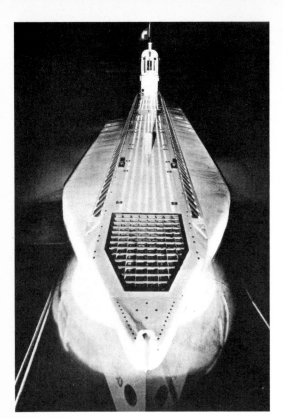

Central ship for Lausanne Exposition

the other involves informing and educating the community. It is not a vain hope that it is precisely around this critical point that the problems may begin to crystallize and find a solution. It is only when a society feels the need to provide itself with a completely new cognitive discipline, which will allow it freedom of choice and initiative (for which the indispensable premise is education — not to be confused with civilization for the sake of industrialization), that it can begin to consider democratic patterns of participation and self-management, which at the moment seem to be the only ones that will permit man to survive the power game.

Having completed this brief analysis of the Italian situation, it seems logical now to take up the problematic questions of pedagogy and of a new economic-political management.

The Pedagogic Problem

Assuming that the debate we have just examined is not lacking in contradictions, but that nevertheless each systematic group of proposals is entitled to be considered to the extent that it is consistent within itself, it may be worthwhile to bring out a few subjects that are in some degree comparable to those discussed, but somewhat different (even if not wholly original), in order to broaden the area of discussion.

Returning to the problem of education, any consideration of it should proceed from a statement of principle: if a political society is inclined toward goals that are not oppressive, it needs must reject — at least in concept — the principle of authority, in favor of its opposite, which is the emergence of the direct voice of the individuals who compose that society, so that they themselves can shape, and not be subject to, their own social system.

To embark on the adventure of nonoppressive design entails rejecting the principle of formalistic composition centered around individual demigods and circumscribed by the binomial set ideology/style, in favor of the principle of extending an awareness of design based on the twin terms ideology/method, to as large a number as possible of technicians and executives who are active, at various levels of responsibility, in the whole process of forming the environment.

The alternatives for design today are: 'free-lance designing,' which is destined to disappear because of exploitation and submission to middle-class ideology, competitiveness, etc. (and this approach is also open to heavy criticism at a time when the ecological system of the whole country, in terms of human survival, is at stake); and 'designing within public agencies,' which is bogged down by the same obstacles as those that hinder the profession itself, complicated by excessively bureaucratic procedures; this might to better purpose be entrusted to the academic world. 'Designing in universities,' and likewise in schools, might thus assume a more fundamental role in this process, but there is a problem that would first have to be solved. That is, how to organize the university as a permanent structure, in which every person would prepare himself to fulfil his function, which would always be within the structure — with the community as the client — and to which he would return whenever the evolution of his own experience led him to didactic explorations. But for design, this criterion, too (aside from the fact that it is unlikely to come to pass for many years), would tend to establish a relationship between the elite and the world of work and production that would end by becoming institutionalized. Among other possible solutions, our own choice (because of its obvious consequences, which it would be superfluous to specify) would be not for groups of planners in the strict sense of the word, but rather for democratic 'planning communities,' working on a nonprofit basis and concerned

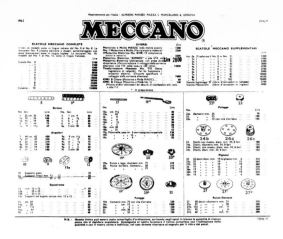

Diagram for the game of 'Meccano'

375

with getting to the root of the problems and contradictions of that very society in which they operate.

It follows, therefore, that the irreversible direction for those who wish to design for construction is that of group work, not in the sense of the usual dormant collaborative practice, but precisely as one of the 'categories' of designing, with full awareness that this activity is in fact collective rather than individual.

On the plane of expressive means, also, if what is sought is widespread and antielitist communication with people at large, creativity should go beyond the traditional practice of the lone architect, who exercises authority even over his own colleagues, to become a community activity in every sense. This should not lead to astylistic neutralization nor to compromise in the statements made, but on the contrary to a kind of unbiased antistyle, based on contradiction and experimentation in method (action designing). From another angle, that of the scientific and technical standpoint, the highly complex nature of problems today requires the organization of work into interspecialized, interdisciplinary groups, in order to ensure the methodological or sociotechnical bases of the project. This last declaration is meant to imply that unification in methodology — which itself should be regarded not as an end but only as a means — must somehow come to pass. This is no simple matter, however, since serious doubts are now arising regarding the universality of methodology, and attempts are being made by certain advocates to reverse any trend in that direction.

'I think that it is a time of our sun on trial,
of all our institutions on trial.

I was brought up when the sunlight was yellow,
and the shadow was blue.
But I see it clearly as being white light, and black shadow.
Yet this is nothing alarming, because I believe that there will come
a fresh yellow, and a beautiful blue,
and that the revolution will bring forth a new sense of wonder.

Only from wonder can come our new institutions...
they certainly cannot come from analysis.

And I said, "You know, Gabor,
if I could think what I would do, other than architecture,
it would be to write the new fairy tale,
because from the fairy tale came the airplane, and the locomotive,
and the wonderful instruments of our minds ...
it all came from wonder!" ' — Louis Kahn (4).

Admittedly, this is the outlook of a poet and cannot be extended to a community, but nevertheless it represents an attitude toward creativity. It therefore seems necessary to overcome certain aspects of methodology in order to recover imagination, fantasy, and meditation — values on which Oriental culture is based, in contradistinction to that of the West, which is founded on logic (and therefore any attempt on our part to reconstruct Eastern thought on a purely logical basis is erroneous). But possibly — apart from the subject of functionalist obsolescence — there may be still another reason for overcoming methodology, and that is the danger it presents of being used as an instrument of colonialization, just to the extent that logic is a typical European tool, which should not be applied in different historical and cultural situations that may demand a different approach. (Cultural anthropology, for one, has had a great deal to say on this subject.) From this it follows that the unification of methodology as a basic tool for interdisciplinary activity cannot yet be proposed on a worldwide scale. The problem arises — or rather

Buddhist Monks, Angkor

takes on a new character — in the contacts, exchanges, and encounters between East and West. This problem fits in perfectly with the present phase of our discussion and allows us to take a calm view of the revolutionary, or pseudorevolutionary, Maoist-inspired phenomena, which have become prevalent among those subject to logical didactic systems. What we are actually facing here is a kind of thought that, although westernized to a large extent, nevertheless has its deepest roots (indispensably so, although this is not always clearly recognized) in Confucianism and in a civilization whose millennia of history give it primacy in continuity, if not in quality.

Conceding the fundamental role of Mao in relation to his people, to whom he has given political identity within a world context, thus avoiding the risk of a nonpresence, the question still remains of how the strictly cultural situation will develop after violent action has been succeeded by reflection; and there is also the problem of the unilateral nature of a struggle carried on over the power of institutions on the basis of an essentially economic logic, with all its inner inconsistency with Chinese anthropological culture.

Contacts among mankind, embracing all the cultures of the world, require the renewal of man's capacity for fantasy and imagination, which should be helped, rather than stifled, by his capacity for logical reasoning (relativism rather than absolutism among cultural systems).

A final problem concerning pedagogy is that of research. In this respect, the situation in Italy is completely chaotic, and there is not the slightest sign of any activity whatsoever. Here the blame does not lie with the behavior of groups or individuals, as regards the cultural debate in question, but with the state and its relations with those who should carry on the research required to raise the standard of quality in every branch of national production. Let us consider the matter of industrialized building, a paradigm for the whole field of design, which is at least underway, however badly; while a similar discussion regarding the production of home furnishings, for example, is still far from being taken into consideration. Two years of union strife in Italy have led the government to recent decisions regarding housing. The goal of a 'state industry for housing' could become a reality rather than an hypothesis if the IRI (Institute for Industrial Reconstruction), comprised of various industries that carry out those social services for which the state itself provides intervention and financial aid, were to involve itself in future building programs. But the level of sophistication in scientific and technological research in this sector is extremely discouraging; judged by the standards of other countries in the European Economic Community, Italy is many years behind. Here, the proposal of a special program of research on the industrialization of building, to be developed into a 'permanent' organ for research, and for the expansion and control of building production, is only now being developed, while other countries have already gone beyond that to a stage of reconsidering and verifying results achieved during at least two decades of actual experience. The general situation of research at the National Research Council (CNR), the culprit at fault here, is as follows. On the one hand, there is still a myriad of little research projects, widely scattered among a great variety of subjects; while on the other hand a quite separate massive research project on the subject of industrialization recently died of inertia. The interconnections among those appointed to responsible positions, the subjects, and the amounts expended certainly yielded no policy of any interest; one can only perceive the usual intention of making a wide distribution of posts among neutral appointees, to give a little support to, and pacify, as many people and institutions as possible. And this brings us to a quite unforeseen project, that of the massive research program. It is important to point out what a large political and financial

commitment was involved in the 'Special Program for Investigation of the Industrialization of Building,' in comparison with the extremely bad use of funds just mentioned.

When it was initiated, a few years ago, the Program set out to 'seek ways and means of achieving the industrialization of building materials and of effecting a structural organization of this sector, better suited to obtaining a product that, in both quality and quantity, would be responsive to the needs of the community, at a cost compatible with the economic and financial policy of this industry.' The entire history of the logic of research has proved that any program whatsoever of this kind is based on a false methodology; and in the present instance, the defect was aggravated by a growing alienation of many individual investigations carried on in the overintellectualized language of initiates, without any power to communicate and serving only to conceal the lack of political commitment, while never succeeding in coming to grips with the real problems at stake.

The Program was boycotted just at the moment in which it might have transformed itself from a temporary scheme into a 'permanent body.' As always, a combination of causes led to this result. The failure of the industrialization program must be seen in the light of the vast crisis affecting all policies regarding science in Italy, which could only be overcome, in our opinion, by a radical revision of the process whereby decisions are reached, and also of power relations, so that scientific research — as a service directed toward all productive forces — is neither broken up into ineffective units nor monopolized.

This incident, which is certainly not unique, demonstrates the backwardness of the policy regarding research on design in Italy. Realizing that even before any connection between school and research can be envisaged, all the problems of school and of research in themselves must be ferreted out, one comes to the conclusion that the all-too-familiar 'professional freedom' situation remains the only possibility, unless radical changes take place; and it follows that the same system of pedagogy will persist, with its vertical structure composed of students/teachers/economic power, for the training of technicians whose only prospects of work lie in their becoming part of private industry (which will welcome them to the extent that they have been 'good students').

Meanwhile research, if one can give that name to what is now being carried on, will remain a technological refinement, hoping to win approval from the deluded purchaser.

Parallel to those questions that concern the elementary aspects of teaching, there are others which to some extent lie outside of pedagogy as a specific field in itself, but which offer the opportunity of an encounter between the stages of teaching and of society. First of all, we must say something about the inclusion of a plan for the study of design within the mental and cultural development of the individual. It goes without saying that a degree conferred in recognition of some special kind of professional attainment, like the baccalaureate, would be ridiculous; diplomas, in fact, tend to set up categories of subordinate classes, resulting in obvious problems of a human and social nature. The continued existence of schools of design — makeshift transformations of stuffy old art schools — which confer an academic title in no way on a par in our market with that conferred by schools of architecture, has created a chronic problem that can only be solved by taking a comprehensive view of the whole question of instruction. But it is only too evident that in Italy, the policy in this respect proceeds — or tries to proceed — by dealing only with partial sectors. As a result, the reform of the university, which does not take into account schools, and which confuses a new kind of

Bechuanaland
377

problem, that of mass education, with an increase in enrollment, still seems after years of weary and contradictory leadership to be incapable of becoming crystallized into a law.

A further task to be performed in the reorganization of teaching is community information and education. Whereas this is at present construed as an attempt to increase demand for the sake of a greater profit and a wider market, it should in reality be altered to create conditions that would form the basis for fostering a process of self-management. This, which we have pointed out as the most serious shortcoming of Italian design, could still be overcome, if a permanent system of instruction could be established that would allow each person to find his own place in it, leaving whenever he felt it to be necessary for himself or others.

A final problem is that of the relation between instruction and employment, a problem that obviously concerns every kind of teaching but is particularly delicate in those areas that have a direct connection with the world of industry. This relationship is being increasingly transformed into one of conflict. The number of students is growing (owing to the longer obligatory term of attendance and the liberalization of academic requirements), while job openings for those with higher levels of education are diminishing. From this comes a contradiction between a system of training that tends to make access to instruction more available, and the system of employment. Furthermore, there still persist the same criteria of selection, first in the schools and then in employment, based on obvious favoritism (because of our closed social system) toward those coming from upper-middle-class families. At this point, any discussion of teaching must necessarily branch out into the politico-economic question.

Precisely as regards design, we would be prepared to assert that the training of the designer has a relation to the needs of the community, and that — besides being formulated on the basis of suitable training — it must actually presuppose a politico-economic system. It is absolutely indispensable that design for the community replace design for a private market geared to the satisfaction of individuals. This would involve controlling induced needs, which require an object to be continually redesigned so that the purchaser feels obliged to get rid of what he already has and buy the new object. He suffers from compulsions that are almost always psychological in nature and related to his economic and social standing, for which the object itself constitutes a status symbol.

If we analyze the relation between the average earnings of an Italian citizen and the expenses to which he is put by his house and design in general, we must conclude that half of what he makes goes for his home and for the satisfaction of the temptations offered by the market. The same industry that gives him a salary for his work reabsorbs that very salary with objects that he himself has been involved in producing. Industry thus only makes a pretense of paying for his labor. What such a system implies is that there must be an end to competition, and that production must be centralized. Essentially, this would be a matter of incorporating the subject of certain kinds of furnishings, other than what is absolutely essential to equip a house, into the legislative procedure governing building as a concern of the state. If one wished to abolish the surplus value on the house-commodity, due either to speculation in land or speculation by the constructor as entrepreneur, there would be no reason why the theme of house furnishings could not also be resolved in these terms. (It is well known that the cost of buying or renting an apartment is equalled by an analogous cost for furnishing it.) Distribution plays a large part in this, since it involves the cost of advertising and agents.

In fact, the number of distributors' licenses went up by 59.7 percent during the 1960s, thanks to an absurd process of fragmentation that reached even greater peaks with respect to domestic electrical appliances. If, on the one hand, small businesses raise costs, on the other hand the major company, in order to counteract the extremely generalized type of sale offered by the traditional retailer, practices a policy of selective selling through a chain of exclusive outlets, or through points of sale that it owns, to differentiate its products from those of others.

In considering economic-political reorganization, we should also take into account the question of employment. The growth in production since the war has not been matched by an equal development in the structure and size of the producing units; and in the sector we are considering, with very few exceptions, not the slightest effort has been made to follow any principle of planning. The reaction to increased demand has consisted in continually increasing the number of small firms at the artisan level, concentrated in the classic fields of production.

Examining the statistics of the 1961 Industry and Trade census, we find that the so-called furniture industry had only two establishments employing about five hundred persons, while the average was represented by firms with no more than sixty employees. Even though these figures are not recent (and the fact that this trade federation has felt no need to update them is significant in itself), the situation has not changed greatly. It is therefore easy to understand why the problem of employment, which is already difficult in other branches of national productivity, offers no guarantee of security in this sector.

It would therefore be appropriate to reexamine all the techniques involved in the system of production and distribution from the point of view of a socialized market. A good example is provided by market surveys. They have been developed to evaluate 'changes in taste,' in order to snare the purchaser with more subtle applications of the induced-need technique; whereas actually such surveys should be among the most useful tools enabling the purchaser to participate democratically in managing the cycle of supply and demand.

In this hypothetical system, the national characteristics of a product, which reflect the difference between one country that produces and another that is forced to buy, would naturally disappear. This kind of game — played not only with refrigerators but with cannons as well — is now well known, and it has been proved that it tends to make producing societies increasingly powerful, while societies unable to compete remain chronically poor. Apart from obvious ethical questions, which cannot be settled by the thinly disguised charity of international funds for the Third World, this results in a permanent inflammation of infection centers in smaller countries in which the technological superpowers play their convenient role of exploitation.

To overcome the narrow confines of national boundaries — which makes the very idea of an 'Italian look' in design seem ridiculous — there would ultimately have to be rigid control of materials, prices, quality, and all the various processes; and the institution of a general system of regulations, whose operation would eventually make the criterion of competition superfluous.

(1). *Talks with Students* (Architecture at Rice, 26). (Houston: Rice University, 1969), pp. 23-24.
(2). Special issue, 'Design,' *Edilizia moderna*, no. 85, 1965.
(3). Pierluigi Spadolini, *Design e società* (Florence: La Nuova Italia, 1969).
(4). *Talks with Students*, p. 2.

RADICAL ARCHITECTURE
Germano Celant

During the 1960s, the 'politics of design and architecture,' the activity of the designer and architect, were enhanced by a new slogan, capable of breathing new life into a sphere of action that for the past few decades had undergone sad vicissitudes and was demonstrably in a state of crisis, owing to its having fulfilled none of the main premises on which its operating strategy was based. The process of mystification has found a new formula with which to conceal the alienated, commercialized aspect of design, still chained to neocapitalist reformism. The troupe of designers and architects has hired new actors who, in Brechtian fashion, hold up new placards that praise the political commitment injected into a television set, a refrigerator, a poster, a Fiat condominium in Mirafiori, or an Italsider or Agip pilot village.

Today, no actions or projects in Italy fail to consider the possibility of assuming a new political aspect. They accomplish this by the deceptive use of a socio-aesthetic concern, manifested either in formal, pseudoaesthetic superstructures, or in speeches and reports that seek to justify the violence done to the natural environment and the acculturization of entire settlements peopled by indigenous inhabitants.

The idea of giving design a new ideological and operational dimension, so that it would no longer be available to serve as an instrument for neocapitalist infiltration, has been transformed into a populist 'socialization' of the same idea. Thus, prime importance has been given to the most advanced technology, which is in fact able to meet the demands of new techniques of production and construction, but not the new concepts and behavior of design and architecture. In the past few years, all that either design or architecture have succeeded in doing is to expand the boundaries of their techniques and production (an expansion that in Italy has of course coincided with the boom in consumption and speculative building), without having changed their goals and attitudes. The function and posture of design, like those of architecture, have remained unaltered, without any ideological or philosophical questions having been raised. Everything has taken place on a superficial level, the level of industrialization, neorevival, styling, populist slogans, and middle-class consumer society.

Having been without any clear and consistent plan of action for many years, design and architecture have looked on, without being able to intervene, while all their would-be social and aesthetic hypotheses gradually collapsed. Meanwhile, they continued to claim responsibility for an intervention so fundamental and revolutionary as to guarantee a real improvement in respect to manufactured articles and architecture. Now that ten years have gone by, they realize that they have attained none of their objectives and seek, a posteriori, to invent an ideological alibi. After having disfigured and destroyed the landscape and caused an urban explosion, after having staked everything on all-out planning and consumption, prefabrication, and the assembly line, they now find themselves enslaved to the very system they had thought to overthrow. Their only recourse is to justify their own actions in a moralistic, reformist, and middle-class way — and continue to produce.

All this is because designers and architects still regard the manufactured object and the constructed building as the sole and inevitable bases for their own activity. Taking refuge in production for production's sake, they alienate themselves and lose sight of their own role as designers and their ideological and conceptual aims. By stressing production and the end product, manufactured or built, they tend to emphasize as the sole aim of their work, not the

Germano Celant is curator at the Experimental Museum of Contemporary Art in Turin and has lectured on radical architecture and *arte povera* at several American and Canadian colleges and universities, including the University of Minnesota, the College of Art and Design in Minneapolis, and the College of Visual Art, Toronto. He is also editor of several series of books dealing with conceptual art and modern architecture. In addition to many articles, his recent published works include *Arte povera + Azioni Povere* (1968), *Art Povera* (1969) and *Conceptual Art, Arte Povera, Land Art* (1971); currently, he is preparing a book on radical architecture.

idea nor the project, but rather the project as carried out, on whatever terms. This means that they have abandoned the idea of playing an active role or fulfilling any ideological and philosophical function with respect to architecture and design, for the sake of a commercialized and deceptive self-expression of their own existence as designers and men of ideas. This esteem for the physical residue of their idea or design — the completed object or building — means that they have negated and abolished any operative or ideological power that their concept and behavior within their profession might have, in order to further a systematic, a priori acceptance of what can be produced or built, without any analysis of the concept and nature of design and architecture.

The overwhelming importance given to the finished object or building has bred an attitude in which activity has supplanted any effort to form a philosophy about design and architecture; these therefore have become redesign and neoarchitecture, more concerned with the decoration and neoformalization of the object and building than with the nature and existence of architecture itself. Accordingly, the formal and decorative aspects, with their physical and formal manifestations, have superseded any consideration of the concepts and manner of behavior that should govern the designer's activity. It is apparent that the revolution begun and carried out by the avant-garde movements of the past are still not understood, for the very nature and conception of design and architecture have changed since the early 1920s, when Russian architects and designers, working within the climate of the Soviet Revolution, replaced the completion of a product or the actual erection of a building with planning and concept. The architectural and urbanistic plans, and the designing of objects, by El Lissitzky, Salijeskaia, Tatlin, Ladovsky, and the Vesnin brothers brought about a crisis, not in the form of the structures, but in the very concept and existence of design and architecture. They sought to deny the importance placed upon completing the construction of a work and stress instead a concept and an attitude regarding a new mode of being and acting in design and architecture, corresponding to a new mode of social existence and action. The Russian idea of making a clean sweep of design and architecture coincided with their conviction that the modes of expression of objects and architecture consistently lose their specific meaning as the history, technology, and behavior of a society evolve, and that this specific meaning is instead inherent in the ideological aims and attitudes of that society.

Developments in design and architecture, therefore, have not always centered around completed objects and constructed buildings (and we must remember that half the new architecture and design is neither building nor object), but have come about through revolutionary changes in the behavior and position taken by architecture and design with respect to existential and social ideas, together with a systematic overturning of the philosophical and ideological significance of the concept of architecture and design. One cannot, therefore, evaluate design and architecture simply by examining buildings and objects; one must also take into consideration all the radical modifications in behavior and concept that have resulted from the emergence of different and disruptive ideological and philosophical attitudes and intentions, affecting both policies and behavior.

Shifting the focus of attention from the building or object to be completed implies an effort to escape from the alienating effect of production and the commercialization of one's ideas, in order to attain an ideological absolute, pertaining alike to philosophy and attitude. Refusing to let oneself become prey to the commercialization and deceptiveness of production could also entail a radical change in the activity of architecture and design; a silence in performance, a nonrealization of one's own ideas and projects. Only this can lead architecture and design out of their institutionalized, commercial phase and give them a place in the true cultural revolution, based on ideas and actions intended to subvert the existing system. Such a silence would not constitute an inclination for lethargy, immobility, and abstention, but rather a desire to clarify, philosophically and ideologically, what design and architecture should be and what they should do. Such a clarification became irrelevant when aesthetics and production undermined the ideology and philosophy of the profession.

Yet design and architecture have never existed solely as evidences of production and the physical residue of ideas and plans; for the most part, they have been considered as the ongoing process of analyzing, ideologically and philosophically, just what architecture and design are and how they behave — and this has always implied questioning their nature, rather than producing objects or buildings. For, if one discusses an object or a building, one cannot discuss the nature of design and architecture; if one produces a building or an object, one accepts tradition and all that it entails, since what has been built or produced is only one type of design or architecture, not design or architecture themselves. If one's purpose is to make objects or buildings, one accepts (and does not question) the prevailing ideas of architecture and design.

At this point, is there any reason why one cannot proceed from design oriented to what can be produced and constructed to design that concerns itself only with the ideology and concept of design and architecture? And if work in the field of design and architecture consists of discussing and questioning the nature of design and architecture, and involves a radicalization of their essential behavior, what prevents us from making this discussion and questioning of the nature of design and architecture, and their essential behavior, a work of design or architecture? If it is the ideological, philosophical, and behavioral stages that are the fundamental ones, why not make of that ideology, philosophy, and behavior a work of design and architecture?

Only by giving renewed importance to the value of philosophical aims in design can the profession go back to being an analysis or definition of the nature of architecture, and commercial values be replaced by the value given to ideology and behavior. Objects, products, and buildings growing out of projects and ideas would become irrelevant to the condition of design. (This does not imply that design or architecture cannot adopt forms of analysis and investigation that make use of concrete, physical schemes, but only points out that intentions, ideas, and attitudes must not be led astray by consumer-oriented, factual information geared to the production system of capitalistic society.)

All the new Italian architecture — Archizoom, Superstudio, etc. — has asserted that its aims are conceptual and behavioral. Proclaiming itself as radical, it no longer wishes to be commercialized or alienated, or to renounce its own ideas and expressive attitudes. This is an architecture that has no intention of being subservient to the client or becoming his tool; it offers nothing but its ideological and behavioral attitudes. It has no desire to produce or complete objects or buildings, but wants rather to function through ideological behavior and actions disruptive of past architecture and design. Its significance lies in the systematic attention that it gives, not to what can be produced, but to an absolute and operative ethic. This consists in making people aware (not by means of commercial products, but by their aim of disengaging themselves from, and breaking with, the present system) of the extent to which design and architecture have been led violently astray by having been made subservient to ideological and behavioral repression. This stress on the ideas, behavior, and ideology of design and architecture has taken on added weight since Marshall McLuhan pointed out that a system of maximum control of communications should include every type of medium involved in communications. Similarly, in the field of architecture and design, this has taken the form of a thoroughgoing examination of all the transactions that take place, from the ideological and planning stage to the final aspect given to the product. This radical interpretation of the system of acting and being in architecture has brought out, in the work of such Italian groups as Archizoom and Superstudio, the importance of the various mediums involved in the creation of buildings and manufactured objects. These mediums have proved to be the raw materials for a new ideological and behavioral position, which can in itself be regarded as design and architecture.

From this has come about an explosion of the mediums used in architecture and the making of objects, and a discovery of the intellectual and attitudinal errors in the ways of acting and being in architecture and design. This explosion has, first and foremost, led to a merging of various concepts of designing, commissioning, and production, which heretofore had been regarded as occurring one after the other but are now seen to be simultaneous phenomena, and therefore highly important factors for meaning and communication.

Planning, Commissioning, and Production as Design and Architecture

With its awareness of the importance that the ideological and behavioral mediums of architecture and design have for these respective fields, radical architecture has expanded the operative and instrumental field of its activity, making use in its plans and conceptions of every possible medium, abstract and concrete, so that the written program has been raised to the status of independent and total architecture. This program is then actually analyzed, not as if it were a plan for or about something, but as an autonomous, tautological medium that may be considered: a) as an abstract-concrete sign, which while lacking any reference to what is real or realizable, has its own significant function, the function of providing informational facts about the nature and future of architecture; b) as an individual or collective expressive sign, signifing the fantastic and imaginative purpose of the group or individual that invented or planned it, with or without reference to what is real or realizable; c) as a concrete sign of hypotheses for actual intervention in real or realizable events or conditions, and therefore to be understood as a metaphor for an article to be produced, an ideological-behavioral work either as architecture, as a work of architecture, or for architecture. In its concrete and physical form, the idea may take the shape of something written, oral, three-dimensional, or visual, according to the various mediums employed. Thus the program takes on a different role in communication and information according to the mediums used; it is, and can be, architecture as photography, as writing, as an oral or written text, as a book, or as a multimedia creation, and its existence as architecture or design will depend upon the mediums. *The mediums are design and architecture.*

The various possibilities for an ideological and behavioral program therefore give rise to innumerable possible combinations of idea and medium, idea and idea, and medium and medium. One can accordingly plan or determine a priori the combining or linguistic possibilities of the various elements, in order to produce an explosion of the program. One can determine in advance the mode of operation, the sequence in which the signs and ideas will be used to end up as design or architecture (e. g., the work of Peter Eisenman in the United States). The table of permutations is infinite. *The table is architecture.*

So, as the program chooses, procedures, mediums (photography, writing, written or spoken words, film, multimedia productions), and ideas can be combined, with permutations and combinations of these factors taking place among them. *Combination and permutation are architecture and design.*

If these mediums, the conceptual and operative program, the ideas, and the multimedia elements are architecture, then the control of these mediums when they come into contact with the organs of information — magazines, books, catalogues, television, and mass media — also comes within the scope of architecture and design. Thus, the group or individual planning or devising the program can either control the ideological and philosophical aspect of architecture by the outright invention of a text, a photograph, writing, or sound, for an architectural concept for the various mass media (e. g., Archizoom); or else the mass media can be considered in the program and work of architecture or design as further mediums to be interpolated into the project as architectonic means (e. g., Superstudio). In this way, all the transactions will be controlled by the designing group or individual, so that information is regarded as architecture. *Information is architecture.*

If information is architecture, it becomes a part of radical architecture: every written or spoken work — book, essay, catalogue, lesson, speech, or lecture — is architecture, unless is is estranged from the reader or listener and forced to become part of some other field of activity, such as the history of architecture or design, or the like. Even so, these, too, may be included. It is the attitude toward the medium which makes it architecture or something else. *The attitude is architecture and design.*

The program, whether written or spoken, is the highest point of abstraction and of disengagement from the commercial and deceptive compromises that have characterized the process of architecture and design in the past. It is the acme of ideological and behaviorial purity in the activity and existence of architecture and design. The radicalization of activity and existence (or doing and being) coincide with nonrealization and silence. *The written or spoken program is architecture at the vanishing point.*

If, instead, one follows the sequence or transaction from idea to production, one encounters the second (so-called) stage, the commission. Commissioning is another medium that has become interpolated into architecture and design. *Commissioning is architecture and design.*

In its continuous control of the means employed, commissioning can take its place as a medium under the direction of the individual designer or group and can therefore assume the role of an operating tool; but its role as information must also be taken into consideration. Commissioning is a message; it has a temporal significance, representing the second stage in the aforementioned sequence.

As the second stage, it comes after the drafting of the program and is therefore retarded with respect to the philosophical and ideological considerations that determined it. When it becomes part of radical architecture, it slows the pace; as a second phase, it shares in the information or the formation of the program. It may therefore be considered among the mass media; the designing body may regard it either as a medium or as a channel of information and will force the patron to acquiesce passively, with no possibility of participating (we are still speaking about the radicalization of the program). The idea is sketched out a priori, without any regard whatsoever for the patron (as a specific and significant entity, a particular message).

In the opposite situation, the patron may become the principal medium of the program for architecture or objects, completely reversing the transaction between the idea and the commission. The patron, a particular message, will assume a primary and fundamental role; he will be architecture, in that he will force the other mediums (idea, attitude, information, program) to conform to his informational requirements.

In the former case, the group or individuals devising the project and idea can prepare or present a series of abstract ideas, a sampling of the programs that have resulted from their own activity and existence in architecture and design, and oblige the patron to accept them strictly as they are, without allowing him to intervene in any way whatsoever, except to provide the traditional financial support (e. g., Superstudio, Archizoom). In the second instance, the patron will dictate his own architecture, and the designing and idea-producing individual or group will be nothing but the medium for fulfilling his commission. In either case, the result will be architecture — in one instance radical, in the other reactionary. The program as conceived and the patron, together, increase the number of possible combinations in architecture and manufactured objects. The combinations can be determined a priori; they are architecture. The patron as economic and ideological medium is message; he may base his demands on money and exploitation, or he may be at the service of the revolution. This message, too, the program must take into consideration in advance, choosing to make its ideological and behavioral content either consistent with, or alienated from, the patron as economic medium. *Predetermination of the ideological and behavioral relationship with the commissioning medium is architecture and design.*

The minimum redundancy results when the concept and ideology of the program coincide with the commissioning medium (radical architecture). The economic medium, the commission, may be the principal message, if the program abdicates in its favor (speculative building). All the mediums that converge in transacting the connections between idea and production are architecture and design. The commission may request an abstract program, leading to architecture at the vanishing point, or it may demand or accept a program for something to be actually produced or built.

Production is a further medium of architecture. It is the most obviously concrete medium and, therefore, the most alienating. As the final medium, it encompasses in itself mediums of production and concrete execution (each medium used naturally augmenting the possible combinations in architecture and the making of objects), and hence introduces into the scheme other elements having a high informational potential. In the case of production, also, the medium — considered abstractly — can enter directly into the ideological and behavioral drafting of the program, functioning as an inferior means controlled in all its informational implications (delays in execution, historicism introduced into the abstract program, new mediums, economics, sociology, industrial materials, ergonomics, the ideological component of production, etc.), and thus introducing physical considerations into the original program. In this case, the initial program may or may not take under advisement the slow pace of production, giving itself an historical aspect or one contingent upon production.

On the other hand, production may regard itself as the primary medium. In this case, its message will distort the ideological and behaviorial aspects of the program and determine the architecture or design to be executed, whether a book or a building, an object or a city plan. As in the preceding cases, the radicalization of acting and being in architecture and design coincides with maximum control of the mediums, among which production must also be included, if one intends to continue the revolutionary sequence; whereas maximum integration will coincide with alienation from this medium also, resulting in a reactionary sequence. The abstract or concrete combinations of the various elements — planning, commissioning, and production — or their independence as informational and formative means, constitute radical or reactionary architecture.

The Architect as Architecture

In this system of mediums used for architecture and objects, the architect and designer also have a part. Ever since Herbert Marcuse substituted sensory for technological mediums, and the physical and organic responsibility of the individual has become an ideological and philosophical concept, the architect or designer, also, with his own behavioral and imaginative ideological office, has become a medium — the architect as architecture.

It is a medium that can either allow itself to be engulfed by the other mediums, or it can establish itself and win respect through its conceptual and ideological programs, giving the maximum importance to its informational role, without conceding anything to the other mediums. Such a degree of self-esteem leads to the minimum ideological dissipation of one's own way of acting and being, and to the greatest degree of osmosis between theory and practice at the initial stage, so that the two constitute a new mystique of architecture. And it is precisely toward a mystique of its own acting and being that the new architecture seems to be heading (including, besides Archizoom and Superstudio, also Ant Farm, the Vienna-Hollein Planet, Abraham, Pichler, Haus-Rucker, Onyx, Archigram, Utopie, and Metabolism). It is a mystique that is by nature concrete and radical in its ideological and behavioral premises, a mystique that refuses to be alienated from its own ideas and its own image, with a sacred implacability regarding its own ideas and concepts of architecture and design, which entails sublimating its own activity and imagination as the primary means of its existence in architecture. These means respect the organic unity of their own thought and feeling, which are no longer directed toward the goals of production and the making of material objects but toward sensory and mental satisfaction.

Nowadays, the architect and designer do not produce more ideas, they rid themselves of ideas, producing ideal programs that are 'less ideas,' mental liberations from one's own acting and being. These minimal projects (Archizoom, Superstudio, Group 9999) can subsequently be adopted by other mediums, such as commissioning or production, but they must be respected in their entirety, precisely because they are free expressions, and hence finished and complete objects and concepts, allowing no other participation except what serves their purposes. They are, in fact, not additional ideas having anything to do with a system of production, but liberating gestures in their own right, hence 'architecture and design in their pure state' — radical architecture.

The repressed imagination of the architect or designer, being an additional medium, bursts forth and expands its power of invention, its involvement with linguistics and behavior. Freed from the restraints of mediums that have one idiom only, it can fulfil itself through many mediums, from photography to film, from writing to books, from concept to design, from listening to projecting. Since the imagination is now freely available for all kinds of fantastic actions, it can include within its scope of action every type of activity or creation, from art to film, literature to philosophy. In its modus operandi, therefore, it can use every expressive means. The higher its level of expressiveness and philosophical and ideological awareness, the greater will be the informational message it conveys about its acting and being.

Thus, the highest degree of ideological, philosophical, and imaginative-behavioral radicalization will still result in the least degree of estrangement from other mediums.

The architect or designer can still, in accordance with theories of his own choice, plan, a priori, the pattern of his acting and being. Following an ideal scheme of behavior and action, he can decide what course to take, what mediums to use, what ideas to control. He can also embrace, a priori, fields of research that particularly interest him (we are still referring to fewer concepts), and he can communicate them through mediums that he controls and directs.

So the existential creative cycle comes to a close.

The architect is architecture in the making, and the physical traces and residue are of no consequence; the tangible work also need no longer exist, and should even be abolished, or at least suspended. For, in fact, as long as alienation in work and activity continues, intellectual radicalization will always have the support of a consumer society and will take physical form in other objects, images, and creations that by their very presence will negate the total intellectual radicalization of work. There must be immobility, then, and not work; silence, and not production; statements, and not construction; concentration on the uses and manner of using one's own thought and action. This means regarding thinking and producing ideas as work, self-expression and invention as work, and the abolition and annulment of work by aphysical and nonproductive intellectual activity.

DESIGN AND TECHNOLOGICAL UTOPIA
Manfredo Tafuri

The essential continuity between Italian design of the prewar period and that of the years from 1945 to 1960 is an historical fact that has not yet been sufficiently recognized. It is surely no mere accident that the formal paradoxes inherent in the decorations by Franco Albini and the firm of BBPR between 1936 and 1940 should have provided a wholly consistent basis for the development of the 'surreal' tendencies of postwar Italian design. I am thinking in particular of the controlled functional distortions evident in Albini's Room for a Man in the housing exhibition at the VI Triennale in 1936, with the bed balanced on a rowing machine and suspended in midair, the clotheshorse used as a ladder, and a wardrobe with a roof that could be walked upon (fig. 1); the disguieting strangeness of the Living Room in a Villa at the VII Triennale in 1940 (fig. 2); and the 'useless machine' in the library of Albini's house (1940), with glass shelves supported by light metal rods suspended from two spars that form a 'V' (fig. 3); I also have in mind the BBPR pavilion at the Paris Exposition of 1937 (1). Some have interpreted such incursions into the dangerous realm of the 'autonomy' of the object as efforts to escape the necessity of coming to grips with the political situation of the time. With greater historical objectivity, one might perhaps see in them something close to the first ironical researches of Bruno Munari. The 'revolt of the object' did not come about, however, because of any such adherence to the poetics of the avant-garde nor because of some kind of 'technological despair,' but because of the alienation of the object from its context. This sense of loss, this forced withdrawal of the object into itself, was the result of the fragmentation of the building trade and the consequent autonomy of its several sectors, which influenced, and was reflected in, the cruel elegance of avant-garde Italian design before the war.

A similar evocation of the magical, manifest in the crystal pavement of the living room Albini showed at the VII Triennale in Milan, pierced by a tree and laid over a flowery meadow, is not part of a comprehensive program. The delicate grills with which Albini broke up the space in his installations for the Hall of Aerodynamics at the Aeronautical Exhibition of 1934 and the exhibition of antique goldsmiths' work at the VI Triennale in 1936, with their reminiscences of works by Walter Gropius and Joost Schmidt, or by Edoardo Persico, indicate divergent tendencies still forcibly held together. On the one hand, these allusive schemes are equivalent to purely hypothetical statements of a modular system longed for as if it were utopia; they are reflections of a scheme tried out on an urban scale in the project for the Viale Argonne in Milan, submitted by Albini, Renato Camus, and Giancarlo Palanti, or on a utopian scale in his project for 'green Milan.' On the other hand, they point the way to a kind of design in which allusion could become an independent instrument of communication, and in which the value of the image would be dissipated in the unreal dematerialization of the object. (And in several works of those years, such as the BBPR pavilion at Paris, already mentioned, the 'anti-twentieth-century' controversy is evident, a revolt directed not only against Giovanni Muzio and Gio Ponti but also against Giovanni Terragni, Giuseppe Pagano, and G. Levi Montalcini).

What I wish to point out, however, is that this kind of unrealistic alienation of design had already taken root in the most advanced outposts of Italian architecture, in a field in which there was an inclination to transform the conditions caused by the backward system of production into an ideology.

Should one wish to give a synopsis of the dominant features of such an ideology, one would have to take into account the quite incredible survivals of the mythologies longed for in their time by the historical avant-garde: the nostalgia for a return to childhood seems an irrepressible fact even in the most progressive research in design.

Manfredo Tafuri, who received his degree in architecture from the University of Rome in 1960, has been a member of the faculties of the architectural schools in Rome, Milan, and Palermo and since 1968 has been professor of architectural history and director of the historical section of the Istituto Universitario di Architettura, Venice. As critic and historian, he is particularly concerned with the social and political aspects of architecture and town planning, from the Age of Humanism to the present day. He is a member of the editorial staff of the periodical *Contropiano* and the author of numerous books and articles; his recent publications include *Teorie e storia dell'architettura* (1968-70), and he is also editor of a two-volume anthology, *Socialism, città, architettura: URSS 1917-1937)* (1971-72).

1. Franco Albini. Man's Room. VI Triennale, Milan, 1936
2. Franco Albini. Living Room in a Villa. VII Triennale, Milan, 1940

One might, of course, compare the experiments previously mentioned to analogous research elsewhere in Europe (Gerrit Rietveld's work of the 1940s comes to mind), in order to give a general picture of a particularly 'difficult' moment in contemporary design. I prefer, however, to stress the paradox inherent in the Italian situation, relating the phenomena of the first postwar years to the atmosphere created by the 'dangerous' survivals already mentioned above. In this sense, one would have to acknowledge that the assertion of a kind of representational quality, heavily laden with allusions, the recovery of a symbolic dimension for the object, the effort to extract 'constructed' space from the laws of automatism in general, constitute only ambiguous responses, all basically inspired by a kind of unconscious realism.

At first glance, it seems surprising, especially after the expectations of 1945 to 1947, that Italian architects should in a very special sense have been subjected to an absolute exclusion of the production of objects from the totality of the building system. Piero Bottoni's efforts to give the 1947 Triennale the character, not of a trade fair, but of a living manifesto within the city of Milan of a program for consolidating design on the object level with design on an urban scale, were doomed to failure because of the completely archaic structure of the building industry and the impediments within it (2). Over and beyond that veritable graveyard of good intentions, the experimental QT 8 quarter, the burial of the themes faced, for better or worse, at the VIII Triennale marks the decisive moment of the internal dissolution in the production methods of building. For all its lack of organic unity, the naïveté of its proposals, the anachronism of the typological research it undertook, and, if you like, the wishful thinking of its plan, QT 8 nevertheless purported to be an experiment continuing the achievements of the European vanguard of the 1930s. Reaffirmation of the principle of unity of design on various scales was not actually of great methodological importance, but it had a certain value as manifesting an inclination to unify the building industry. It can surely be said that a program of this sort was not in the slightest degree avant-garde, in comparison with the historical achievements of modern city planning. The Siedlungen (housing developments) of Ernst May and his collaborators in Frankfurt, between 1925 and 1930, were the fullest realization of direct inclusion of the small-scale production of objects within the parameters controlled by the vaster scale of an urban undertaking. Thus it might be claimed that the famous 'Frankfurt kitchen' precisely because of its situation within a context of production directly in proportion to, and controlled by, the urban scale, was the most convincing example of 'radical' European design in the period between the two wars (3). Leaving aside the intrinsic deficiencies of the urban and building policies of the German Social-Democratic regime (4), it is certainly at Frankfurt, rather than at the Bauhaus between 1924 and 1928, that the principle of including design within the whole complex of building found coherent expression (5).

QT 8, therefore, should be regarded as a plan rather than an actuality. And it is with respect to its program that the Milanese quarter planned by Bottoni and his associates is of historical significance as having initiated a comprehensive plan for overall production, within which the various sectors of the building process would find their rightful place. In his Obus plan for Algiers, which was on the most comprehensive scale possible, Le Corbusier made extremely clear the extent and margins of freedom allowable for the various participating sectors. QT8, being a reactionary plan presented solely by intellectuals, within an economic context in which a productive and technological reorganization of the building industry played no part,

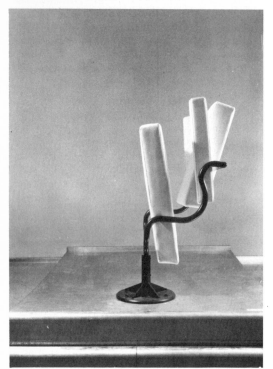

3. Franco Albini. Living room in the architect's house, Milan. 1940
4. Franco Albini and Franca Helg. Consigliere folding chair. Original design made 1960 for the Communal Offices, Genoa. Iron tubing and foam rubber

had the defect of having chosen to pursue a shortsighted policy of realism, for which, in the long run, the whole complex had to pay the penalty. There is probably no other project of the 'era of reconstruction' in Milan so utopian in its estrangement from the city, and at the same time so nostalgically backward looking, as QT 8.

It was just this moment, that is, the moment when Italian capital decided what strategy it would adopt during the period of reconstruction, that determined the conditions whereby the ideological program of design became utopian, deciding to make of its qualitative contribution to industrial production a privileged field for mass education. In order to explain this, we must resort to an analysis of the system.

First of all, it should be noted that there had to be a process of adjustment between the production system of object design and the inadequate, fragmented productive capacity of the building trade, its levels of craftsmanship, and the impediments caused by an archaic system of public intervention. At a time of crisis early in the 1950s, caused by the impossibility of finding enough capital to provide the necessary housing, by the trend toward ownership of dwellings, by the isolation of the market for new houses (with a consequent immobility of the market and a scale of rents determined according to the estimated costs of new construction), and by the concept of state support for building materials as a kind of charitable undertaking, the production of furniture underwent an adjustment to bring it into line with that sector of building least subject to planning — that dominated by privately owned luxury property. In this sector, also, the complexity of production constituted a further impediment to any likelihood of unified organization and any possibility of attracting the interest of industries having a high level of capitalization (6).

The inadequacy of state support and of the credit system furthered the fragmentation of production and led to a building boom based on very low, and constant, technical factors. This is not the place to demonstrate how this backwardness in organization was intimately bound up with the concentration of capital in other more developed sectors. It is nevertheless a fact that the aggravation of the urban problem and the crises within the various sectors (already noticeable in the early 1960s and still unresolved today) paralleled the situation in the production of objects for use in the home (7).

Finding no support for a policy oriented toward the concept of the house as a product for mass consumption, Italian industry logically turned to a market that required no fundamental technological innovations, and that corresponded in a positive way to the traditional values inherent in the 'quality object' produced by craftsmen.

This caused a decisive break between design for the home and industrial design, which was concerned with creating forms for mass-produced objects manufactured by sophisticated technology. Not surprisingly, this very identification with a consumer elite opened up for Italian design an international market for distribution of its products. The lack of those restrictions necessitated by advanced production methods or by a mass market aided the quest for a way in which to retrieve what official criticism has defined as the 'freely-arrived-at combinations' in the mainstream of modern movements (8). It is precisely these conditions that impelled Italian design to resume once more the researches into form that had been undertaken between 1930 and 1940. We have, on the one hand, Marcello Nizzoli's technological designs for Olivetti, and the styling of Pinin Farina; on the other hand, we have the exotic elegance of Marco Zanuso and Paolo Chessa, the hermetic objects of Bruno Munari, the

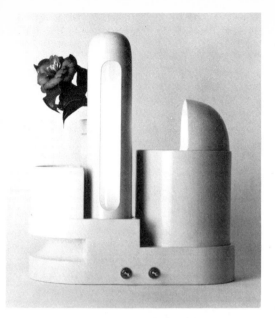

5. Gae Aulenti. Rimorchiatore ('Tower') lamp.
6. Gae Aulenti. Showroom of Knoll International, Milan. 1971
7. Gae Aulenti. House of a Collector, Milan

'neomodernism' of Gio Ponti, the expressionist distortions of Carlo Mollino, and the 'representational' design of Ignazio Gardella, Albini, and the BBPR firm.

Given an industrial program that allows a considerable margin for 'quality' production, the solution of Italian design was to accept these backward conditions and attempt to overcome them by a magical sublimation. Thus the goal became to make the object 'eloquent' and, frequently, to have it reveal the level of craftsmanship that produced it.

There was a reaction of both amazement and scandalized shock on observing, between 1957 and 1961, the appearance of tendencies seeking to reconcile the ambiguous oscillations of Italian design in its nostalgic flirtations with a snobbishly longed-for historicism with the frowned-upon declarations of adherence to the 'orthodox heterodoxy' of the modern movement. The intolerance, complacency, shrewdness, and restlessness of the new generation in Milan and Turin, which have given birth to the pseudo-movement wrongly called 'neoliberty,' were actually only the results of an attitude deeply rooted in the ambiguous development of modern Italian architecture (9). BBPR's installation of the Castello Sforzesco in Milan is a particularly outstanding example of the trend toward representation and the compromise between technological elegance and a return to the realm of 'magic.' It is in the same direction as the frank reversion to the past of Guido Canella, Roberto Gabetti, and Aimero Isola, with no undue half-measures in regard to consistency with the 'tradition of the new,' and it has the merit of having brought to light, without any bias, the extent to which those engaged in cultural controversy in Italy have remained firmly buried under layers of inhibitions and guilt complexes. The themes of play, magic, and the mystique of the eloquence and ambiguity of the object: these are the themes with which the new generation in the North came to grips from the late 1950s up to the time of their group exhibition organized in Milan in 1960 by the *Osservatore delle arti industriali.* It is precisely the theme of the object, moreover, that allows the new tendencies to declare their own position most precisely, and their statements in the exhibition catalogues are in themselves evidence of this.

For Vittorio Gregotti, the problem was to introduce into design 'the uncertainty of shadow rather than the brilliance of light... Who, while seated in one of my chairs, could ever weep? To make an interior design has become as impossible as to paint a picture to decorate a room — impossible, at least, for an honest person to do so without a sense of shame' (10). Guido Canella called for 'absolute openmindedness' as the psychological attitude most capable of 'provoking the recreation of existing images, as allusions that enable one to penetrate beyond reality, or the degradation of stylistic schemes so that they can be adapted to contemporary life. That is to say, the *dolce vita*, or the *Adelchi* performed in a circus' (11). Highly significant is this identification of the suspension of judgment called for here (we should not forget the manner in which the phenomenological 'vogue' infiltrated Italian culture during the 1950s) with the intention of immersing oneself in the dense complexities of reality through a planned program of an ambiguously desecrating nature. What neither Italian nor foreign critics have grasped in the phenomenon of 'neoliberty' is just this process of a contemporary rediscovery and ridiculing of 'values.'

Apart from the subjective eruption of deeply rooted Oedipus complexes regarding traditions of the modern movement (12), such a sadistic evocation of emblematic forms, immediately subject to degradation by

a necessarily 'heedless' use, betrayed an awareness of crisis that is still felt by Italian architects and designers. The furniture shown at the *Osservatore delle arti industriali* exhibition has only one common characteristic — its delight in vague allusions. The constructivism of the freestanding bookshelf by Gae Aulenti, reminiscent of Frank Lloyd Wright and Neo-plasticism, the ironic craftsmanship of the cabinet and little table in turned wood by Michele Achilli, Daniele Brigidini, and Guido Canella, the monumentality of the teak and foam-rubber chair by Leonardo Fiori and Carlo Segre, the complacent cabinet with flap by Roberto Gabetti and Aimaro Isola (a true nostalgic synthesis of nineteenth-century style), the ostentatious purism of Leonardo Ferrari and Aldo Rossi — all revealed a split within the actual situation of design, rather than offering any fruitful suggestions for development (notwithstanding the insistence of the designers on the value of their method).

The scandal they caused was really quite unjustified. Had not this beaten track of irony, the recovery of representation, and the exaltation of the qualities of craftsmanship already been taken, perhaps, by the prewar Italian masters and the staler experiments of the 1950s? That equivocal historicism, laden with nostalgia and prone to equate autobiographical confession with the opening of a road to salvation from alienation (Fellini's *8 1/2* is somewhat in this vein), is indicative of a crisis within the discipline, concealed precisely by the desperate course of taking refuge in the limbo of individual reactions. This autobiographical and revivalistic tendency also has its own proclamatory content. Like the esoteric acrobatics of Albini or the BPR, this neurotic quest for 'authenticity' highlights the deficiencies in the relations of design itself with the system of production.

In this sense, the individual works by the young designers of Milan and Turin represented a farewell, even though a retrogressive one, to utopia (figs. 4-8). Moreover, this clear detachment from the mainstream of the modern movement signified a refusal to continue perpetuating a world of illusions that are promptly belied by reality, in order to take on the role either of spectators inclined to mockery and irony, or, inversely, of actors engaged in contradicting reality precisely in order to assume, like a chameleon, its outward semblance.

It might be said that between 1958 and 1965, Italian architecture and design attempted a 'trial of strength'; but not one of those involved in the architectural controversy of those years summoned up the necessary courage with which to define the objective of his own work as an intellectual technician (13). And that is why — over and beyond strictly commercial reasons — the adherents of the *nouvelle vague* in the North almost immediately saw their plans commercialized. The revival of the representational character of the object was certainly the most 'realistic' proposal that could have been advanced. By imparting an ironic tone to their historical revivals, always concealed under the veil of vague allusion, they could only show their compliance — with the 'shrewdness of doves' — with a situation in which a backward industrial system had repudiated the designer's activity. It suffices to compare the objects at the *Osservatore delle arti industriali* exhibition, so clamorously at peace in their surreal isolation, with the schism in design observable at the XII Triennale in Milan (14).

Outwardly, we seem to be confronted here with two contrasting tendencies that are mirror images of one another. In actuality, both reveal the same impotent state of design, which is also made clear in the reports of the judges for the Golden Compass awards offered by the Rinascente in 1959 and 1962. At the Rinascente Center in Milan, on January 23, 1960, Giulio Carlo Argan read the following statement (15): 'After a careful consideration of the whole situation, the Commission

8. Ignazio Gardella and Anna Castelli. Bed in polyester resin.

for Grand National and International Prizes offered by the Rinascente Golden Compass has unanimously recognized a state of crisis in the field of design: a crisis that ... concerns every country that has given special attention to this problem. It is apparent in a stasis, and even an abandonment of certain positions already attained, and it leads to an inadequate correlation between designers and producers, and between these and the consumer; and among its primary causes is the lack of organizational and didactic systems. This crisis is especially manifest in the tendency of design to concentrate on peripheral and superficial themes, such as interior decoration, rather than on major methodological problems, such as the industrialization of the building trade, which are the only themes that would allow design to develop within a wider context of social aims. This same tendency of design to veer away from concentration on methodology results in a tendency of commercial firms to specialize in the sporadic production of high-quality goods, as if this were an indication or guarantee of technical ability, instead of laying down the principles and lines for a systematic development of a kind that would ensure the same high level of quality for the whole cycle of production' (16).

Once again, ideology has turned the world topsy-turvy. According to the statement quoted above, it was design that determined the orientation of economic policies in this branch of industry, and not the other way around. Thus, the withholding of the national Golden Compass award that year can be understood as a punishment inflicted on the servant in order to intimidate the master. This is highly indicative of the illuminism that pervaded Italian culture in the early 1960s. We have, on the one hand, Vittorini's invitation to intellectuals to bring their specialized instruments into the technological world; and on the other, an analogous invitation from the Golden Compass jury for design to act as a pressure group for a cultural policy (17).

Meanwhile, a Triennale on the theme 'Home and School' served to underline the futility of any proposal not geared to the level of the production program. It also led to a concentration of critics — including studies by Gillo Dorfles. Emilio Garroni, Umberto Eco, et al. — on the theme of the relationship among communications, language, and consumption. There was in these studies a special emphasis that does not appear in the works of Max Bense, the foremost international scholar in the field of the aesthetics of technology (18).

The introduction of a relation between entropy and the semantic ambiguity of aesthetic information (which is the basis for conceiving the form of a product as a channel of communication that is, and can be, regulated by a system of binary values like those of a binary computer) was welcomed as a fountainhead for new programs of formal exercises. In these studies, time after time, the redundancies or essential points dealt with the desire for design to overcome the crisis independently, by communicating it through objects, or by a strict control over the effective promotion of the object itself.

Analytical studies on the theory of communication, meanwhile, avoided complete elucidation of the indissoluble links between technological aesthetics, the theory of symbols, and the capitalistic theory of development, to take on the role of an ideology of compensation. Analyses of the relations between communications and consumption, and between the theory of technological innovation and the theory of linguistic innovation, almost always contain a suggestion that, given the premises, is wholly gratuitous. As an extensive information system directly involved with the world of advertising, design stands out as one type of activity in which indeterminate efforts at semantic restructuring could successfully regain for the discipline itself a 'social,' 'humane,' and even

revolutionary role, to counteract 'distortions in consumption.' It is hard to establish precisely the extent to which Italian design was influenced by the massive wave of semiological and structuralist literature. It is certain, however, that the designers didn't let slip such a convenient alibi for their intellectual work, responding to repeated invitations to 'resemanticize' the object and recover its myths by veering increasingly toward the surreal, toward the field of willfully unresolved formal ambiguities, and toward the gratuitous (19). Thus we witness in Italy a true historical paradox. Impelled by the challenge to an ideological 'frenzy' of 'radical' design, the esoteric and anguished ironies of Canella, Gregotti, Aulenti, Ettore Sottsass Jr., et al., opened the way for a return to the same atmosphere in which 'radical' design had first originated. The antiutopian regression was therefore fated to give birth to new utopias — even though here it is a question of a 'theater of utopia' in which pure 'plays of anticipation' are performed with conscious detachment.

It is, after all, wholly logical that design should have become the focal point for the neo-avant-garde in the field of planning. For, in fact, it is only in the composition of architectural microcosms that it is possible today to hold forth the promise (without any pretense at credibility) of subjective liberation through the reconciliation of 'man' with 'the soul of things,' and with the unfathomable depths of his own repressed impulses.

That design and environmental planning should have approached the poetics of Surrealism or Pop is therefore not the result (or not solely the result) of a superficial assimilation of imported figurative modes. To be within the confines of production and be lured into escaping its laws by accepting the invitation to make a subjective, private leap into the sublimated universe of 'artificial paradises': just such a liberating utopia lies at the core of Surrealist ethics, just such an imaginary prospect of a 'reign of liberty' which must follow the 'reign of necessity,' just such a pillaging of the individual unconscious in the effort to recover the archetypes — now finally laid bare — of human values.

This tendency, discernible in some of the more recent works of Ettore Sottsass, Jr., or Afra and Tobia Scarpa, runs through several of the stylistic paradoxes of Fabio De Sanctis and Ugo Sterpini (typical examples are the chair in iron and oiled leather by the Officina Undici of Rome, and the triangular cupboard with its clear reminiscences of Futurism), to reach its culmination in the refined eclecticism of Gae Aulenti (figs. 5-7), and the (increasingly commercialized) irony of Archizoom (fig. 10), Ugo La Pietra, and the Superstudio group — which have obvious connections with such foreign experiments as those of Archigram, the Austrian 'Salz der Erde' group, etc. (20). Thus, the very branch of institutionalized design that is engaged in small-scale production has taken on the task of trying to fulfil all the hopes for a thoroughgoing planned renovation ot the whole human environment. It can in fact be said that the ideological responsibility that Italian discussions of 1945-1960 had entrusted to urban planning has gradually, from 1960 on, been transferred to the field of design.

The reason for this shift is evident. As planning policy, with all its obstructions and contradictions, descended from the Olympian heights of ideological behests to the solid ground of practice, gradually escaping from the checks imposed by hypothetical forecasts, it became impossible to reconcile hopes and commitments of a purely intellectual nature with the techniques of a purely scientific evaluation of the future. When planning became one of the basic institutions for capitalist development, it was obliged to burn

9. Aulenti, Aymonino, Paciello, Bonfanti, Gardella, and Macchi Cassia. 'Transformation of the Landscape: The Italian Coasts.' XIII Triennale, Milan, 1964
10. Archizoom Associati. Homogeneous residential quarters. 1970

11-12. Enzo Mari. Two views of 'Module 856.' VI Biennale of San Marino, 1967

its bridges behind it and abandon the 'highest principles' on which its beliefs rested. Thus, time after time, technicians have had to come to terms with individual sectors for which forecasts have clearly been made within the context of political and economic choices determined at a higher level. It is no accident that for the past several years (apart from the inevitable fringe of incorrigible ideologists), it has been painters, sculptors, designers, and critics of design who have formulated the hypotheses for a comprehensive 'alternative' new order (21).

Ideology has thus found the most appropriate place in which to carry out its own retrogressive activity. It is not by chance that it took refuge in the very sector of production which was least involved in the economic cycle, so that design saw its own shortcomings transformed into a doubtful privilege. The isolation and autonomy of the planning sectors thus played a double role. On the one hand, they provoked a permanent state of crisis and widespread symptoms of frustration among designers; on the other hand, these crises and frustrations were sublimated into a frame of reference that always looked to the future, with the expectation of seeing the utopian phoenix arise from its own ashes.

Still more should be said. The more design is used to redeem, a posteriori, urban or building systems dominated by production methods that are completely out of phase with the present stage of the development of capital in Italy, the more its theorists tend to regard as permanent what is obviously a transient state of affairs. Or rather, they tend to distort the situation and see in it a crisis within the production of objects.

Thus, unconsciously, they reject the only prospect that could restore a context to design and give it a basis rooted in reality and the material conditions of production, but which would at the same time require them to restrain every desire to make design itself the new horizon of a mankind freed from its own contradictions. The integration of 'product design' within an economic cycle, revolutionized in its production system and integrated within the total dynamics of capitalistic development, is what has come to be both hoped for and feared. The independence now exploited by Italian designers in order to fill the institutional vacuum of their own discipline with cloudy ideologies can thus be seen as a two-faced mirror. It offers opportunities for finding a sphere of action (or of theorizing) that would allow avant-garde ethics to survive, yet at the same time it would cast that sphere of action into the limbo of the 'abstract frenzies' of planning, redeemed, according to Tomás Maldonado's recent book (22), only by vague and generalized 'hopes.'

This is confirmed by the designers' ambivalent attitude toward the world of technology. The quest for a proper relation between the object and production techniques, and for a faithful interpretation of the properties of the material or of molding processes (such as we find in Albini's little Adriana armchair of 1951, or his portable Cicognino table (1954); Mangiarotti's production for Cassina; Tobia Scarpa's lamps for Flos and his furniture for MIM; and in general in all Italian design that tends to revaluate the structure of the object), is countered by a tendency to experimentation that seeks to dissociate itself from industrial management, giving rise to attempts at a 'planning revolution,' exemplified by Mari and Archizoom (fig. 10).

When a designer who is as well equipped as Mari — by virtue of his researches into the 'exact' use of semifinished products, his design of objects completely fabricated by machine (either molded in a single piece or by successive stages of manufacture), and by his

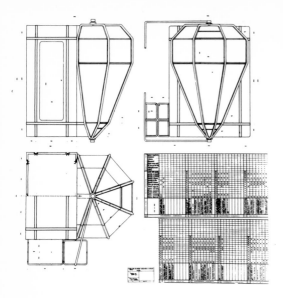

researches on the use of the module and components, and the elements of form — finds it necessary to justify his own work through an all-too-ingenuous 'revolutionary' phraseology, one cannot fail to note a symptomatic rift between his avowed intentions in design and his actual practice.

The elimination of the superfluous, and the identification of the object by its pure value as 'sign,' which have allowed Mari — and, in general, all research in 'programmed art' — to translate into their products their yearning for an historical link with the founders of the avant-garde, have already assumed ironic overtones in the 'machine for processions' exhibited in the 'Interventi nel paesaggio' ('Participation in the Landscape') exhibition of 1967, and in the 'Module 856' made for the VI Biennale at San Marino (figs. 11-12). In the latter, the visitor gazed into a container shaped like the figure '7,' which he found completely empty except for the reflection of his own image in a mirror on the opposite wall. The theory behind such a 'shock machine' was that the spectator's supposed uneasiness would evoke in him a chain of recollections, contrasting with the uneasiness aroused in him by the assortment of goods and images assembled in the group exhibition — from superabundance to the sudden isolation of the pure sign-man (23).

In actuality, confronted with the complete strangeness of the object in which he encounterd himself, the spectator could only come to a realization that he himself was the real 'superfluity' to be eliminated; the superfluity of the image was caused by the 'useless' intrusion of the human figure. Mari's 'Module 856' thus differs from other visual and kinetic games of Op and Pop experiments; its intention of establishing antialienating communications succeeds instead in creating the absolute 'model' of alienation.

The silence of the object, meant to propose an alternative to the superfluity of communications, leads to the discovery that its own 'void' is the only mute message it has to convey. In the face of such nihilism, devoid of any utopian remnants, it is hard to comprehend what meaning (aside from the indication of a subjective disquietude) the confused aspirations of Mari can signify for a new 'revolution of technicians,' for an 'alternative commitment,' or for a design either opposing, or accelerating, the process of social development (24).

I have cited Mari merely as an example. What the neoobjective tendencies of design reveal is, in actuality, only the urgent necessity of wiping out every *mémoire involontaire*, every semantic residue. To present the 'sign as sign,' estranged from its physical context, and with all its coordinates referring solely to the all-encompassing horizon of the technological universe, means — all denials to the contrary notwithstanding — a direct adherence to the theory of communications, coinciding with what Max Bense has defined as the tension between the 'form-product' and the horizon of 'technological determinism.' We are dealing here with a program, that is, which in its 'reconciliation' between technics and aesthetics, ethics and cybernetics, and its elision of contradictions, once more sets itself up as mass education, seeking in its comprehensive scheme to control everything 'irrational' and use it for the purpose of an overall development of the 'given' reality (25).

In this sense, Mari's researches have the same value as those of Joe Colombo or the projects and pedagogy of Max Bill, Tomás Maldonado, and Andries van Onck (fig. 13). All their work comes under the heading of a veritable 'theory of technological integration,' a synthetic kind of design in which the control of the irrational becomes 'productive' on the basis of its systematic adoption of the principles of the cybernetic

13. Sketches for cabins for cranes, ergonomically correct, made by students of the Istituto Statale d'Arte, Rome, in advanced course in industrial design and visual communication taught by Andries Van Onck and Ernesto Rampelli. 1970-71

universe. Thus, according to J. Habermas's metaphor, the objectified *homo faber* becomes equated with *homo fabricatus*; moreover, the essential statement of programmed art is 'our purpose is to make you a partner' (26).

It is doubtful whether on the basis of such a program, design could escape from the limitations of its own scope, the object, and embrace the entire sphere of human relationships. Rather, a systematic control of the links between production and information, with an entire program for reintegrating communications and technology, could operate as the surest means of incorporating the domestic environment within the urban landscape. The most important consequence of the discovery of the extent to which communications can be controlled lies precisely in the production of forms contained within the world of self-regulating systems. By leading experimentation in form and its uses back into the sphere of a process of collecting multiple information, design has found a suitable, independent field for development, closely intertwined with those forms of 'repatriation' of subjectivity in that realm of artificiality par excellence, the city.

It is here that we discover the meaning underlying the anguished and protesting literature of the 'radical' designers. The institutional crisis of design is actually merely a pretext. They realize that what is now at stake is the role of general control, at the planning level, of this kind of extension of their responsibilities. The true 'hope for design' lies in seeing the ideologist (in the case of the designer, an ideologist who has translated into real fact and a usable product the aspiration toward banality inherent in the whole historic evolution of the avant-garde) once more assigned the role of planner, in an increasingly highly developed integration design within the dynamics of the city.

Is this anguished aspect of the controversy over design in any way in conflict with the tendencies revolving around what is commonly defined as the resemanticization of the object — whose origins we have traced back to the history of Italian architecture of the 1930s and '60s? There seems to be no doubt that the avant-garde tendencies of Italian design have chosen for themselves the field of an ongoing and planned 'semantic distortion.' This has, however, an aim that goes far beyond that naively recognized in official criticism, which up to now has usually been prone to lament, in a moralistic way, the present mechanisms for inflating the consumption of consumer goods by the process of 'styling.'

It is hardly worth mentioning here that, in a capitalistic system, there is no break between production, distribution, and consumption. All the intellectual anticonsumer utopias that seek to redress the ethical 'distortions' of the technological world by modifyng the system of production or the channels of distribution only reveal the complete inadequacy of their theories, in the face of the actual structure of the capitalist economic cycle (27).

And that is why every judgment of contemporary design — and of Italian design in particular — should disregard the self-justifications of the designers, in order to explore without prejudice beyond that ambiguous incline over which glide their allusive metaphors, as full of ill-concealed autobiographical references as of an uncontrollable sense of guilt. Even the most recent works of Gae Aulenti, Sottsass, Jr., or Archizoom connect them, in spite of themselves, with the tradition founded by Albini, Zanuso, and the BBPR, which proposed design as an authentic kind of 'shock therapy.' The production of Poltronova, the 'center for eclectic conspiracy' installed by Archizoom at the XIV Triennale in Milan in 1968; but we might equally well cite, in the same Triennale, the space arranged by Giulio Confalonieri, or the metamorphic environment with mirrored walls and fluorescent wings installed by Cesare M. Casati, Gino Marotta, and Carlo Emilio Ponzio (28), are undoubtedly related to preceding and parallel experiments in 'Pop environment' projected into a utopian dimension. But at the core of such a leap into a willfully *autre* dimension, we find an atmosphere like that proposed in 1964 for the XIII Triennale (fig. 9). The interruption of every flow or environmental relationship, the obstructing of any specification of the particular use of space of objects, the suspension of space itself within the dimension of a universe of open, indefinite possibilities, even the negation of any specific quality of space, which is transformed simply into a place in which to 'plot cruel conspiracies': these are certainly objectives that young Italian designers have borrowed from Öyvind Fahlström and certain 'minimal art' experiments. It is not accidental that the theme of the XIII Triennale should have been 'Leisure,' nor is it accidental that the Italian designers response to this theme took the form of essays in complex and multidimensional communications techniques, poised midway between absolute autonomy and the effort to transmit messages that would evoke a completely 'liberated' environment (29).

It is precisely this evocation of streams of communication, filled with a desire for destruction, that characterizes the association between design and the 'technique of the ambiguous image' so dear to Aulenti and Gregotti. Here, too, this technique expresses itself in the form of 'shock therapy.' The sequence of images brought together within a space diluted with objects (we have in mind the ironic anthropomorphism of Archizoom and some of the experiments of Sottsass, Jr.) alludes to a veritable utopia of liberation, attained through a succession of shocks to perception. The playful potential values are charged with a utopian straining toward a world of complete liberation, like that evident in the first Dada manifestos or experiments: that ideal of man freed from work, and of Eros reinstated as the totality of experience for oneself and the world (30). Emilio Garroni has rightly interpreted the tendencies observable at the XXIII Venice Biennale as the conversion of playful art into a perspective focused entirely on the future (31).

Marcuse + Fourier + Dada: the designer absorbs all the ingredients for a systematic reconnoitring of techniques whereby the spectator can be reconciled with the future — since the present is condemned. Utopian space, often constructed without any irony whatsoever, leads directly back to the urban environment, sublimating its chaos, its multiplicity of dimensions, the constant mutability of its structures. These new *Merzbauten* offer the promise of a nonwork continuum, guaranteed by the most advanced forms of technology and, consequently, by the world of development.

But together with the steady proliferation of such 'artificial paradises,' private and public, the phenomena of profound structural changes within the productive sector are laying the groundwork for a new mass design, destined to dispel the prevailing ideological mists.
There seems to be no doubt — especially in view of the inevitable changes in its institutions that design will be compelled to make because of the restructuring of building activity — that planning methods and the role of the designer will come to be differentiated, to correspond to different levels of production, and in accordance with the automation of machine-driven equipment. The introduction of techniques in which the planning and control of production will be rigorously programmed will obviously entail entirely new activities for the designer, different from those that he has traditionally carried out heretofore. The designers' insistent demand to be given a leading role in the planning of production — a demand refused by

Italian industry while it was still at a low level of industrialization —
has now become totally out of step, if not absurdly willful, in the light
of the present development of automation.

Up to the time when the industries producing durable goods began
to use techniques of fabrication that demanded highly complex
machinery and highly skilled operators, the processes of production,
with a low level of capitalization and of productivity, always offered
the designer some scope for arbitrarily transforming the product
'qualitatively.' But recent technological developments, the necessity
for international consolidation of capital, and the ever-increasing
concentration of capital in those highly developed industries that have
faced the problem of planned and permanent modernization, tend to
limit that 'arbitrary scope.' Today, the planning of production cycles is
being entrusted to managerial systems controlled by computer
programming.

In a report made more than five years ago, Andries van Onck already
pointed out that such a restructuring was already in progress, citing
as an example the Olivetti system for continuous numerical control of
manufacturing equipment, which is worth considering here (32).
The Olivetti system proposes the elaboration of geometrical and
technological information regarding the work to be developed, and
the translation of these data into the motion of the machine that
carries out the operation. 'Programming translates into simple and
conventional form the geometric information contained in the design
and supplies the machine with the technical information needed to
execute it. The result, transcribed on a perforated tape, is sent to an
Elaboration Center, where elaboration of the data is carried out in a
Computor Center by electronic devices that produce a magnetic tape,
which is sent to the workshop. The fabrication of the object by
machinery is governed by the magnetic tape through a control
unit (33).'

In order to evaluate correctly what consequences the introduction of
numerical control will have for the production process, we must place
any projected analysis in the right perspective in relation to
capitalist restructuring of the economic cycle, which directly
involves the new roles assigned to the various phases of programming
and elaboration.

Despite all the many contradictions, inconsistencies, and delays due to
the problems raised by a situation that had remained stagnant for
far too long and the deficiences of the political class, one can
nevertheless foresee the conditions for a total restructuring of the
building trade in Italy, with the participation of economic experts in
the whole cycle, or with the application of state capital as in the
concerns under the aegis of the IRI (Institute for Industrial
Reconstruction) and others. This is not the place in which to undertake
a detailed study of these phenomena, which are only in their initial
stages; but it is certain that all the processes of design will have to
cope with such prospects in the future.

In other words, the groundwork has already been laid for the object to
lose its position of isolation, with respect to both production and form.
A building program that overcomes the impediments of land
speculation, and which plans the production of houses as if they were
consumer goods, involves unification of the hitherto divided operations
of building and design. Obviously, this will not necessarily lead to the
realization of the urban utopias propounded at the highest ideological
levels of the modern movement. But it is equally obvious that the
difficult process of capitalistic restructuring now underway, obliged to
disentangle a sector so completely bogged down in backward
systems as building has been, cannot fail to have irreversible

consequences in the field of 'design' as well. Design planned as a product for mass consumption and united with the processes of renewal in the building trade — that is the prospect opening up before us today, a prospect to which Italian industry, in particular, seems called upon to respond in a concrete way. To gc any further in predictions would be to enter upon the slippery terrain of ideology. It is certain, nevertheless, that this is the road that this branch of capitalistic development must inevitably take, and that these are the new systems which architects and designers will soon have to take into account.

This does not mean that there will still not be a wide margin for the production of objects and environments that will allow designers bent on 'saving their souls' to carry out their solitary rites of exorcism undisturbed. The nostalgic longing for magic, for the golden age of the bourgeois mystique, still continues to be cherished, even at the most highly developed levels of capitalistic integration, as a typical method of compensation. And this will be the case, as long as the magicians, already transformed into acrobats (as Le Corbusier himself finally realized) (34), agree to the ultimate transformation of themselves into clowns, completely absorbed in their 'artful game' of tightrope-walking.

(1). For these subjects, see the interpretation offered by Maurizio Fagiolo dell'Arco, 'Albini e la favola dell'arredamento' (*La Botte e il violino*, no. 5, 1965), pp. 30 ff. Interesting, from an historical point of view, is C. Zanini's negative judgment of the Living Room in a Villa at the VII Triennale of Milan, which he considered a decadent expression of 'an almost dangerous elegance' ('A proposito di un arredamento esposto alla VII Triennale, Costruzioni,' *Casabella*, Vol. XIV, no. 157, January 1941, pp. 34-39). See also Giuseppe Samonà, 'Franco Albini e la cultura architettonica in Italia' (*Zodiac*, 1958, no. 3), pp. 83 ff.

(2). Immediately after the Liberation, Piero Bottoni was named Special Commissioner for the Milan Triennale. Those appointed as a council to the Commissariat (Albini, Belgiojoso, Della Rocca, Gardella, Rogers, Pollini, Peressutti, and Clerici) called upon about a hundred painters, architects, and technicians to assume responsibility for the 1947 Triennale. Thus, the first plan for the Triennale Study Center took shape, but shortly thereafter, in 1949, it went through a crisis when the management of the Commissariat was dismissed. This crisis became clearly apparent in the gap between the project for QT 8 and the 'Spontaneous Architecture' exhibition organized for the IX Triennale. Cf.: *Catalogo del QT 8*, Milan, 1947; Franco Buzzi-Ceriani and Vittorio Gregotti, 'Contributo alla storia delle Triennali. 2. Dall'VIII Triennale del 1947 alla XI del 1957' (*Casabella continuità*, 1957, no. 216), pp. 7-12. See also Carlo Doglio, 'Accademia e formalismo a base della Nona Triennale,' and the self-critical articles of Gentili and Albini (*Metro*, 1951, no. 43).

(3). The term 'radical' is used here in its specifically political sense, in accordance with the essay by Giulio Carlo Argan, 'Architettura e ideologia' (*Zodiac*, 1957, no. 1).

(4). This subject has been developed by Barbara Miller-Lane, *Architecture and Politics in Germany 1918-1945* (Cambridge, Mass.: Harvard University Press, 1968); Carlo Aymonino, 'L'Abitazione razionale' (*Atti dei congressi CIAM 1929-1930*, Padua: Marsilio, 1971); Manfredo Tafuri, 'Socialdemocrazia e città nella Germania di Weimar' (*Contropiano*, 1971, no. 1).

(5). On the structural crisis within the Bauhaus, important as indicating the first moment of a rift between the ideology of design and the planned organization of production, see Walter Dexel, 'Why Is Gropius Leaving? (On the Bauhaus Situation)' (originally published in German in the *Frankfurter Zeitung*, 1928, no. 209, March 11.; reprinted in English translation in Hans M. Wingler, *The Bauhaus, Weimar, Dessau, Berlin, Chicago* (Cambridge, Mass.: MIT Press, 1969), pp. 136-37. The historic crisis at the Bauhaus, which Dexel at the time recognized as caused by the failure of this center for experimentation to become incorporated within the national planning of industries producing consumer goods (owing to a fortunate nationalization of this sector), can be seen as the first indication of a crisis that still affects international schools of design, from Ulm to those in Italy.

(6). Because of the importance that the theme of the restructuring of housing policy has for our analysis, we give here a brief bibliography, selected from the most interesting recent contributions: Bernardo Secchi, *Analisi delle strutture territoriali* (Milan: Angeli, 1965); idem, 'Il territorio come sistema di relazioni umane, naturali e tecnologiche' (*Parametro*, 1970, no. 3/4, pp. 10-15); Franco Corsico, 'Le Riforme in una nuova strategia: la casa' (ibid., pp. 16-19; but see the entire issue); Enrico Fattinnanzi and Salvatore Petralia, 'Il Disagio urbano in Italia' (*Zodiac*, 1970, no. 20, pp. 188-201); A. Villani, *La Politica dell'abitazione* (Milan: Angeli, 1970); Francesco Dal Co, 'Sviluppo capitalistico e riforme, il problema della casa' (*Contropiano*, 1970, no. 3, pp. 615-33); Giorgio Ciucci and Mario Manieri Elia, 'Ciclo frutture e composizione di classe dell'edilizia' (*Contropiano*, 1971, no. 1, pp. 173-204); Paolo Cacciari and Stefania Potenza, 'Posizioni del movimento operaio organizzato sul problema della casa' (ibid., pp. 223-34).

The approach of the last three articles, in particular, accords with the hypotheses on which the present article is based.

(7). Cf., among other works: A. D'Angelo, 'L'Industrial design in Italia' (*Siprauno*, 1966, vol. I, no. 3, pp. 90-100); Enzo Frateili, 'Fortuna e crisi del design italiano' (*Zodiac*, no. 20, 1970, pp. 116-18); the entire issue of *Edilizia moderna*, 1965, no. 85, devoted to design (and in particular the investigations of Italian designers). It is also interesting to reread, as documenting the situation concerning the controversy over design at the beginning of the 1960s, Attilio Marcolli's article 'Disegno industriale' (*Superfici*, 1961, no. 7/3, pp. 26-46). See also note 15, below. Also worth noting are the articles on design included in Attilio Marconi, *Topos, Khora e Architettura* (Rome: Silva, 1969), pp. 217 ff.

(8). We refer particularly to the various pages devoted to Ernesto Rogers in *Casabella continuità*.

(9). It seems superfluous to document further here the polemics on neo-liberty that have appeared in *Casabella, Architectural Review, Architettura cantiere*, and *Comunità*. See, however, the historical perspective given by the present author in *Teorie e storia dell'architettura* (Bari: Laterza, 1970², pp. 72-77) and the debate between Roberto Gabetti and Paolo Portoghesi, with a note by Carlo Guenzi, published under the title 'Revivals e storicismo nell'architettura italiana contemporanea' (*Casabella*, 1967, no. 318, pp. 12-19).

(10). 'Arredamento per i senzatetto' (*Nuovi disegni per il mobile italiano*, catalogue of the exhibition, *Osservatore delle arti industriali*, Milan, March 12-27, 1960). 'In furnishing a house', Gregotti continues, 'either cne provides instruments (an easy-chair with a whisk, a table, a refrigerator, a bed, a television set) or one tries to represent a fact; and if the representation ends up as a farce, we should not forget, after all, that clowns have always acted as mediators between mortals and the forces of destiny.'

(11). 'La Prova del nove', in ibid.

(12). Canella (ibid.) declared: 'The Italian masters, who with the passage of time and events have become parents of a young generation, do not deserve children as loving, grateful, and understanding as we. Our affection (if not our devotion) has not satisfied them, ever since their sins became an experience that we, too, had to suffer. Our having followed and then surpassed them (after a brief interval at the beginning of our formation, when we went tc the sacred texts of the Modern Movement to check their deviations), when after the war they abandoned the blind path of uncommunicativeness and tried, on their own initiative, the less aristocratic road of representation (which, like it or not, meant a retraction), has today caused us to be severely reproached for atheism with respect to method.' In this passage there can already be found an acknowledgment, however intolerant, of the continuity with the lessons of the preceding generation that the 'dissident' generation feels it necessary to preserve as a device. These confessions of Canella's should be compared with the article by Aldo Rossi, 'Ventiquattro per cento' (in the same catalogue of the 'Nuovi disegni' exhibition), which is devoted to defining new programs for mass design, with polemical barbs directed at the 'neoliberty' and formalist tendencies. The 1960 exhibition was sharply criticized, with his usual superficiality, by Bruno Zevi in an editorial 'La Storia non volge a ritroso' (*L'Architettura: cronache e storia*, vol. 6, 1960, no. 56, pp. 74-75.

(13). The reference is to a study by Franco Fortini, *Verifica dei poteri* (Milan: Il Saggiatore, 1965¹, 1969²), which marks a precipitating moment in the turmoil that had existed in Italy since the early 1960s, with a renewed criticism of ideology and intellectual work. It may be worth recalling that in the 1950s, Fortini himself had made one of the most interesting contributions to the controversy over industrial design, in which he was extremely critical both of Gerratana's idealism and the position take by Argan. Cf. Franco Fortini, 'Disegno industriale (1954)' in *Dieci inverni, 1947-1957* (Milan: Feltrinelli: 1957), pp. 125-27.

(14). A harsh criticism of the ideas behind the XII Triennale was contained in the editorial by Bruno Zevi, 'La XII Triennale di Milano: dodici punti fermi per la XIII' (*L'Architettura: cronache e storia*, vol. 6, 1960, no. 59), pp. 290-91. See also the summary, 'L'Anti-catalogo della dodicesima Triennale di Milano' (ibid., vol. 6, no. 61), pp. 438-76. These criticisms, however, merely touch the surface of the problem and avoid any consideration of the structural aspects of the situation in Italy.

(15). See the report of the jury, 'Commissione dei Grandi Premi nazionale e internazionale "Compasso d'oro" - La Rinascente 1959'.

(16). A few years later, the jury for the 1964 Golden Compass awards pointed out a series of shortcomings in the structures of production and distribution, as well as in the cultural commitment of the designers, and summarized them: '... in the infrequent organic integration of design, understood as an indispensable and irreversible phase of the production process; in the constant employment of the designer as merely a consultant extraneous to the production process itself; in the slight acount of cultural discussion on design in Italy by those responsible; in the almost total absence of institutions capable of forming future design; in the scarcity of exchanges and scantness of information; in the slipping of many sectors of design into a neostylistic phase of modern formalism'. The jury — Dante Giacosa, Vittorio Gregotti (himself, it may be observed, one of those engaged in that 'slipping'), Augusto Novello, Bruno Munari, and Gino Valle — cited 'the great imbalances in production and distribution throughout Italy's realm and society,' which designers should take into account, at the same time emphasizing the importance of the aesthetic link between production and communication.

(17). In this regard, a great part of the literature on industrial design after 1960 is symptomatic. As typical examples, apart from the well-known contributions of Argan and Dorfles, we may cite: Enrico Crispolti, 'Contropiede al disegno industriale' (*Marcatré*, 1964, nos. 8/9/10, pp. 122-26); Rosario Assunto, 'La Crisi del Design' (*La Botte e il violino*, 1965, vol. 2, no. 2, pp. 3-9); Filiberto Menna, 'Design, comunicazione estetica e mass-media' (*Edilizia moderna*, 1965, no. 85, pp. 32-37); Augusto Morello, 'Pragmatica del design' (*Edilizia moderna*, ibid., pp. 44-45); Tomás Maldonado, 'Riflessioni profane intorno all'architettura e al suo insegnamento,' ibid., pp. 96-101); and the articles by A. D'Angelo and E. Frateili, cited in note 7, above. Perhaps the most significant of these articles is that by Maldonado, in which his proclamation of the crisis in the ideology of design, instead of redeeming itself by simple indications of method, directs itself to the 'professionalism' of the designer and the architect.

(18). Cf. Max Bense, *Aesthetica* (Baden-Baden: Agis-Verlag, 1965; idem, *Semiotik, allgemeine Theorie der Zeichen* (Baden-Baden: Agis-Verlag, 1967); idem, *Einführung in die informationstheoretische Aesthetik* (Reinek bei Hamburg: Rowohlt Verlag, 1969). We should also refer, however, to the fundamental work by Giangiorgio Pasqualotto, *Avanguardia e tecnologia. Walter Benjamin, Max Bense e i problemi dell'estetica tecnologica* (Rome. Officina Edizioni, 1971), in which Bense's contribution to the demystification of a utopia that makes 'political' use of science and technology — propounded by Benjamin as well as by current supporters of 'alternative' sciences and techniques — is roundly attacked.

(19). See, for example, the final pages of Emilio Garroni's book *La Crisi semantica delle arti* (Rome: Officina Edizioni, 1964).

(20). One of the most significant documents of 'protest design' is the recent issue of the periodical *IN: Argomenti e immagini di design* (vol. 2, March-June 1971), devoted to the theme of 'the destruction of the object.' The articles therein by Archizoom, the Superstudio group, Ugo La Pietra, and others are among the most symptomatic recent texts on the literary mystification regarding their own objective place within the sphere of intellectual work. Beginning with analyses that invariably end up with the all-embracing character of capitalistic development, and the impotence of the intellectual either to oppose it or determine its results, they end by excluding any possibility of a leap into utopia, and finally wind up by justifying the most banal forms of the avant-garde: 'The act of designing, seen as a prefiguration of the future, is coming to an end; the only active force is becoming mass intellectual production. If the elimination of work implies that the sole human production will be intellectual, in order to ensure that such a condition does not become the most absolute degree of alienation or 'freedom of consumption,' it is necessary to ensure collective management of automation ... Today, everyone clearly feels himself becoming a member of the proletariat; nevertheless, he is unable to face this problem directly because he is blocked by the inhibiting mechanisms that society has instilled in him ... [Therefore] one should methodically combat those repressive structures — both moral and formal — which would reinstate for the individual the environment that surrounds him as a situation consecrated by 'values and meanings' that are natural and rational, the direct projection of his 'own' nature itself' (Archizoom). 'If, as we have seen, the city is the temple of objects, then taking them away would mean its destruction. The meaning of the bourgeois city is lost when only what is necessary is produced' [!?] (Ugo La Pietra - Gian Luigi Pieruzzi). 'The metamorphoses that the object must undergo are those charged with the values of myth, sacredness, and magic by the reestablishment of relations between production and use, going beyond the fictitious links between production and consumption ... The alternative image (which is, then, the hope of an image) is that of a more serene, relaxed world, in which actions find their full meaning, and in which life is possible with only a few more or less magical implements. Objects like mirrors — that is, reflection and measure' (Superstudio). Apart from their folkloristic aspect, such accumulated 'consumer ideologies,' offered to those nostalgic for the worst of the neo-bourgeois myths immanent in some of the 1968 movements (but see also the inquiry undertaken by Enzo Mari, published in *NAC*, 1971, nos 8/9), serve as justification for an 'environmental design' that is often on a high formal level (as in the projects of Archizoom), poised between a game and a technological utopia — yet not capable of escaping from the morass of pure ideology.

(21). One might cite numerous pertinent references; here we shall mention only three, which are highly symptomatic: Pierre Restany, 'Le Livre blanc de l'art totale, pour une esthétique prospective' (*Domus*, 1968, no. 469, pp. 41-50); Filiberto Menna, *Profezia di una società estetica* (Milan: Lerici, 1968); Pietro Consagra, *La Città frontale* (Bari: De Donato, 1969).

(22). *La Speranza progettuale* (Turin: Einaudi, 1970).

(23). Mari had already used the theme of a container for images in his setting for RAI (Italian Broadcasting Company) in Milan, 1965, in collaboration with A. and P.G. Castiglioni.

(24). Cf. the interview with Enzo Mari conducted in spring 1970 and published in *Zodiac*, 1970, no. 20, pp. 10-31, and his book *Funzione della ricerca estetica* (Milan: Comunità, 1970).

(25). Cf. Max Bense, *op. cit.*

(26). *Teoria e prassi nella società tecnologica* (Bari: Laterza, 1969), p. 219.

(27). In this connection, see the criticisms directed by Enzo Frateili ('Fortuna e crisi del design italiano,' cited in note 7) against the tendencies leading to 'illustration' of the challenge to the very forms of design (although his argument is quite alien to the premises underlying our own analysis).

(28). See *Domus*, 1968, no. 466.

(29). At the XIII Triennale, dedicated to 'Leisure,' one witnessed the collective discovery of the possibility of using the most complex techniques of information as a direct means of enabling the 'urban spectator' to experience directly the language of technological development, in the wake of the most contrived semantic and polyglot experimentation. Significant, also, is the attempt to relate such linguistic ambiguity — on the level of the 'cybernetic surreal' (a path already taken by Le Corbusier and Edgard Varèse in the Philips Pavilion at the Brussels Exposition of 1958) — to the 'positive protest' against 'consumer distortions' of capitalistic development. On this subject, see *Casabella continuità* (no. 290, 1964), with articles by Gillo Dorfles, 'La XIII Triennale,' pp. 2-17, and by Guido Canella, Enrico Mantero, and Luciano Semerani, 'La Triennale dei giovani e l'ora della verità,' pp. 44-47. The article by Gianugo Polesello, 'Questa Triennale e l'architettura discoperta' (ibid., pp. 33-42), in which he turns his analysis into a pure linguistic discussion, appears the most objective in laying bare the fundamental schism between formal experimentation and the tendency toward increased content.

(30). The division of urban space into 'heterodirectional' places (those for work) and those for recovering one's individual creativity (or nonwork places) is a theory that has been explicitly advanced by such students of social communications as Abraham A. Moles and Edward T. Wall. Moles, in particular, proceeds to discuss the utopia of the 'reprivatization of society,' on the basis of a split, in time and space, between two modes of associated life: one characterized by intense social interchanges, 'the other, on the contrary, empty and isolated by law ... made for an autonomous structure, in which autonomy increases through such techniques as a metered subscription to a municipal distributor of intelligence, like that which "time-sharing" offers us' (Abraham A. Moles, 'Le Conchiglie dell'uomo,' *Casabella*, 1970, nos. 351/352, pp. 53-57). The split hypothesized by Moles is not even a utopia, strictly speaking, since it corresponds precisely to situations that can be fully verified in reality. On p. 57 of the article cited, he goes on to say: 'The cybernetics of general systems offers still another doctrine to urbanistic theory, the "theory of aggregates": human beings have two parts in their lives, in space-time; one is subdued, colorless, and socialized, and skilfully restrained; the other is free and creative within the isolation cell of leisure — the human desert after 6:00 p.m. We already know that when we find systems haphazardly combined in such aggregates, they tend to decompose into parts isolated from one another but strictly related to one another in their behavior, even if they apportion among themselves no more than the use of a substratum of those links: the circuits of fluids and services.' On the ideology of design as the prophecy of a play society, see again the work by Menna cited in note 21. The ideological pole of Moles's studies is only another aspect of researches directly involved with the productive sector. See the article by Tomás Maldonado and Gui Bonsiepe, 'Scienza e Design' (*Ulm*, 1964, no. 10/11), which describes how methods developed by Moles with respect to components and usable functions have been used to study the structural benavior of a product.

(31). See Emilio Garroni, '33ª Biennale di Venezia: "jeu" et sérieux' (*Op cit*, 1966, no. 7), pp. 25-48. After having rightly pointed out the desecratory value of the kind of provocation induced by the new experiments in 'play' art — 'The game itself testifies to the fact that it is a type of discourse reduced to minimal terms' (p. 36), Garroni goes on to observe that 'It has two faces: one turned toward the present, in which one consumes and is consumed, without any appreciable cultural residue; the other turned to the future ... where it is filled with seriousness and new meanings, where in short a "positive discourse" takes place, capable of being not only used, but even carried further, in subsequent studies' (ibid., pp. 44-45).

(32). 'Metadesign' (*Edilizia moderna*, 1965, no. 85), pp. 52-57.

(33). Olivetti, *Sistema per il controllo numerico continuo di macchine utensili*, edited by the Direzione Pubblicità e Stampa della Ing. C. Olivetti & C., S.p.A. Ivrea. Such new systems for the automation of manufacturing machinery are gaining increasing use, thanks to the development of methods for computing between men and machines by way of so-called 'programming language.' (One such Machining Center is illustrated on page 337 of this catalogue.) See the proceedings of a conference sponsored by Olivetti (Turin, October 14-17, 1968), published in *Linguaggi nella società e nella tecnica* (Milan, 1970), and in particular the contributions by Mauro M. Pacelli, "I Linguaggi conversazionali e la loro influenza nella ricerca e nell'industria' (ibid., pp. 433 ff.), and A. Caracciolo di Forino, 'I Linguaggi speciali di programmazione' (ibid., pp. 381 fl.). This level of technical development is also discussed by Maldonado — in the least ideological part of his studies — as well as by Van Onck. The latter, in particular, in his didactic experiments, which are among the most notable among current researches, has tried many applications of operative research and mathematical models related to design. Particularly interesting is his suggestion to interpolate among the variables in linear programming a measurement of the 'quantity of redundancy,' in the communicative sense, in an object or series of objects. (Even if one is obliged to admit that 'actually one can assert that consumers are rather indifferent to the *quality* of redundancy.') See E. Frateili, P.L. Spadolini, A. van Onck, 'Situazione del design' (*Marcatré*, 1967, nos. 34/35/36, pp. 126-34). Although there exists a large quantity of literature on this subject, there are few examples of its practical incorporation within the didactic system of schools of industrial design in Italy, which today are suffering from a crisis that threatens to become permanent.

(34). Cf. 'Parlons de Paris' (*Zodiac*, 1960, no. 7), p. 53: 'A bon entendeur salut' ... / Un acrobate n'est pas un pantin ...,' etc.

A DESIGN FOR NEW BEHAVIORS
Filiberto Menna

The purpose of this essay is to pinpoint a crisis situation in design (understood as referring to articles for consumption), and the ideological about-face that has occurred in certain sectors of design; and to discuss the criticism, both theoretical and operative, that has been voiced on many sides against the structure of modern industrial society. In the sphere of design, also, there has been a new climate of opinion, marked by growing suspicion of the object for consumption and by the design of community structures and goods, and even by a utopian prefiguring of totally different situations, proposed as criticisms of, and alternatives to, the present condition. The turning point in design is connected with events that have occurred in other areas of aesthetic investigation, in which the artist has abandoned the plastically autonomous work, an object that is an end in itself, and turned his attention to creating ambiential structures (environments), to the use of the body as a means of relating and communicating (body art), and to the mental processes that preside over artistic activity (conceptual art).

The new trend gravely threatens the emerging aspects of design, still preoccupied with producing refined objects endowed with an aesthetic appearance that will make them welcome both at home and on the international market. The alternative concept of design includes a decisive ideological factor that seeks to unmask the reasons for a system that has increasingly taken on the characteristics of a situation in which the individual is generally manipulated. There is also a more acute awareness that the space for intellectual action has been diminishing, and particularly in the case of aesthetic design; and there is a growing temptation to abandon the specific field of design to devote oneself exclusively to the arena of political commitment.

This essay, however, takes as its point of departure an assumption of the essential function that can be performed by individual creativity and the aesthetic dimension in the difficult task of liberating mankind. In short, it is intended to counter the claims of technology and politics that they can do everything themselves, with an affirmation of the necessity for a joint discourse of several voices, precluding the return to any sort of hegemony.

The Motion of the Horse
Once more, political rationale preempts the entire space of action. It again proposes itself as the hegemonous force for every possible transformation of man and society. Once more, it whittles down the space for intellectual action, again disregarding the proposal for specific discourse and replacing it unreservedly with political action in the strictest sense. And once more, it whittles down the space for aesthetics and displays an inordinate suspicion of any possible manifestation of artistic activity. Suspicions are concentrated above all on design, which is accused of sharing responsibility for having created the society of mass production and consumption. 'Design nihilism' (1) is born as the sole alternative to the 'technical intellect,' which has reached the point of total reification and manipulation of man, reducing him to an entity that can be calculated, counted, and manipulated. The system of technological manipulation is founded on these bases. As the Czech philosopher Karel Kosìk has written, 'The technical intellect designs reality not only as an object of domination — utilitarian, calculable, and disposable — like a field that lies before us, fundamentally predictable, maneuverable, and manipulable, but also as perfectible toward an evil infinity.' (2) The infinite-series nature of industrial production is the horizontal

Filiberto Menna is Professor of the History of Contemporary Art at the University of Salerno; he also lectures and conducts specialized courses in several schools and institutions, among them a course on industrial and visual design at the University of Rome. His writings have focused on historical and critical problems of art, architecture, town planning, and industrial design in relation to society. Among his recent works are *Profezia di una società estetica* (1968) and *La Regola e il caso: Architettura e società* (1970).

equivalent of this vertical, temporal fugue of the infinite chain of events created by technological 'progress.'

But questioning of technology and planning has now come to be accompanied by the questioning of politics, especially since it has become evident that, within the context of the two major spheres of influence in the present-day world, technology and politics have led, by different routes, to the total manipulation of mankind. It is necessary to take note of this and understand at last (really understand) the ineluctable relation between means and ends. The objectives may change, but if the same old methods are proposed all over again, if we still put our trust in rigid, authoritarian organizational structures, we shall fall back again into the same illusion that we can transform man unbeknown to him. In short, we must finally realize that our situation has reached point zero, now that technology and politics have failed to achieve their designs for mankind.

If technical manipulation transforms the individual into a machine for consumption, political manipulation deprives him of the right of consent and self-determination. 'Power is not omnipotent,' wrote Kosìk, 'and its possibilities, however great, are limited. Power can create the conditions that allow man to move freely (and thus develop and approach his own humanity), but it cannot perform this movement in his stead. In other words, freedom can be established with the help of power, but each individual creates his own independence, without any intermediary' (3).

Design can make an effective contribution to the creation of a space and place propitious for this movement of the individual toward self-realization. Without claiming to be able to change everything, and before intervening with his own specific activity, without, that is, hoping vainly to pull himself up by his own bootstraps, the designer can still accept certain assumptions without necessarily being led to accept the generally anticipated consequences: from design to the market, along the road marked out by industrial offices. There is a 'feedback.' The course can be slowed down, it can be led into unexpected bypaths, return to the starting point, and become modified. For the straight line that links design to consumption without any deviation, the designer can substitute a line that takes a zigzag course, like the 'motion of the horse.' And who knows whether he won't succeed in a checkmate? There are some who seem to have well understood Sklovsky's predilection for the 'winding pathway' of the horse, which seems to 'move sideways,' in distinction to 'Pedestrians and Kings who because of official duties rely on one faith only' (4). The architects of the Superstudio group resort to processes of 'lateral thinking,' asserting their right to counter one-track efficiency by 'non-functioning' and 'making mistakes.' 'Architecture today,' they declare, 'seems stabilized within self-sufficient, perfect catégories and types... organized according to technological and artistic patterns and conditioned by formal power structures. Its processes are logically linked together and susceptible of proof by simple checks aimed at controlling the return of all the data fed in. Institutionalized architecture apes power... every action is planned, developed, and carried out in precise sequences. To upset this tempo and break the chain of cause and effect involves disqualifying the profession, that is, not abiding by its rules, therefore not "functioning" and hence "making mistakes" (5).

Design still has a range of moves in the game that it plays daily with production and consumption. The outcome, nevertheless, is dubious. Design might end up the loser if it does not suceed in effecting a radical revision of its relations with production. But it could also win some tactical successes, which might allow others to enjoy a strategic advantage. Technology can play an important role in this game, offering design the possibility of overcoming traditional patterns, rigidly bound by static definitions of a formal-figurative nature. But design must urge technology on without timidity, freeing to the maximum its future potential, which at present is held back by having to serve the market. In short, it is necessary to create objects and buildings that stimulate the freest and most vital behavior possible, pure and simple items for coping in general with daily existence; items that will not decrease but increase the individual's possibilities for choice and foster his maximum direct participation in the process of shaping the environment. Only thus will it be possible for design to contribute to the overthrow of the helter-skelter system in which we live today, in the East as well as in the West, a system in which, as Jacques Famery has written, 'consumers are completely shut out of the decisions regarding the creation of their objects and their environment, and when all is said and done, they have no possibility of self-fulfillment in their daily lives' (6).

The Crisis of the Object

Hypotheses of the possible rejection, destruction, or surmounting of the object have become increasingly frequent in the last few years in various fields of artistic endeavor. Criticism, for its part, has repeatedly dealt with this phenomenon, which we may define as the 'crisis of the object' (7). The trend is well known. The artist's attention shifts from the work to the processes of its formation. What matters is no longer the finished object, enclosed within the absoluteness of its form, but the process that led to the result. Preference for the process rather than the result may be regarded as a special aspect of the more complex dialectic relationship between aesthetic intention and its concrete expression. Overturning the more ancient and widespread tendency to attribute to the work itself a certain margin of privilege, preference for the intention appears as the preference for a broader and less limited range of possibilities, whose locus is entirely within the realm of the 'imaginary,' and which from that vantage point can lead to a multiplicity of directions and alternatives — something that the finished work (the object), with its definitive solution, can only to some degree inhibit. Obviously, an aesthetic based on the ideology of success will always find in this disparity between intention and completed work a clear superiority of the latter, of the concrete as against the imagined, adopting as its criterion of judgment the adage that the road to hell is paved with good intentions.

The phenomenon involves various kinds of problems, from those of a strictly linguistic nature to those that are sociological. The latter approach is the one most frequently adopted, in the sense that the crisis of the object, the rejection or destruction of the product, is regarded as related to the process of commercialization of the work, and the consequent effort to remove aesthetic activity and the products that stem from it from this uncomfortable situation. Rather than in the field of design, which only recently seemed to face this problem squarely, a verification of the means of aesthetic communication and their possible impact on the process of transforming and liberating society was first accomplished in the field of the figurative arts, beginning about 1960, through an analysis of the essential components of aesthetic communication (the producer, the vehicle, the recipient). The first of these terms to be considered was the vehicle, either two-dimensional (the picture, the painting) or three-dimensional (the sculpture, the object). On the basis of this process of structural verification, it was possible to draw a distinction between a 'picture' and a 'painting,' in that the former

407

exists essentially on a narrative plane, as imagery, and as emotional expression, whereas the second, on the contrary, operates primarily on the plane of systematic analysis of its own constituent elements, and thus on a predominantly structural level. Applying this distinction to three-dimensional works, the 'sculpture' always entails metaphorical references or narrative content, whereas the 'object' presents itself as an ensemble, a pure structure, or as 'an autonomous entity made up of internal dependencies.' If we consider the work of art, or, rather, the artistic communication, as a function of the dual polarity metaphor-metonymy, we may say that to proceed from 'picture' to 'painting' and from 'sculpture' to 'object' also indicates a shift in communication toward the second of these rhetorical terms. In either case, the work as an object (irrespective of whether it is two- or three-dimensional) is not (yet) destroyed. It appears, nevertheless, as a structure that derives its value and significance from the working process immanent within it, which it reveals with the greatest possible clarity — whether it be an achromatic work by Manzoni, a monochrome by Schifano, a spatial structure by Lo Savio, or programmed objects by the Milanese artists of Group T. Naturally, the viewer, or recipient, is directly involved in this process of analyzing and verifying the communication, to the extent to which the work shifts his attention from merely enjoying its effect to analyzing the process that formed it and the resulting structure, or induces him to abandon his purely passive attitude, so in keeping with the psychology of consumption, and transform himself from consumer to producer, from 'consumer to technician,' in the apt phrase of Gianni Colombo. Furthermore, the entire process entails a reinterpretation of the character of the producer, or, rather, a modification in the point of departure of artistic activity; shifting attention to the formative processes is, in fact, indicative of an effort to lead the artist's private and privileged experience back into the context of a more generalized and everyday experience. The artist rids himself of the aura of the conjuror who, with a casual 'Hey presto!' draws strange objects from the secret recesses of his marvelous jacket; instead, he shows others the paths he has taken and the various stages of the process. Lastly, there has been a change in the factor of communication, which is the result of the first three; or, rather, there has been a change in the interpretation of the place in which the aesthetic event occurs. The notion of the space that the work establishes is equated with the notion of the reality in which we are all inextricably bound up, which we can experience by moving around in it, and by establishing ever-changing relationships with the objects that define it.

If we analyze the meaning that these works have for us at the moment in which we encounter them, we note that the most significant fact is their physical presence, their proximity that impinges upon our senses besides involving us mentally. At the same time, we observe that the object, precisely because it rejects the role of traditional sculpture, contributes to our realization of a vital space, transforming itself into a kind of stage property for the mise-en-scène of our daily existence. Accordingly, the object finally locates itself not only within the real and inhabited space, but incorporates and keeps within itself a portion of space. The work transcends the discontinuous boundaries of the object and transforms itself into environment, into a space that encloses it. Theater and architecture are the end points of this spatial concept, which moves from the denial of a hierarchically constructed perspective space to a final rejection of the traditional interpretation of theater and architecture themselves, understood as rigidly predetermined decor. But the transformation of the work and its escape from the limits of an object with its points determined within a total space entail a further modification on the part of the viewer, who now becomes an actor in the aesthetic event. For now, for the first time, total space assigns man the role of protagonist.

But what man is in question? And further, can total aesthetic space contribute (and if so, to what extent) to transforming the individual who lives in a generally manipulative society, and raise him to the condition of a man who is capable of self-determination and the full exercise of freedom? And finally: is the object destined to disappear once and for all in this new kind of space? If we analyze the human component in this new spatial concept from the same point of view as heretofore, we note that we are no longing dealing with man as undergoing a purely rational experience, nor yet as man understood as host to a purely perceptive process. Instead, the subject seems to be involved at every level, participating in the event as an intellectual and psychosensory totality, as a total corporeal being. The answers to the other questions seem more difficult, for they would require a more extensive discussion of the function of art itself and of the aesthetic dimension, apart from a consideration of the possibility that the functional object, too, may acquire other, more complex, and more remote meanings.

The Imaginary Object and the Real Object

These questions might be clarified in psychoanalytical terms, especially with the aid of the relationship that Freud established between play, daydreaming, and artistic activity, all brought together under the common denominator of the pleasure principle (8). Certainly the artist, as opposed to the child at play and the adolescent building castles in the air and reshaping the world according to his wishes, does not remain in the pure realm of the imaginary but comes back to earth, bringing back from his adventurous excursion certain observations that become the concrete works, the objects that take their place among the other objects of our daily life. But the work of art, the object with an artistic intention, wears a very particular guise, because it presents itself as an object neither real nor unreal, endowed with the same sensory concreteness as the things that surround us, and at the same time, because of its emergence from deep and distant realms, enveloped in a kind of halo, a strange and permanent phosphorescence.

The art object therefore dwells in an intermediate zone between the imaginary and the real, as the tangible result of an experience sustained by the pleasure principle and yet clashing with the reality principle. Proceeding from these Freudian premises, analytical research has sought to define in more precise terms that no-man's-land, introducing the concept of 'transitional experiences' and deriving from this, inter alia, the aesthetic experience (9). Transitional moments include a whole series of experiences, such as play and artistic activity, which so to speak lie halfway between the pleasure principle and the reality principle, insofar as the examination of reality (which would allow the experience to be placed in one domain or the other) is not denied (as happens in hallucinations, for instance), but simply held in abeyance.

If, then, we admit that the art object is a transitional object, since it does not have the pure transparency of the images of daydreams and yet appears separate and different from real objects and the realities of man's life, the shift of attention to the formative processes and the diminished emphasis on the object in favor of total aesthetic space become significant as efforts to lead the privileged and separate experience of the artist back to the context of everyday experience, and to relate aesthetic experience as closely as possible to life experience.

But to what extent is it possible to abolish the space that separates the art object (and the transitional area in which it is located) from the real object, and from the concrete reality of life? Admittedly, mistrust of the object, fostered by a legitimate suspicion of its ultimate commercialized destiny in the times in which we live, also stems from a recognition of the real-unreal condition of the object itself, and hence of its only partial state in comparison with the total dimensions of life. On the other hand, the perfect, harmonious universe glimpsed by the artist in his imagination and rendered tangible in his work does not consent to an acceptance of a separate, divided, unhappy life, but demands a radical change in real existence. The art object, interpreted psychoanalytically as anamnesis and resurrection of the original love object, can offer basic suggestions for such a transformation but also reveals itself as a not wholly adequate instrument. The artist desires something more than a transitional experience, separated by its very nature (even if by an extremely narrow margin) from the concrete leading of life. What he wants now is no longer a substitute or alternative experience that could save him temporarily from the recurrent anxiety of abandonment (the daily tragedy), but the transformation of living into a continuous, daily love object. It is André Breton's effort at total desublimation: if Eros is the basis of artistic activity, and poetry a kind of (transposed) lovemaking, one must reduce 'art to its simplest form of expression: love' (10).

But is it possible to confirm the 'simplicity' of this operation? The releasing of the object into total space and life as it is lived is thus a sign of an effort to transpose the imaginary state (in which everything comes out right) to the state of life (in which everything, or nearly everything, has to be remade according to that pattern), without passing through or stopping in the intermediate, transitional state, that of the object. It is an attempt at a complete overturning of the imaginary into the real, to overcome the neurotic dichotomy between the pleasure principle and the reality principle, and, finally, to prefigure a utopia with 'a qualitatively different reality principle' (Marcuse).

But, in bypassing the dimension of the object, one also bypasses (or tries to) the dimension of language, forming a close alliance against it of two provinces that cannot be expressed in words — the province of imagination and that of living, both of which are immediate and primary, and situated on either side of linguistic mediation, or, rather, the mediation of a language that expresses itself as metaphor, serving as a bridgehead in the real-unreal space of transitional experience. The ecstasy of daydreaming turns into an ecstasy of living, imaginary space into the real space of life. The silence of new mystical experiences, or the silence of life lived in the artificial paradise of drugs or the voluntary confines of community minorities, represent the first tangible evidence of this process.

The leap beyond language, then, can only take the form of a leap beyond time and history — at least of this time and this history, as Schiller divined when he indicated that play and the aesthetic experience were ways in which to combine being and becoming, to abolish time within time. The same suspension of the metaphorical language of art, which many propose in favor of an exclusively political commitment, is another symptom of this urge to abandon the intermediate domain of aesthetic experience and to live among things and transform them. We do not deny a similar possibility: fatigue, divisiveness, anguish, do not make us love this time and this history, and if art is just a partial, fragmentary, and precarious sign of that 'eternity of pleasure' of which Nietzsche spoke, we can even look forward to its 'death,' as long as that death signifies the establishment of the rule of pleasure. The quantitative progress of civilization will certainly contribute to creating the objective conditions that favor such a radical transformation, but this will not be possible without some kind of mutation that definitively converts aggressiveness into Eros.

Design and Utopia

Utopian imagination gains fresh impetus from these ideological premises. This upsurge has two motivations, apparently contradictory but actually strictly connected, as complementary aspects of the same fundamental necessity. To those who wish to propose a radical alternative situation to modern industrial society, utopia offers the model of an integrated, harmonious condition, a 'perfect space' (11). It is this model that has largely inspired the international youth movement; its models belong to the realm of utopia, and to it belong those 'perfect spaces' proposed from time to time — the Cuban Revolution, the aesthetic ideology of Marcuse, Maoist China. But utopia does not imply escaping from the present, but rather proposing an alternative pattern of life that, by the processes of negative thought, can unmask the contradictions, conflicts, injustices, and repressions of the present and further the task of constructing a new society. The radical nature of the alternative derives from the fact that its pattern is situated in another dimension from that of today, in a distant space-time (and in the absolute present of the imaginary), of which it reveals to us 'perfection in the luminous intensity of the absolute' (12).

In Italy, the most extreme hypotheses of this kind arose precisely within the field of planning and design, but they came into being in a region lying outside the localities typically given over to industrial production. The Florentine architects of the Archizoom and Superstudio groups have made a decisive contribution to the protests against design regarded as purely the execution of objects for consumption, and to the invention of a new typology for the equipment needed for daily life. But their most significant contribution has probably been the utopian aspirations that underlie their designing. They are negative utopias in that they are not yet aimed at the building of ideal cities, but rather at an eradication of architecture and city planning, in order to liberate man 'from all formal and moral structures preventing him from being in a position to pass judgment freely on his own condition and design history' (13).

For the Superstudio group, the refusal to design (or to design in a certain way) appears in the guise of a millennarian utopia, an urbanistic *parusia*, which regards the object, architecture, and the city as *ornamentum mundi*, an initiatory pathway leading to wisdom and harmony, with a definitive leap outside of time and history: 'Behold the vision of the twelve Ideal Cities,' wrote these Florentine architects, 'the supreme goal of twenty thousand years of humanity's blood, sweat, and tears; the final haven of Man who possesses Truth, free at last of contradictions, doubts, lies, indecisions; definitively, totally, immovably filled with his own PERFECTION' (14).

In these proposals, we encounter a provocatively ironic tone and a strong, conscious element of play. But in this designing game, the architects show their understanding of what is happening and their wish to have an impact on reality, in the sense indicated by Marshall McLuhan, namely by producing models of situations that have not yet come to fruition in society and by building 'Noah's arks with which to meet the coming change' (15).

The situation that has not yet come to fruition in society is the pattern for an individual and collective existence, in which the rights of the imagination are recognized, and in which a new aspect of nature, understood as a new, deeper, and broader aspect of the vital, is recovered. Going back to nature means freeing the deep psychic structures, reinstating them in the round of daily experience, and preventing the exploitative use that industrial society and the work ethic have made of them for the sake of production and consumption. The reasoning we can deduce from this proposal is simple. If progress and technical civilization had to sacrifice the deep psychic structures of the individual for the sake of necessity and biological survival, sublimating them into a 'principle of performance,' the time has come to move in the opposite direction, from 'civilization' to 'nature,' recovering the latter as a specifically autonomous and originally positive stage of the human being.

A Design for New Behaviors

The negative utopia has a concrete, reverse plan that seeks to operate in the present: to destroy the object, architecture, and the city means to propose a new philosophy of design, a kind of design that refuses, insofar as possible, to provide rigid, authoritarian structures in which the individual has no possibility for the independent exercise of his own choices. Creating an environment conceived as a 'neutral parking lot' or as 'furnished-empty-space,' of a kind to restore to each individual 'the right to develop privately his own habitat' (16); or to think of the environment no longer in terms of an ensemble of closed, self-sufficient objects, but as an 'object-structure' and a 'system of flux' (17), means opposing one of the prevailing trends in Italian design of the past few years, that which has attained its eminence on purely stylistic grounds, seeking an aesthetic quality in the product as a status symbol. Design has in fact shut itself up within a kind of aristocratic Apartheid, abetted by a production policy that has not faced decisively the problem of collective needs but has been preoccupied above all with gratifying a limited social class, moderately up-to-date in matters of culture and taste. This has also been confirmed by the way in which the consumer has used certain foreign products, beginning with the Scandinavian, doing away with their originally collective intention and assimilating them within a sophisticated environment whose forms, paradoxically, border on the realm of kitsch. The very success of Italian design in foreign countries is to a large extent bound up with this stylistic approach, which has met the aspirations, conscious and unconscious, of a society swiftly evolving toward a condition of affluence.

The new philosophy of design seeks to avoid this mortgage to the production-consumption binomial, the pivot of the present economic system in highly industrialized countries. It wants to offer a possibility of overcoming the ironclad conditioned reflex that has been established between the quantitative spread of production and the atomization of consumption. Finally, it seeks to achieve eminently communitarian ends, by constructing environments that stimulate the active participation of every individual through the creation of spaces endowed with strong mental, psychological, and sensory appeal. In this activity, the designer takes a different point of view toward production, and he looks with interest at new modes of community life, such as those adopted by a minority of youths in various countries of the world; he begins to execute (precisely for these minorities) ambiences and habitats in which a participatory type of outlook is already becoming evident. Examples of this are the large dance halls where modern youth now congregates, patterned after locales that have already become legendary, like the Savoy, which Le Corbusier understood so well, precisely because it offered the possibility of a communal, ritual, and intensely unanimous life (18); or drugstores, where the function of consumption is likewise exalted — but one has only to compare the kind of activity in such places to that in a large department store, to understand the profound difference between the two.

Viewed in this light, design seeks to liberate individual behavior, willingly sacrificing the execution of finished objects that are plastically autonomous, that can be bought, carried home, and enjoyed in jealous bourgeois privacy. If these designers really have to build something, they prefer to make structures and environments that will enlist the active participation of the user, in the first person, in the shaping and enjoyment of his own surroundings.

Design for domestic life cuts out its own slice of life within this kind of design, whose aim is a different sort of daily existence. It is always a design for behavior, for moments of pause, rest, reading, study, conversation, and games. The domestic round moves between poles of contraction and expansion, isolation and involvement, private and community life, even if it be only the restricted community of one's friends and relations. Here design can play an important role. It can make the domestic pattern a pattern for behavior that is more expansive, more committed in a social sense. But it has to do away not only with the lofty and the solemn, but also the rigid and predetermined — in short, everything that has pretensions to endure as long as a masterpiece or encourages the sentimental languishings of the kitsch-man. Objects must definitely abandon their claim to direct the behavior of the user in an unvarying fashion, but without necessarily following ephemeral vogues. The problem is rather to root the future history of function in basic, relatively unchanging structures, in fundamental patterns of behavior, a sort of archetype of daily life. The recovery of these patterns can be left to the memory, which operates not only on the conscious level of history and culture, but also on the unwitting level of the individual and collective unconscious, nourished, if need be, even by anthropological situations remote in time and space.

Thus, function can reacquire the repetitive form of the rite. Entering the house once more becomes crossing a threshold, a boundary that separates the exterior from the interior, which are different psychical spaces. The house is a space of concentration; going through the doorway marks a significant event, breaking a static equilibrium and setting in motion the lines of force latent in space, as a pebble breaks the surface of water. The potential energy of space becomes condensed and thins out, it contracts and expands; with its systoles and diastoles, it scans the rhythm of daily existence. Various psychic and perceptive fields intersect, creating vital nuclei in the inhabited space; every function takes place in relation to others, and at the same time isolates itself within the circle of its own field. The continuous space of historic Rationalism presupposed an equivalence of gestures and modes of behavior. Here, on the contrary, space is concentrated and coagulated in denser nuclei, reserved for functions that to some extent may be considered privileged, more charged with meaning: actions that could, and should, regain anew their original symbolic, ritual, and mythic significance.

(1). For an analysis of this phenomenon, see Tomás Maldonado, *La Speranza progettuale* (Turin, 1970), pp. 36 f.
(2). *La Nostra crisi attuale* (Italian translation of original Czech, Rome, (1969), p. 94.
(3). Ibid., p. 72.
(4). *La Mossa del cavallo* (translated into Italian from the Russian *Chod — Kónia*, Moscow and Berlin, 1923; Bari, 1967), pp. 7-8.
(5). Superstudio, 'Distruzione, metamorfosi e ricostruzione degli oggetti' (*IN, Argomenti e immagini di design*, March-June 1971), p. 15.

(6). 'Dall'oggetto subito all'ambiento vissuto' (*IN*, op. cit.), p. 83.

(7). See the papers by Giulio Claudio Argan, Alberto Boatto, Maurizio Calvesi, and Filiberto Menna in the symposium on 'The Crisis of the Object,' sponsored by the Istituto Nazionale di Architettura in Rome, November 25, 1968. See also the issue devoted to 'L'Oggetto e il comportamento' ('The Object and Behavior'), *Marcatré*, nos. 6-7, Spring-Summer 1971. Finally, see the series of articles in the issue on the theme of 'La Distruzione degli oggetti' ('The Destruction of Objects'), *IN*, March-June 1971.

(8). 'Il Poeta e la fantasia' (Italian translation of *Der Dichter und das Phantasieren*, Vienna, 1907; in *Saggi sull'arte, la letteratura e il linguaggio*, Turin, 1960), pp. 49-59.

(9). D. W. Winnicott, 'Transitional Object and Transitional Phenomena' (*International Journal of Psychoanalysis*, vol. 34, n. 2, 1953), and F. Fornari, 'Principio del piacere e principio della realtà nel fenomeno Beat' (*Psicanalisi della situazione atomica*, Milan, 1970), pp. 267-89.

(10). *Poisson soluble* (Paris, 1924), p. 100.

(11). Martin Buber, *Sentieri in Utopia* (Italian translation of *Pfade in Utopia*, Heidelberg: Lambert Schneider, 1950; Milan, 1967), p. 17.

(12). Ibid., p. 17.

(13). Archizoom Associati, 'Residential parking lots' (in catalogue of the VII Paris Biennale, September-November 1971).

(14). Superstudio, *Premonizioni della parusia urbanistica* (mimeographed text, Florence, 1971).

(15). *Gli strumenti del comunicare* (Italian translation of *Understanding Media*, New York, 1965; Milan, 1967), p. 381.

(16). Archizoom Associati, 'La Distruzione degli oggetti' (*IN*, op. cit.) p. 11.

(17). Superstudio, 'Distruzione, metamorfosi e ricostruzione degli oggetti' (*In*, op. cit.), p. 15.

(18). *Quand les cathédrales étaient blanches* (Paris, 1937).

SUMMARY
Emilio Ambasz

The intention of the exhibition 'Italy: The New Domestic Landscape' and of this publication has been to recognize the cultural achievements of modern Italian design, to honor the accomplishments of Italy's gifted designers, and, by presenting a selection of the most outstanding examples of their works during the last decade, to illustrate the diversity of their approaches to design. We have also wanted to stress that Italian design is important not solely because of its remarkable formal production, but also because of the high level of critical consciousness with which its proponents — designers and critics, individual and communal users — are now questioning the sociocultural meanings and implications that the phenomenon of design has for Italy.

The concerns of these designers and critics are so wide in their range, and their perception of problems so acute, as far to transcend their local circumstances. Thus, evaluating the situation in Italy may lead to a better understanding of the recipricocal relationships that exist in general between design and society today.

Up to now, as many of the foregoing articles point out, Italian design has been limited to the production of single items and small environments. Only rarely has it extended to the creation of large environments or communities of objects. In Italy, as in most industrialized societies, these beautiful, isolated objects and microenvironments usually exist in the midst of deteriorating towns and urban areas, the evidences of whose history are being rapidly eroded, while the surrounding countryside is being laid waste by ecological neglect.

This emphasis on consumer products might seem to show the designers' lack of interest in social problems and a kind of design that has the community as patron, rather than private individuals or businesses. In reality, however, it is a response to a number of interrelated industrial and cultural factors. Although most Italian designers are highly aware of the needs of both the community and industry, an intricate set of market conditions has led to their concentration on the designing of single articles and small environments that have no real connection with a 'culture of the dwelling.'

There is no organic coordination between the furniture industry and what might be called — if it existed — the building industry, which to be truly industrialized would require long-term capital investments that only a comprehensive housing policy could ensure. At present, however, the government's policy and programs in this sector exist primarily as aspects of the employment problem rather than as an integral part of urban planning. This has left architects and designers a very small, and at that very rarefied, field in which to operate. With few exceptions, opportunities for town planning and large-scale architecture in Italy are frozen, thus forcing most architects, designers, and would-be planners to take refuge in the peripheral areas of product design. The success of their products on the international market has been a temporary boon to the national economy but has aggravated the situation by concentrating attention on short-term gains rather than farsighted goals.

Moreover, within the province of product design, the designers have been further alienated by the fact that the Italian furniture industry has not yet fully resolved the problems inherent in switching from an artisan type of fabrication to industrial processes. Lacking state support for a policy oriented toward conceiving the home in terms of mass production, Italian industry has quite understandably turned to an area that has required no basic technological innovations and is chiefly responsive to the traditional values represented by the high-quality object. This has caused a continuing widening of the gulf between the design of the home itself and the design of mass-produced objects for its decor. Coupled with the lack of any ongoing controversy about urban planning or the industrialization of the building trade such as might have served to crystallize the issues involved, this situation is responsible for the pervasive sense of crisis and diffuse symptoms of frustration among Italian designers in the past several years.

The roots of that crisis, however, are not grounded solely in socioeconomic factors. Much of the cultural debate regarding Italian design originates within the psyches of the designers themselves, their inhibitions, and their guilt complexes. The disintegration of the traditions that generated the modern movement has brought about an identity crisis, compounded by guilt feelings regarding their own complicity in reenforcing the prevailing values of a consumer society. The consequence has been the Italian designers' keen awareness of social questions and a disquieting self-doubt regarding their own role. This accounts for the lucidity, as well as the rhetoric, with which they articulate their predicament.

The environments presented in the exhibition manifest two opposite attitudes to design currently prevalent in Italy. The first involves a commitment to the designing of physical objects as a problem-solving activity, capable of formulating solutions to the problems presented by the natural and sociocultural environments. By contrast, the attitude we have characterized as that of 'counterdesign' prefers to emphasize the need for a renewal of philosophical discourse and for social and political involvement as the way in which to bring about structural changes in society.

The environments proposed by designers in the first category reveal many facets of the present industrial and ideological controversy. As far as possible within the scope of an exhibition, they explore the possibilities for change that the prevailing situation in Italy might allow. No clearly defined line separates those who seek to change established conditions by means of technology from those who attempt to use design as a symbolic means for reforming the present. We find among the latter group, for example, some who assign to the object an architectural and representational value, and others whose heavily ironic symbolism is protected by a layer of ambiguous allusions to individual and collective protest. There are also those who design deliberately unattractive objects with gloomy colors and redundant formal details, for the purpose of declaring the ultimate futility of their protest.

Designers who explore the possibilities that technology may offer for overcoming existing shortcomings seek to develop such domestic service elements as kitchen, bathroom, and storage ensembles, which in the absence of any industrialized housing system might for the time being act as surrogates. Other designers try to solve the same problem, at least partially, by recycling and adapting as environments structures derived from totally different industrial processes — such as aluminum freight containers — in order to make them serve as habitats. Still others, despairing of ever being able to change the backward managerial methods and building techniques of the existing construction industry, propose the concept of the mobile home as a strategy for directing the attention of Italy's highly qualified automobile industry toward an effort to attack the problem of mass-produced dwellings.

Those who believe that only sociopolitical involvement and concentration on urban problems can provide substantial solutions bring into question the entire concept of physical design as the designer's prime task. In conflict with the present, the avant-garde in

Italy quite logically looks to the future as offering the opportunity for all reconciliations and understandably chooses counterdesign as its mode of operation.

Yet even these avant-garde groups, too, are located within, and restricted by, the present social structure and production system. As a metaphor for collective action, some of them attempt to recover the archetypes of human conduct, proposing that the individual withdraw into an image of paradise in which Eros and freedom from work are postulated as man's total experience of self and the universe. But the traditional signs of utopia have been changed, for what the counterdesign groups set forth are negative utopias. They do not aim at ideal cities, but rather at the eradication of architecture and city planning, in order to free man from all the formal and moral strictures that prevent him from passing free judgment on his own condition and history. The negative utopia, however, is not intended for the future, but for the present. Destroying the notion of the object and the city means opposing the dominant trend now prevailing in Italian design, which cultivates the aesthetic quality of the product as status symbol and imposes middle-class standards on the urban environment. The approach of these radical designers also implies an alternative to the closed cycle of dispersed production and atomized individual consumption; it purports to recover design for communal ends. The goal is the creation of spaces wherein the active participation of every individual in his own daily life may be stimulated, and mental, psychological, and sensorial functions regain the quality of continuously renewed rituals.

Italian design today displays as wide a spectrum of conflicting schools of thought and action as can be found anywhere. By inviting the exponents of these diverse philosophies to present their points of view, this exhibition has sought to bring into focus the most significant current positions on design, applicable not to Italy alone. Thus, we have on the one hand designers who thoughtfully and imaginatively attempt to offer alternative, concrete solutions to problems of the present; and on the other hand, those who elaborately explain their conviction that no solution at all is possible until the present structure of society is changed.

Their statements make it possible to draw some specific conclusions and venture some observations regarding the present state of the design endeavor in industrialized societies. Generally speaking, the international rebellion against the object came about not only because of the doctrines of cultural vanguards, nor the hysteria induced by 'technological despair' in the face of an increasingly mechanized civilization, but also because the object lacks a suitable sociocultural context. In the case of Italian design, specifically, despite many observable contradictions, one can sense the circumstances that could lead to a change in its scale, methods, and goals.

Italian designers now have a choice between at least two possible courses. The first alternative is to remain limited to the production of isolated objects and small environments, intended for only one segment of the social strata, and for the foreign as much as for the home market. The second choice would require facing the need to build at least two million rooms a year, without spoiling the urban and rural landscape. This would demand an enormous effort that could not be confined to the designing of isolated objects or buildings but would embrace, as a complex whole, the task of building entire urban environments. Such a development could establish the favorable circumstances in which the object might lose its isolation, both in production and use.

A merging of the divided sectors — construction and design — will not automatically lead to urban utopias. It is nevertheless clear that the disentangling of so backward a sector as the building industry from its present obstructions, and the projection and realization of a comprehensive housing program, could have beneficial effects for an understanding of design in its role as a social service.

Obviously, the issues raised by this exhibition, the designers' statements, and the articles in this catalogue transcend the boundaries of Italy. They reflect, in great part, the growing distrust of objects of consumption emerging in all industrialized countries and undermining the traditional notion of design as being solely concerned with producing cultural objects. There is a more acute awareness of the ways in which the individual today is manipulated, of the diminishing scope for intellectual action and, in particular, for aesthetic concerns. In reaction to this situation, there is an increasing temptation to abandon the specialized field of design itself, in order to concentrate on the political front to the exclusion of all others. Thus, political action is put forward as the dominant force for bringing about any transformation of man and society.

But the task is far too complex (and human beings too complicated) to entrust to political action alone. What is needed is a discourse of many voices, excluding any form of hegemony. A resort only to politics tends to restrict intellectual freedom, and by concentrating on reaction against the established, disregards the possibilities for action that aesthetic invention and imagination may reveal.

In addition to being a political stance, the rejection or destruction of the object has become a well-known form of aesthetic activity and criticism, widely observable throughout many fields of artistic endeavor. In design, as in several avant-garde tendencies in other arts, attention is increasingly shifting from an exclusive concentration on the form of an artifact to encompass the processes that generated it, and the processes which it in turn generates. This preference for process over end product is an aspect of the conflict between aesthetic intention and the concrete work that results from it. Together with a shift of concern from object to environment, it has brought about a redefinition of the task of design.

From this redefinition, two converging and complementary interpretations emerge. The first sees design as functional and symbolic formalization, capable of effecting changes from within by the power that the designed object or environment exerts on the user and his behavior. The second, a political definition, sees design essentially as conflict, which must effect changes from without. In reality, however, the two definitions are aspects of one system of thought. An aesthetic of design founded not only on the concrete object, but also on its sociocultural context, entails a redefinition of the user as well, as enactor of an aesthetic event. Hence, the concept of environment presupposes for man the role of active protagonist rather than of mere passive spectator.

Thus, design ultimately transcends both object-making and conflict, to encompass all the processes whereby man gives meaning and order to his surroundings and his daily patterns of life. Without claiming to solve everything, design can nevertheless move man toward an authentic realization of himself.

Credits

EXHIBITION CREDITS

Exhibition directed and installed by Emilio Ambasz
Coordination: Thomas Czarnowki
Assembly and documentation of objects: Anna Tucci

Objects: Designers
(numbers refer to pages on which the works are illustrated)

Archizoom: 101, 103, 108
Arioli, Roberto: 82, 85, 87
Aroldi, Danilo and Corrado: 62
Asti, Sergio: 48, 64, 88, 109
Aulenti, Gae: 43, 109

Bartolini, Dario and Lucia: see Archizoom
Bassi, Giampiero and Giovanni: 64
Becchi, Alessandro: 119
Bellini, Dario: 68
Bellini, Mario: 31, 48, 68, 70, 120
Bicocchi, Giancarlo and Luigi: 130
Bimbi, Carlo: see Internotredici
Boccato, Marilena: 84
Boeri, Cini: 29, 61, 121
Bonetto, Rodolfo: 28, 33, 49, 71-72
Branzi, Andrea: see Archizoom

Casati, Cesare: 96, 118
Castelli Ferrieri, Anna: 47, 52
Castiglioni, Achille: 65-66, 102, 118, 123
Castiglioni, Livio: 67
Castiglioni, Pier Giacomo: 65-66, 102, 123
Catalano, Umberto: 114
Cattelan, Franco: 53
Ceretti, Giorgio: 101 (Gruppo Strum), 103
Colombo, Joe: 30, 40, 45, 53, 62, 116-17, 123
Coppola, Silvio: 42
Corretti, Gilberto: see Archizoom
Cuneo, Marcello: 63

D'Aniello, Pierangelo: 43
Decursu, Giorgio: 44
Deganello, Paolo: see Archizoom
De Pas, Jonathan: 34, 44, 57, 95, 114
Derossi, Piero: 101 (Gruppo Strum), 103
D'Urbino, Donato: 34, 44, 57, 95, 114

Facchetti, Gianfranco: see Group G 14
Ferrara, Gianni: see Internotredici
Frassinelli, Piero: see Superstudio
Frattini, Gianfranco: 67

Gardella, Ignazio: 47
Gatti, Piero: 113
Gigante, Gian Nicola: 84
Gilardi, Piero: 99
Gioacchini, Nilo: see Internotredici
Gramigna, Giuliana: 35, 77
Gregotti, Vittorio: 60
Group G 14: 32
Gruppo Architetti Urbanisti Città Nuova: 63
Gruppo Strum: 101

Iliprandi, Giancarlo: 124
Internotredici: 133

Jacober, Angelo: 43

Klier, Hans von: 107

La Pietra, Ugo: 55
Lenci, Fabio: 122
Leonardi, Cesare: 27
Lomazzi, Paolo: 34, 44, 57, 95, 114
Lucci, Roberto: 101
Lucini, Ennio: 82

Macchi Cassia, Antonio: 60
Magistretti, Vico: 33, 39, 41, 59, 65, 109
Magris, Alessandro and Roberto: see Superstudio
Mangiarotti, Angelo: 46, 82, 128
Mango, Roberto: 78
Manzú, Pio: 72
Mari, Enzo: 54, 76-77, 83, 89-91
Marotta, Gino: 101
Masi, Gianfranco: 114
Massoni Luigi (Studio BMP): 129
Matta, Sebastiano: 115
Mattioli, Giancarlo: 64
Mazza, Sergio: 35, 56
Meneghetti, Lodovico: 60
Monsani, Roberto: 130
Morozzi, Massimo: see Archizoom
Munari, Bruno: 131

Natalini, Adolfo: see Superstudio

Orsoni, Umberto: see Group G 14

Paolini, Cesare: 113
Pareschi, Gianni: see Group G 14
Peduzzi-Riva, Eleonore: 64, 86-87
Pensotti, Pino: see Group G 14
Pesce, Gaetano: 35, 97-98
Pietrantoni, Marcello: 101
Piretti, Giancarlo: 34, 36-37, 51
Pizzo Greco, Alfredo: 98
Ponzio, Emanuele: 96, 118

Raimondi, Giuseppe: 100
Rosselli, Alberto: 33, 38
Rosso, Riccardo: 101 (Gruppo Strum), 103

Salvati, Alberto: 126-27
Sambonet, Roberto: 79
Sapper, Richard: 44, 69, 71, 73-74
Sarfatti, Gino: 61
Scarpa, Tobia and Afra: 28-30
Seassaro, Alberto: 132
Siard, Marcello: 57
Soavi, Giorgio: 83
Sottsass, Ettore, Jr.: 50, 75, 104-6
Stagi, Franca: 27
Stoppino, Giotto: 48, 60
Studio BMP: see Massoni
Studio OPI: 80-81, 84
Studio TG: 80
Superstudio: 100

Teodoro, Franco: 113
Toraldo di Francia, Cristiano: see Superstudio
Tresoldi, Ambrogio: 126-27

Ubaldi, Roberto: see Group G 14
Ufficio Tecnico Snaidero: 125

Valle, Gino: 72
Vignelli, Massimo: 77

Vigo, Nanda: 58
Zambusi, Antonio: 84
Zanuso, Marco: 35, 42, 44, 69, 71, 73-74

Objects: Manufacturers
(numbers refer to pages on which the works are illustrated)

ABET-Print, Cuneo: 105, 130
Acerbis, Bergamo: 98
Anonima Castelli, Bologna: 34, 36-37, 51
Arflex, Milan: 29, 35, 118, 121
Arredoluce, Monza-Milan: 58
Arteluce, Milan: 60-61
Artemide, Milan: 33, 39, 41, 56, 59, 63-65, 67
Autovox, Rome: 71

Bazzani, Milan: 43
BBB Bonacina, Milan: 44, 114
Bernini, Milan: 42, 49, 122-23
Boffi, Milan: 123, 129
Bracciodiferro, Milan: 97
Brionvega, Milan: 68-69, 71
Busnelli, Milan: 32

Campeggi, Milan: 126-27
Candle, Milan: 64
Cassina, Milan: 28-31
C & B Italia, Como: 35, 48, 98, 120
Cedit, Milan: 88
Cini & Nils, Milan: 80-81, 84
Cinova, Milan: 35
Cristal Art, Turin: 100

Danese, Milan: 76-77, 82-83, 89-91
Driade, Milan: 33

Elco, Venice: 27, 53

Flexform, Milan: 28, 116
Flos, Brescia: 65-66, 102

Gabbianelli, Milan: 63, 77, 82, 85, 87
Gavina, Milan: 42, 54, 115
Giovanetti, Pistoia: 119
Giovenzana (for Heller Designs): 77
Gufram, Turin: 99, 101, 103
Guzzini, Macerata: 80

Heller Designs: see Giovenzana

I.C.F. De Padova, Milan: 130
Italora, Milan: 72

Kartell, Milan: 30, 40, 44, 47-48, 52, 57

Longato, Padua: 57

Martinelli-Luce, Lucca: 109
Minerva, Milan: 70

Necchi, Pavia: 73
NY Form, Bologna: 114

Olivetti, Ivrea-Turin: 75, 83
O-Luce, Milan: 62

Planula, Pistoia: 107
Poggi, Pavia: 55, 109
Poltronova, Pistoia: 50, 95, 100-1, 103, 106, 108, 128
Ponteur, Bergamo: 96

RB, Bergamo: 124
Reed & Barton, Taunton, Massachussetts: 78
Robots, Milan: 131

Salviati, Venice: 88
Sambonet, Vercelli: 79
Saporiti, Varese: 33, 38
SICART, Vicenza: 84
Sintesis, Varese: 48
Sirrah, Imola: 64
Sit-Siemens, Milan: 74
Snaidero, Udine: 125
Solari, Udine: 72
Sormani, Como: 116-17
Stilnovo, Milan: 62, 101
Studioluce, Pistoia: 64

Terraillon, Pianezza-Turin: 73
Tisettanta, Milan: 46
Turri, Giosuè, Milan: 133

Veglia Borletti, Milan: 72
Venini, Murano: 109
Vistosi, Venice: 86-87

Xilema, Vicenza: 53

Zanotta, Milan: 34, 43, 45, 102, 109, 113, 118

Objects: Containers

Designers: Emilio Ambasz, in collaboration with Thomas Czarnowski
Tecnical consultants: initial phase, Giancarlo Piretti; construction phase, Justin Henshell. Construction: Eckol Containers Systems, Glenside, Pennsylvania. Lamps: Artemide

Patron: Anonima Castelli, Bologna, manufacturers of wood, metal, and plastic furniture

Photography: All black-and-white enlargements by Aldo Ballo, Milan, except: Serpentone, by Giancarlo Iliprandi; Cub 8, by Foto Masera; Monoblocco and Domusricerca, Foto Casali-Domus; A 1, by Studio Professionali, Milan; Safari, by Archizoom. All transparencies by Valerio Castelli, Milan, except: Sottsass case, by Ettore Sottsass, Jr.; Archizoom case, by Archizoom

Environments

Designer: Gae Aulenti
Patrons: ANIC-Lanerossi; Kartell
Producers: Kartell; with the assistance of Zanotta

Designer: Joe Colombo; collaborator, Ignazia Favata
Patrons: ANIC-Lanerossi, Elco-FIARM, Boffi, Ideal-Standard
Producers: Elco-FIARM, Boffi, Ideal-Standard; with the assistance of Sormani
Film: directed by Gianni Colombo and Livio Castiglioni

Designer: Ettore Sottsass, Jr.
Patrons: ANIC-Lanerossi, Kartell, Boffi, Ideal-Standard
Producers: Kartell, Boffi, Ideal-Standard; with the assistance of Tecno
Film: directed by Massimo Magri

Designer: Alberto Rosselli
Patron: FIAT
Producers: Carrozzeria Renzo Orlandi, Carrozzeria Boneschi, Industria Arredamenti Saporiti, Boffi; with the assistance of Valenti, Nonwoven, Rexedil
Film: CINEFIAT (Ernesto Prever and Osvaldo Marini)

Designers: Studio Zanuso — Marco Zanuso and Richard Sapper
Patrons: ANIC-Lanerossi, FIAT, Kartell, Boffi
Producers: FIAT, with the participation of Boffi; Kartell
Film: directed by Giacomo Battiato

Designer: Mario Bellini; collaborators, Dario Bellini, Francesco Binfaré,

Giorgio Origlia; collaborators for technical development, Centro Cassina
Patron/producers: Cassina, C & B Italia; with the contributions of Citroën, Pirelli
Film: directed by Davide Mosconi; visual ideas by Mario Bellini, Francesco Binfaré, Davide Mosconi, Giorgio Origlia

Designer: Ugo Pesce
Patrons: Cassina, C & B Italia, Sleeping International System Italia
Producers: Centro Cassina, with the assistance of Sleeping International System Italia
Film: directed by Klaus Zaugg

Designer: Ugo La Pietra
Patron: ABET-Print, with the collaboration of Silcon and Moro
Audio-visual program: Ugo La Pietra with Piero Castiglioni

Designers: Archizoom Associati (Andrea Branzi, Gilberto Corretti, Paolo Deganello, Dario Bartolini, Lucia Bartolini, Massimo Morozzi)
Patron: ABET-Print
Audio-score: Giuseppe Chiari

Designers: Superstudio (Piero Frassinelli, Alessandro Magris, Roberto Magris, Adolfo Natalini, Alessandro Poli, Cristiano Toraldo di Francia)
Patron: ANIC-Lanerossi
Film: Superstudio

Designers: Gruppo Strum (Piero Derossi, Giorgio Ceretti, Carlo Giammarco, Riccardo Rosso, Maurizio Vogliazzo)
Patrons: Gufram, Casabella
Consultant for photography: Paolo Mussat Sartor

Designer: Enzo Mari

Designers: Gianantonio Mari and Studio Tecnico G. Mari
Planning: Gianantonio Mari
Project line and texts: Ezio Mari
Graphics: Ornella Selvafolta
Collaborators: S. Ando, M. Matsukaze
Photography: S. Pazzi

Designers: Group 9999 (Giorgio Birelli, Carlo Caldini, Fabrizio Fiumi, Paolo Galli)

Graphic and Audio-Visual Information Systems
Production and realization: Olivetti Corporate Advertising — Audio Visual Department

Orientation Gallery and leaflet: Art Director, Franco Bassi; Designers, Adriana Balzonella and Giacomo Sala

Introductory Film, Objects Section
Text and visual story: Emilio Ambasz. Director: Giacomo Battiato. Producer: Sergio Lentati (Politecnico Cinematografica, Milano) Cameraman: Alberto Spagnoli. Sets: Oliva di Collobiano. Photography: Dido Mariani. Music: Tito Fontana

Closed-Circuit-TV Introduction to Environments Section
Text: Emilio Ambasz. Technical development: Olivetti Audio-Visual Department

Critical Commentary on the Exhibition (Implicor Olivetti)
Text: Emilio Ambasz. Art Director and designer: Umberto Bignardi Hardware designer: Hans von Klier, collaborator, Malcolm Alum Graphic designers: Roberto Pieraccini, Poppi Ranchetti, Mizio Turchet, Natasha Poblete. Audio-visual technology: Giancesare Rainaldi Photographers: Adolfo Fogli, Ennio Canziani

PERMISSIONS

We are grateful to the sources listed below for the use of quotations on the following pages:

p. 18. Antoine de Saint-Exupéry. *The Little Prince*. Translated by Katherine Woods. New York: Harcourt Brace Jovanovich, copyright 1943.

p. 139. William Empson. *Seven Types of Ambiguity*. 3rd edition, revised. New York: New Directions Publishing Corporation, 1953.

Upheaval. New York: Monthly Review Press, 1969.

p. 144. David Cooper. *The Death of the Family*. New York: Pantheon Books, 1970.

p. 145. Harold F. Searles. *The Non-Human Environment*. New York: International Universities Press, 1960.

George Steiner. Review of *Beyond Reductionism*, edited by Arthur Koestler and J. R. Smythies. *The New Yorker*, March 6, 1971.

p. 331. 'Il Tempo libero.' Catalogue. XIII Triennale, Milan, 1964.

pp. 337-38. Paolo Chessa, in *IN: Argomenti di architettura*, no. 3, September 1961.

p. 346. Alberoni. *Consuma e società*. Bologna: Il Mulino, 1964.

p. 348. Ivor De Wolfe. *The Italian Townscape*. London: The Architectural Press, 1963.

pp. 355-57. From script by Ugo Pirro and Elio Petri for *La Classe operaia va in paradiso*, produced by Euro International Films, 1971.

pp. 370, 375. Louis Kahn. *Talks with Students*. (Architecture at Rice, 26). Houston: Rice University, 1969.

p. 372. Vittorio Gregotti. Special issue, 'Design,' *Edilizia moderna*, no. 85, 1965.

pp. 372-73. Pierluigi Spadolini. *Design e società*. Florence: La Nuova Italia, 1969.

pp. 402-3. Special issue, 'La Distribuzione degli oggetti,' *IN: Argomenti e immagini di design*, vol. 2, March-June 1971.